ADVANCED SIMULATION AND TEST METHODOLOGIES FOR VLSI DESIGN

T0332897

ADVANCED SIMULATION AND TEST METHODOLOGIES FOR VLSI DESIGN

Gordon Russell and Ian L. Sayers

Department of Electrical and Electronic Engineering
University of Newcastle upon Tyne

Van Nostrand Reinhold (International)

First published in 1989 by
Van Nostrand Reinhold (International) Co. Ltd
11 New Fetter Lane, London EC4P 4EE

Set in 10/12 pt Times by Keyset Compostion, Colchester
Printed in Great Britain by St Edmundsbury Press,
Bury St Edmunds, Suffolk

ISBN 0 7476 0001 5

British Library Cataloguing in Publication Data

Russell, G. (Gordon)
    Advanced simulation and test methodologies
    for VLSI design.
    1. Very large scale integrated circuits.
    Design and construction. Simulations
    I. Title   II. Sayers, Ian L.
    621.381'73
    ISBN 0 7476 0001-5

CONTENTS

PREFACE

With the advent of LSI and subsequently VLSI, technology has never witnessed such an unprecedented growth in the use of electronic systems in a wide range of applications. Although the ability to integrate several hundreds of thousands of transistors into a monolithic integrated circuit offers many advantages, a major concern in the design of these complex devices has been the ability to verify and in some instances guarantee their fault free operation. The testing of digital circuits has been problematic since time immemorial; however with the introduction of VLSI the problems have been aggravated by several factors in addition to those associated with the increase in complexity and reduction in the pin to gate ratio.

- Market trends have encouraged the prolific use of VLSI in many applications ranging from consumer products to critical commercial controllers, consequently reliability is of paramount importance.
- The increased requirement for ASICs, has necessitated the development of more sophisticated CAD tools. However, the greatest advances have been in layout and simulation with little improvement in area of test. Consequently, inexperienced designers can now produce extremely complex but untestable chips.
- Again, with the introduction of ASICs the high costs of testing cannot be amortized over a large number of components since ASICs are inherently low volume products.
- Traditionally the design and test of a circuit were considered as two distinct phases in the development of an electronic system. Designers must now be educated that in order to produce testable VLSI circuits testing must be considered as an integral part of the design process.

Hence the objectives of this text are, first, to emphasize that testing is an integral part of the design process and must not be considered as an afterthought. Secondly, it summarizes the new developments in simulation, test and design for testability about which a vast amount of research work has been produced and which is disseminated over a large number of technical journals.

Chapters 1 and 2 comprise the introductory material required in the book. Chapter 1 describes the essential components of a design verification system emphasizing the need to consider testing as an integral part of the design process. Chapter 2 describes the faults and fault models considered in testing VLSI circuits.

Chapter 3 discusses the topic of simulation at different levels of abstraction and describes the basic algorithms used in the simulation process. A brief description is also given of the architecture of hardware accelerators used for simulation. The topic of simulation is discussed further in Chapter 4; however in this instance it is the behaviour of the circuit under fault conditions which is of interest. In view of the high cost of fault simulation several alternative approaches are also described.

Chapter 5 describes the current techniques for test pattern generation at functional and gate levels of abstraction. At gate level, the algorithms discussed are for generating tests in combinational circuits since many of the Design for Testability techniques, discussed in Chapter 6, reconfigure sequential circuits into combinational circuits for the purposes of testing.

In view of the vast amount of research effort directed at endeavouring to facilitate the testing of complex VLSI circuits, Chapters 6 and 7 are concerned with design for testability techniques. Chapter 6 essentially summarizes the methods currently used, covering *ad hoc* techniques, structured approaches and built-in self test. Chapter 7 continues with this theme and describes a range of hybrid designs for testability techniques which combined the structured and built-in self test approaches.

With the introduction of VLSI and the subsequent reduction in device size researchers have observed an increase in the occurrence of intermittent faults. The prolific use of VLSI in a wide range of applications has dictated that the reliability of circuit operation is of paramount importance necessitating the continuous testing of circuits during normal operation. Consequently, Chapters 8 and 9 are concerned with topics related to fault tolerant design. Chapter 8 describes the general techniques used, for example, hardware and information redundancy and the design of total self checking circuits. Chapter 9 discusses a particular information redundancy technique employing residue codes and describes its application to a wide range of arithmetic and logic functions; the design of an eight-bit data path circuit incorporating residue codes for concurrent error detection is then described in detail.

Chapter 10 continues with the topic of testing and design for testability; however, in this instance emphasis is placed on its applications to regular structures, for example, RAMs and PLAs.

The final Chapter discusses the application of expert systems to the testing problem. The objective of employing expert systems to test circuits is to try to incorporate the macroscopic view that the test engineer has of the circuit into automatic test generation programs and hence make them more efficient. Expert systems have also been developed to assist designers to choose the most appropriate design for testability techniques for a given circuit, in view of the wide range of methods available each with its own advantages and disadvantages.

The treatment and scope of this text is such that it can be used by students on taught MSc courses involved with IC design and research students involved with the design of VLSI circuits, as well as professional engineers and managers in industry who are involved with the design, test and reliability of VLSI circuits.

GR
ILS

ACKNOWLEDGEMENTS

The authors wish to acknowledge Mrs K. Kidd for her assistance in typing the manuscript and also for her prowess in deciphering the hieroglyphics in the original text.

Due acknowledgement is also given to all the authors and co-authors of the work discussed and cited in the text. Apologies are offered to anyone whose work may have been discussed but not cited, although every effort has been made to ensure that this has not occurred.

The authors also wish to thank Professor D. J. Kinniment for the facilities used, within the department, in the preparation of the manuscript.

1
INTRODUCTION

1.1 PROBLEMS OF TESTING VLSI CIRCUITS [1]–[4]

Over the past decade a vast arsenal of CAD tools has been developed to assist engineers in all phases of the design process for VLSI circuits and systems. However, the most significant advances have been within the physical layout phase, which in the past was an extremely time consuming and error prone task. The advances in IC layout tools, in addition to reducing design times, have further highlighted the testing phase of a design, both before and after fabrication, as the major issues in the design of VLSI circuits. The objectives in testing a design are twofold:

1. to ensure that, before fabrication, the circuit behaviour satisfies the intent of the designer, that is, it is free from functional or logical design errors; and
2. to detect faulty devices, after fabrication, either on the wafer, as packaged ICs or as a system component in the field.

However, the testing of devices after fabrication is, by far, the most pressing problem currently confronting both designers and test engineers, and is considered to be, potentially, the major obstacle to the full exploitation of the benefits to be obtained from realizing extremely complex systems as VLSI circuits.

The symptoms of the problems encountered in testing VLSI circuits are

1. increased testing costs, which are proportional to testing time, and are increasing rapidly with circuit complexity;
2. test pattern generation and evaluation times are increasing with circuit complexity; however, this is offset to some extent by improved performance and reduction in costs of computer systems; and
3. increase in the volume of test data, that is, the number of input stimulus–output response pairs required to test the circuit.

The question to be asked, subsequently, is 'Why does the increase in circuit complexity create such enormous testing problems?' An insight into the cause of the testing problem can be gained by first considering some aspects related to the testing of small and medium scale integrated circuits. The testing of these low complexity devices was a relatively simple process since

1. Internal nodes in the devices were readily controlled and observed from the terminals of the device. Problems of testing are directly related to the degree of difficulty encountered in controlling/observing internal nodes in the circuit;
2. The relative simplicity of the functions realized in an IC permitted exhaustive testing to be performed economically; and
3. Complex systems were subsequently constructed from well tested devices.

With the advent of LSI and subsequently VLSI, the gate to pin ratios increased, hence internal circuit nodes became less controllable and observable. The increase in circuit complexity also precluded the use of exhaustive testing techniques; furthermore the increase in complexity has invariably been associated with an increase in the sequential depth of the circuit. The problems of automatically generating test patterns for sequential circuits has only been solved for circuits with a limited degree of sequentiality. The difficulties encountered in testing highly sequential circuits have been solved, to some extent, by using design styles which permit the circuit to be reconfigured into a combinational circuit for the purposes of testing; however, the reconfiguration techniques have the side effect of extending test application times. Consequently, the reduced accessibility, the increase in sequential depth, the need to determine fault coverage of test patterns since exhaustive testing cannot be used, and the increase in the number of faults to be processed have all become factors contributing to the problem of testing VLSI circuits.

The problems of testing VLSI circuits have an obvious 'knock-on' effect to the systems designers who must now use components which are not 100% tested, unlike using small and medium complexity devices where this could be guaranteed. Thus, when the prototype system does not perform the intended function the system designer is confronted with the problem of deciding whether the fault is due to a design error or a faulty component. It may be considered that the system designer's problem could be alleviated by screening the incoming VLSI parts for faults. This approach, however, would require the use of expensive VLSI testers, furthermore the circuit vendors do not usually supply sufficient information about the details of the circuit to permit the system designer to generate a comprehensive set of tests for the circuits.

More recently, the problems of testing VLSI circuits have been further aggravated by

1. rapid changes in technology
2. increase in the requirements for Application Specific Integrated Circuits (ASIC)

3. increase in the use of Customer Own Tooling (COT).

When the semiconductor technologies used to manufacture the circuits change rapidly, there is insufficient time to adequately study the way in which defects from the process manifest themselves as faults. If it is assumed naïvely that the effects of defects in a new process can be adequately represented by the standard fault model, it is possible that devices, which have not been properly tested, are sent to customers.

The ability to integrate a large number of logic functions onto a single VLSI circuit has increased the requirement for ASICs. However the ASIC market has certain characteristics which add to the problems of testing VLSI circuits, for example

1. ASICs usually require a fast design time if the products in which they are used are to maintain a competitive edge, hence they have a short product lifetime;
2. Due to the unique functions performed by ASICs, production volumes are low, hence test costs cannot be amortized over a large number of components;
3. The uniqueness of ASIC functions requires new test programs to be developed for each design.

In order to offset some of the cost of ASICs, due to the low volumes, the users of the ASIC components are designing the circuits themselves (COT) and having them fabricated in silicon foundries. Consequently the onus of ensuring that a circuit can be adequately tested falls upon the designer, who in the past would design circuits with little thought about testing and assume that the test engineer would produce an adequate test program for the circuit. Since designers have had little experience in testing VLSI circuits, the outcome of this situation is that the test engineers in the silicon foundry are presented with long test sequences containing many redundant and ineffective tests, the formats of the test sequences are unstructured and incompatible with the ATE equipment and the test requirements, for example in terms of speed, stipulated by the designer are outside the capabilities of the test equipment. Invariably when the 'tested' components are returned to the customer and are subsequently found to be defective an argument ensues between the designer and the silicon foundry; however, since testability is an intrinsic characteristic of a design it is ultimately the responsibility of the designer to ensure that the device can be tested adequately.

To alleviate some of these problems several proposals have been made:

1. Test considerations must become a major theme in the design of a circuit and not an afterthought;

2. Test engineers must become involved with a design at the outset and work in conjunction with the designer; the designer through his knowledge of the circuit must provide the test vectors, the test engineer with his knowledge of ATE must integrate them into an efficient test program within the capabilities of the test equipment;

3. Circuit designs must be governed by rules to ensure that the final circuits can be readily tested. Software aids should be available to check if a circuit complies with the design rules;

4. Higher quality in fabrication should be sought to reduce manufacturing defects; and

5. Testing standards should be developed so that test programs are readily transportable across a range of test equipment.

The current climate in the semiconductor industry indicates that it is a buyer's market, hence whether the IC manufacturer is a high volume producer or a silicon foundry it is important to ensure that quality products are delivered to the customer, because once a company has a bad reputation for delivering defective components, it is extremely difficult, in the present climate, for the company to redeem itself.

1.2 A COMPUTER AIDED DESIGN VERIFICATION AND TEST GENERATION SYSTEM

With the increase in the use of ASICs and the adoption of COT to reduce design costs, the roles of designer and test engineer are merging; furthermore, since the problems of test must be considered at the outset of a design there is a growing requirement for an integrated set of design verification and test generation tools. The major components of such a system and their interaction are shown in Fig. 1.1, and comprise the following CAD tools and knowledge bases:

1. simulators to perform empirical design verification
2. fault simulators to determine test coverage
3. automatic test pattern generation programs
4. information on available design-for-testability techniques
5. information on fault models and embryonic tests.

With the increase in the use of expert systems in the design process, Fig. 1.1 also illustrates where these systems could be incorporated to assist designers in overcoming the problems of testing VLSI circuits.

Each of the components in the system is described briefly below, although a deeper discussion on each is given in the appropriate chapters in the book.

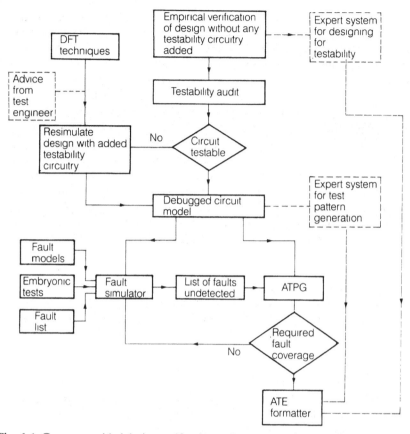

Fig. 1.1 Computer-aided design verification and test generation system.

1.2.1 Empirical design verification and fault simulation [5]–[8]

A major problem confronting designers of complex integrated circuits is that of ensuring the correctness of the logical design of a function prior to manufacture. In the past design verification was performed using the technique of 'breadboarding'; however present day circuit complexities and design techniques, e.g. dynamic logic using charge storage, render this technique impractical. Computer simulation techniques have replaced breadboarding for design verification and can be used at various levels of abstraction. The technique comprises creating a model of the circuit, selecting a set of input patterns which will potentially exercise the circuit through its various functions, and observing the predicted results. The accuracy of the simulation results and subsequently the amount of information obtained about the behaviour of the circuit can be varied by altering the complexity of the models used to represent gates and other functions in the circuit. The continual increase in circuit complexity has been the driving force for further developments in simulation tools to improve speed and accuracy of the technique. Improvements in speed have been obtained by either developing more efficient algorithms, using special purpose hardware to run the simulations or performing the simulation process at higher levels of abstraction. The ability to simulate at higher levels of abstraction is also compatible with top-down design philosophy. However, a problem with this philosophy which has troubled designers is that of checking the integrity of the decomposition from one level of abstraction to another. The difficulty can be resolved to some extent by using the technique of regression testing and mixed-mode simulation, which permits parts of a circuit described at different levels of abstraction to be simulated concurrently. To check the integrity of a decomposition the complete circuit is simulated, first at the higher level of abstraction; a given function is then replaced by a representation of the function at a lower level of abstraction and a mixed-mode simulation is performed with the same inputs applied to the circuit; the output responses from both simulations are subsequently compared.

The types of errors identified during the simulation process can be categorized as database errors and design errors:

(a) Database errors

This class of error includes schematic entry errors, model errors and input stimulus errors. Schematic entry errors are typically the mis-spelling of signal names. The occurrence of simulation model errors, in which the model does not mimic the correct behaviour of a function, is quite high initially but becomes less frequent as the simulator matures. Stimulus or waveform description errors are probably the most common type of database error; this may be a simple typographical error, or may be more serious in that the input waveforms neither fully exercise the circuit nor set up the necessary conditions to perform a given operation although the circuit design is correct.

(b) Design errors

This class of errors includes logic errors and timing errors. Logic design errors occur when a particular connection of gates fail to realize a given logic function; the identification of these errors is the most common use of simulation. Timing errors are the result of incorrect delays in the circuit producing race or hazard conditions; these errors tend to be input dependent and in some instances do not exist in the actual circuit but have been generated by the simulation models. More recently timing verifiers have been developed to analyse critical paths in a circuit; the analysis performed is totally independent of any input conditions.

The simulation process, however, is an empirical method of design verification, since the only valid comment which can be made about the results obtained from a simulation run is that for a given set of inputs the simulation model predicted the correct output response, which does not mean that the design is necessarily fault free. Consequently simulation as a verification tool can only be considered as a first order approximation. In an attempt to improve the quality of the process of proving the logical correctness of a design several techniques, used by computer scientist to prove the correctness of computer programs, have been adopted, namely, symbolic simulation and formal verification techniques. In symbolic simulation the output is a symbolic expression of the input variables and essentially represents the simulation results of a large number of test cases all of which would produce a similar outcome. The requirements of symbolic simulation make the designer think more deeply about a design, hence many conceptual design errors are removed before the simulation is run. Formal verification techniques use a mathematical approach to prove the equivalence between the description of a circuit at two different levels of abstraction, usually the initial specification and the gate level implementation. However, the use of formal verification techniques require the designer to become skilled in the use of theorem provers and the art of inventing 'assertions' or facts about the state of a circuit at various control points, so that comparisons can be made between the different representations. However, these more rigorous approaches to hardware verification are still in their infancy, although it is an active area of research.

A necessary adjunct to the process of test pattern generation is fault simulation, where the output response of the circuit is predicted under fault conditions. Fault simulation permits the designer to determine the fault coverage of sets of test patterns or discover what other faults can be detected by a given test during the test generation process, as a means of reducing test generation costs. Thus fault simulation essentially answers the question of whether, if a given fault occurs in the circuit, the test vectors are sufficiently comprehensive to detect it. Again, with the increase in the use of COT, fault simulation is the only tool generally available to ensure the quality and integrity of the test vectors sent to the silicon foundry to test devices after fabrication. Although fault simulation is closely associated with test pattern generation, the fault simulator is also a useful tool for identifying logical design errors, since if a set of patterns developed to exercise a section of circuitry is run through a fault simulator and a large number of faults in the section of circuitry are flagged as undetectable it may indicate that a logic design error may exist in that part of the circuit or that part of the circuit is redundant; the ability to identify redundant parts of a circuit is particularly

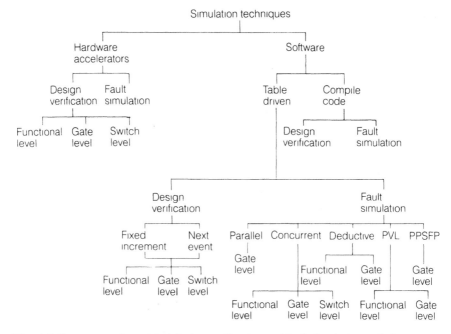

Fig. 1.2 Taxonomy of empirical design verification and fault simulation techniques.

useful in gate array design. Some fault simulators also have a 'fault blocking trace' mechanism which essentially provides the information why a given fault in a circuit cannot be detected by the applied test patterns. However, with the continual increase in circuit complexity, gate level fault simulation, as a means of determining fault coverage, is becoming extremely expensive and alternative approaches are being developed, such as using data extracted from testability analysis, critical path analysers, probabilistic fault simulation techniques etc.

A taxonomy of simulation techniques is shown in Fig. 1.2.

1.2.2 Fault modelling [9]–[11]

Fault models are a means of describing the effects of defects in a circuit which can cause an error in the output of the circuit. The effectiveness of test patterns in detecting faulty devices depends upon the accuracy of the fault models in reflecting the operation of the circuit under fault conditions. A defect in a circuit may cause either a permanent fault or an intermittent fault or may affect the performance of the circuit. However, in screening the circuits after fabrication, testing is performed primarily to detect circuits with permanent fault conditions. In generating test patterns to achieve this objective the fault model used is the stuck-at-fault model, in which it is assumed that the effect of a defect in the circuit is to hold either the input or output of a logic gate permanently at a logic 1 or logic 0. The most identified failures in MOS circuits are open connections, shorts and pin-holes in the gate oxide. A taxonomy of the fault conditions which can cause a circuit to malfunction, both during the design and fabrication phases of a circuit, is shown in Fig. 1.3.

The development of a good fault model is a complex task and requires a knowledge of circuit design, logic design and stuck-at-fault testing

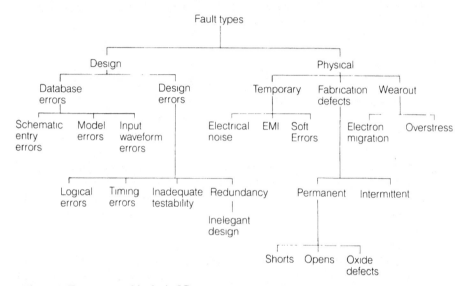

Fig. 1.3 Taxonomy of faults in ICs.

techniques. The modelling process starts by gaining a thorough under-standing at the circuit level of how a given logic function realized, say in CMOS technology, operates and then selecting a set of manufacturing defects which would adversely affect the operation of the circuit. Thereafter, for each defect an exhaustive simulation of the input states, at circuit level is performed and the subsequent output responses tabulated. The defective circuit responses are then compared with the response of a logic model of the faulty circuit in order to identify the undetectable faults and also establish a relationship between the physical defects and logical fault models. The undetected faults may then be analysed; subsequently it may be decided

1. to alter the model to permit the detection of a fault (model iteration)
2. to disregard the defect since the circuit operation is insensitive to it
3. to alter the fabrication process or circuit layout to remove the possibility of its occurrence.

Until recently the stuck-at-fault model was considered to be an adequate representation of a large number of failure modes in integrated circuits, particularly in the bipolar technologies; however defects in the MOS technologies and in particular CMOS can create fault conditions which cannot be represented by the stuck-at-fault model and require a more complex model. However, the use of a more complex model will subsequently have an adverse effect in increasing the cost of fault simulation and test pattern generation. In this respect the requirements for a good fault model are contradictory in that they should be accurate, i.e. model realistically faults in the circuit, and at the same time be tractable, i.e. the model should be simple in order to be applied to complex circuits.

1.2.3 Test pattern generation [12]–[16]

The objective of test pattern generation is to derive input patterns to the circuit which will exercise the circuit in such a way that if any faults are present in the fabricated circuit the output response of the circuit will differ from that of the fault-free circuit. The algorithms used for test pattern generation, in general, produce non-functional tests, in that they exercise individual gates in the circuit to ensure that there are no stuck-at-faults on the input/output of a gate or short circuits, without any regard to the global function performed by the circuit; this type of testing is called *fault oriented*. Inherent in this testing strategy is the assumption that the circuit, via simulation, is shown to be free of logical design errors. Circuit testing can be performed, either explicitly or concurrently. The explicit testing method, which uses specifically generated input patterns, is the most commonly adopted method of detecting faulty devices immediately after fabrication and is run totally independent of the normal operation of the device. However, in concurrent testing, which was introduced primarily as a means of detecting intermittent faults (Chapter 2) during normal field operation, the normal data inputs function as test vectors and the testing and normal functional processes are run simultaneously; concurrent testing is not normally used for fabrication testing.

The test generation process is intimately associated with the fault modelling and fault simulation processes; fault simulation enables the effectiveness of the test patterns to be ascertained, whereas the fault models are the means of simulating the effects of defects in the circuit. The test generation process comprises setting up the necessary conditions in a circuit to provoke a fault condition and subsequently propagate the effect of the fault to an observable output; thus if a given input sequence can neither provoke a particular fault nor propagate the effect of a fault to an output, the input sequence will not detect that fault. A taxonomy of the techniques used in test pattern generation is shown in Fig. 1.4 and these will be discussed in more detail in Chapter 5.

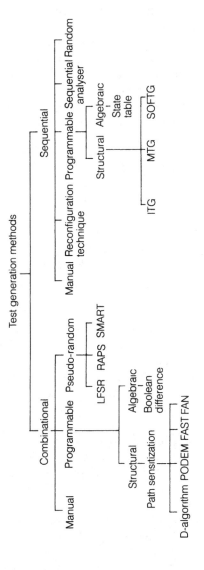

Fig. 1.4 Taxonomy of test pattern generation techniques.

At present most automatic test pattern generation programs use a gate level representation of a circuit, consequently as circuit complexities increase, present test generation methods are becoming inefficient. To improve the efficiency of test generation attempts have been made to perform this process at higher levels of abstraction, in particular functional level. However, functional level testing is considered not to have a good fault coverage [17] and automatic test generation programs operating on functional level descriptions are not readily available. Other solutions which have been proposed as a means of alleviating the problems of testing complex circuits include the use of self-testing techniques, improvements to the fabrication process to eliminate certain defects and also abandoning the requirement to obtain high fault coverages of say 90–95%. The underlying philosophy [2] of this last proposal is that tests should be generated to test about 70% of the nodes in circuit, but the nodes must be scattered about the circuit uniformly, in this way about 70% of the faults in the circuit would be detected. However, fabrication defects tend to create several defects, of which at least one will be detected by the test set; thus the actual number of faulty circuits identified may be greater than that predicted from the value of fault coverage for single faults. Using this philosophy the fault coverage readily obtained using functional tests may be adequate for the fabrication testing of VLSI circuits.

1.2.4 Design for testability [18]–[22]

In an attempt to curtail the increase in the costs of testing VLSI circuits a design style has evolved which eases, to some extent, the problems not only of generating test patterns but also applying the test patterns to the circuit. The design style is classified, broadly, as designing for testability and embraces a number of techniques which range from a simple set of guidelines to formal structured design methods governed by a set of design rules which can subsequently be checked for violations. In general the essence of making a circuit more testable is to eliminate uncontrollable and unobservable nodes in a circuit and also to avoid physical circuit structures which may result in faults which are difficult to test; this later approach is called *physical design for testability*. It is significant that the main driving force for the development of structured design for testability techniques originated from the computer manufacturers, who have access to a vast amount of computing resource and hence can use the sledge-hammer approach to solve the testing problem, indicating a degree of foresight about the problems to be incurred by testing unstructured designs of VLSI complexity. It is also significant that the computer manufacturers proposing structured design techniques had captive fabrication facilities. However, system houses producing complex designs using high volume parts could not gain access to components incorporating structures to enhance testability since the semiconductor manufacturers considered that for a high volume product the cost of the increase in area incurred by incorporating structures to enhance the testability of their circuits was far greater than the reduction in test program generation effort through the use of these structures. However, with the increase in the use of silicon foundries for the production of low volume ASICs, structured design for testability techniques will have to be used since design time and costs are more important than area overheads.

In testing VLSI circuits, the increase in circuit complexity is accompanied by an increase in the volume of test data in terms of input patterns and output responses to be stored and processed. A solution to this problem has been found in the use of built-in test systems which permit the basic elements of a test system to be incorporated into a chip. The use of built-in test systems have the advantage, also, of permitting the circuit to be tested at normal operating clock rates. A taxonomy of the basic design for testability techniques used is shown in Fig. 1.5 and a detailed description of some of the approaches is given in Chapter 6.

A major problem in using the current design for testability techniques is that they are implemented at gate level; however, many VLSI systems are realized using large regular structures and functional blocks – for example ROMs, RAMs, PLAs and microcontrollers – hence there is a requirement to consider design for testability at a higher level of abstraction, which has resulted in the development of testable functions, for example ROMs which have on-line error detection and self-testing PLAs.

A natural extension of built-in self-test techniques is the design of VLSI circuits, which are essentially systems, with a concurrent testing capability; the built-in self-test techniques can only be used to test the circuit off-line (explicitly). Interest in the use of concurrent testing techniques in VLSI circuits has resulted first from the increase in the occurrence of intermittent faults as VLSI circuit densities increase, and second from the increase in the use of VLSI circuits in critical applications. Concurrent testing techniques employ information redundancy (coding techniques) and checking circuits to continuously monitor the circuit during its normal operation; some coding techniques also have the capability not only of error detection but also error correction which enables the circuit to be fault tolerant.

A general taxonomy of basic fault tolerant techniques is shown in Fig. 1.6 and a more detailed taxonomy of self-testing and fault tolerance techniques used with regular functions is shown in Fig. 1.7.

Although it is recognized that some kind of design for testability techniques must be used in testing VLSI circuits there is still some opposition [22] to its use, for example:

1. Some designers consider that testing is not a design problem, any problem in this area must be solved by the test engineer;

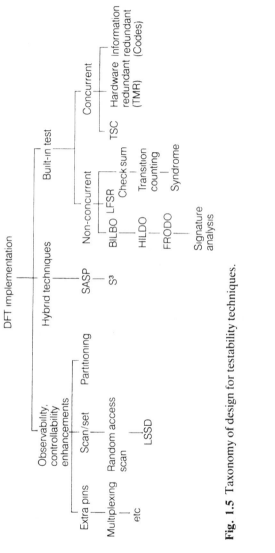

Fig. 1.5 Taxonomy of design for testability techniques.

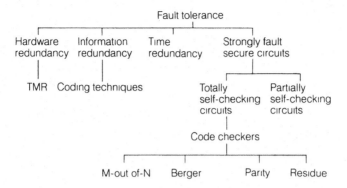

Fig. 1.6 Taxonomy of basic fault tolerance techniques.

2. The use of design for testability techniques is always perceived in terms of its disadvantages – namely increase in circuit area, increase in I/O pin count etc., since these are easily quantified – and not in terms of its advantages, for example, reduced test program generation times, increase in fault coverage etc., which are more difficult to quantify; and
3. There is also a naïve management attitude which assumes that since engineers have managed to adequately test complex systems in the past without recourse to design for testability techniques, they will continue to do so in the future.

Some semiconductor manufacturers, although using design for testability techniques in their devices, will not document this information for use by the customers, since the legal departments in companies have not found a way to adequately describe the function of the testability enhancement structures to avoid possible prosecution over product liability in the event of the device failing. Furthermore supplying information on how the device is tested to a customer essentially involves the manufacturer in an unwanted support role, since information will have to be given to the customer on the adequacy of existing test data when applied to a redesigned part.

1.3 ECONOMICS OF TEST [23]–[26]

A major obstacle to the pervasive use of VLSI is the cost of testing these devices; consequently several surveys have been carried out to assess the various cost factors. The analyses performed assume, in general, that the circuits are combinational or can be reconfigured into a combinational circuit for the purpose of test generation. Results have shown that test generation times increase as the square of circuit complexity, assuming that a path

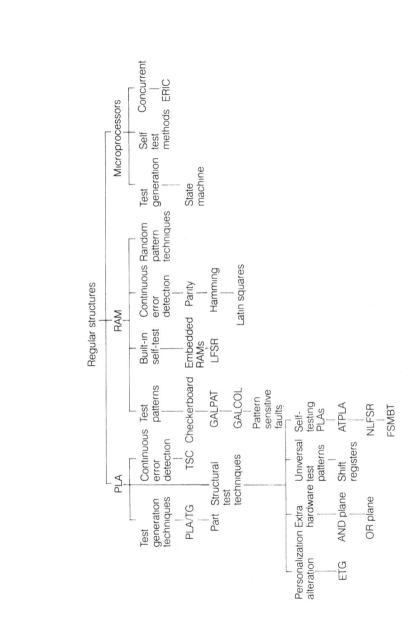

Fig. 1.7 Taxonomy of self-test and fault tolerance techniques used on regular structures.

sensitization algorithm is used and that the amount of backtracking to resolve inconsistencies is negligible; fault simulation time, which is a necessary adjunct to test pattern generation, increases as either the square or the cube of the circuit complexity depending upon the technique used. It was also noted that a small number of patterns produced at the start of a test generation process have a high fault coverage (65–85%), the tests generated subsequently to detect specific undetected faults have little success in detecting any other remaining faults. Consequently, most test pattern generation programs adopt a two-phase test strategy comprising a set of 'global' tests and a set of 'clean up' tests. The global tests, which initially detect a large number of faults, comprise pseudo-random binary sequences; the cost of generating these sequences is negligible although fault simulation must be used to determine the fault coverage; a 'clean-up' test set, using some deterministic test generation technique, is then invoked to produce patterns to detect the remaining faults.

The actual cost of testing a device is a function not only of the cost of developing the test program but also the cost of the Automatic Test Equipment (ATE), which includes tester operation costs, ownership costs, maintenance etc. Thus the cost of testing a device is obtained from the cost of generating the test program divided by the number of devices, plus the cost of the ATE divided by the tester throughput; the tester throughput is an average value since some devices will be rejected after a few tests whilst others will undergo the complete test sequence.

The cost of test program development is not simply the cost of test generation and fault simulation, but also includes the cost of device characterization tests, particularly if the device under test is a new design being fabricated on a new process. From the results derived from the characterization tests, limits (i.e. go/no go values) are set up for the parametric tests used on a full production test run.

A major cost factor in testing devices is the initial capital outlay on test equipment, a common mistake is to purchase overly expensive equipment for a given test environment. In order to avoid buying a test system which is more expensive than required, the tester is divided into its major component parts, the costs of performing individual sections of a test program are then determined and the sections of the tester used in each instance are identified; the degree of sophistication required by each section of the tester is then decided upon with respect to the types and numbers of devices to be tested.

Another factor to be considered in evaluating testing costs is the effect of incorporating design for testability techniques in a circuit. A number of test cases have been analysed in which the cost of testing a circuit without any testability enhancements was compared with a circuit incorporating a structured design for testability technique and another with self-test structures. In the analysis the variables considered were the volume of components manufactured and the cost of the area overheads incurred by the testability enhancement circuitry. It was concluded that for high volume components it was uneconomical to incorporate testability enhancement structures, in general, since the reduction in yield resulting from the increase in circuit area significantly increased manufacturing costs. Furthermore the additional test circuitry increased the occurrence of early life failures of devices in the field, subsequently increasing field repair costs. However, for low volume production it was concluded that self-test methods were highly cost effective, particularly when the testing costs at component level, board level, system level and in an operational environment are considered, even when area overheads are high; this is encouraging with the increase in requirement for low volume ASICs.

It was also concluded that, provided extensive burn-in testing was performed, self-testing techniques could be used economically in high volume devices which were used in applications where 'down-time' is expensive. Furthermore, since the current processing capability of semiconductor manufacturers can produce more complex circuits than those submitted by designers for fabrication, some of this excess capability may be utilized in incorporating additional functions to either explicitly or concurrently test the chip.

1.4 REFERENCES

1. Eichelberger, E. B. and Lindbloom, E. (1983) Trends in VLSI testing. VLSI '83, Elsevier Science Publishers.
2. Gutrel, F. (1984) In pursuit of the one month chip: testing. *IEEE Spectrum*, September, 40–6.
3. Hnatek, E. R. and Wilson, B. R. (1985) Practical considerations in testing semicustom and custom ICs. *VLSI Design*, March, 20–42.
4. Wilson, B. R. and Hnatek, E. R. (1984) Problems encountered in developing VLSI test programs for COT. *Digest of Papers, 1984 International Test Conference*, October, 778–88.
5. Breuer, M. A. and Parker, A. C. (1981) Digital system simulation: current status and future trends. *18th Design Automation Conference Proceedings*, June, 269–275.
6. Henckels, L. P. and Haas, R. M. (1975) Hardware simulation or software simulation – A comparison between the two techniques for digital testing. *IEEE Proceedings ISCAS' 75*. June, 355–8.
7. Levendel, Y. H. and Schwartz, W. C. (1978) Impact of LSI on logic simulation. *Digest of Papers, COMPCON '78*, Spring, 102–19.
8. Babitz, A. and Lender, K. (1984) Using simulation in the design process – a case study. *Digest of Papers, 1984 International Test Conference*, October, 229–36.
9. Beh, C. C., Arya, K. H., Radka, C. E. and Torku, K. E. (1982) Do stuck fault models reflect manufacturing defects? *Digest of Papers, 1982 International Test Conference*, November, 35–42.

10. Timoc, C., Buehler, M., Griswold, T. *et al.* (1983) Logical models of physical failures. *Digest of Papers, 1983 International Test Conference*, October, 546–53.
11. Abraham, J. A. and Fuchs, W. K. (1986) Fault and error models for VLSI. *Proc. IEEE*, **74**(5), 639–54.
12. Stieglitz, C. B. (1975) An LSI test overview. *Digest of Papers, COMPCON '78*, Spring, 97–101.
13. Muehldorf, E. I. and Savkar, A. D. (1981) LSI logic testing – an overview, *IEEE Trans. Computers*, **C-30**(1), 1–17.
14. Siewiorek, D. P. and Lai, L. K-W. (1981) Testing of digital systems. *Proc. IEEE*, **69**(10), 1321–33.
15. Rasmussen, R. A. (1982) Automated testing of LSI. *Computer*, March, 69–78.
16. Abadir, M. S. and Reghbati, H. K. (1983) LSI testing techniques. *IEEE Micro*, February, 34–51.
17. Bottorff, P. S. (1981) Functional testing folklore and fact. *Digest of Papers, 1981 International Test Conference*, October, 463–4.
18. Grason, J. and Nagle, A. W. (1980) Digital test generation and design for test-ability. *17th Design Automation Conference Proceedings*, June, 175–89.
19. Segers, M. T. M. (1982) Impact of testing on VLSI design methods, *IEEE J. Solid State Circuits*, **SC-17**(3), 481–6.
20. Williams, T. W. and Parker, K. P. (1983) Design for testability – a survey, *Proc. IEEE*, **71**(1), 98–112.
21. Abraham, J. A. (1983) Design for testability, *1983 Custom Integrated Circuits Conference Proceedings*, September, 278–83.
22. Parker, K. P. (1986) Testability: barriers to acceptance, *IEEE Design and Test of Computers*, **3**(5), October, 11–15.
23. Goel, P. (1980) Test generation costs analysis and projections. *17th Design Automation Conference Proceedings*, June, 77–84.
24. Huston, R. E. (1983) An analysis of ATE testing costs, *Digest of Papers, 1983 International Test Conference*, October, 396–411.
25. McMinn, C. (1983) The impact of a VLSI test system on test throughput equation, *Digest of Papers, 1983 International Test Conference*, October, 354–61.
26. Varma, P., Ambler, A. P. and Barker, K. (1984) An analysis of the economics of self-test, *Digest of Papers, 1984 International Test Conference*, October, 20–36.

2
FAULTS AND FAULT MODELS IN VLSI CIRCUITS

2.1 INTRODUCTION

The testing problem is one of identifying those chips that do not meet their functional specifications. A chip can fail a test for numerous reasons, such as

1. *Physical faults caused by the manufacturing process.* These types of faults may include such problems as a break in a conductor or a short between conductors;
2. *Problems due to the ageing of a component.* This may occur due to mechanical overstressing i.e. excessive temperatures or electrically i.e. excessive voltages. These effects can lead to corrosion of the aluminium conductors or charge build up on the gate oxide.

Once the above faults have occurred they are generally permanent. However, there is another set of faults which are only temporary in nature.

The temporary faults are far more difficult to detect since they occur randomly. Consequently models for these faults are more complex than the permanent fault models. Typical temporary faults include

1. intermittent faults due to environmental conditions such as electrical noise
2. transient faults due to electromagnetic interference
3. soft errors caused by the interaction of charged particles with the silicon substrate.

In order to generate a set of input patterns to detect a fault it is necessary to have a representation of the fault, i.e. a fault model, which can be used to simulate its effect. The output response of a circuit to a given fault can subsequently be predicted by incorporating these models into a class of simulators called fault simulators. The fault simulator can be used on a circuit described at various levels of abstraction i.e. transistor, gate or functional level. Fault simulation was developed to predict the fault coverage of a set of test patterns i.e. the ratio of detected to undetected faults in a circuit. The accuracy of this prediction is dependent upon the accuracy with which the fault models represent actual failure modes within a circuit. Several of the fault models currently used in practice will be described.

However, before discussing these fault mechanisms and fault models in detail the effect of scaling on VLSI device characteristics will be described. This is important as many of the failure mechanisms will become more prevalent as device dimensions are reduced.

2.2 THE EFFECT OF SCALING ON VLSI CIRCUITS

In order to sustain the growth in complexity of VLSI circuits it is necessary to reduce the dimensions of the active devices used in current technologies. This reduction in device dimensions is referred to as scaling down and can be implemented in three ways [1]. Figure 2.1 shows the layout of a typical MOS transistor with its important physical dimensions marked. Each of the techniques to be described generally involves the reduction of certain transistor parameters by a scaling factor; Table 2.1 details the three scaling techniques and their effect on the physical characteristics of the transistor.

The first scaling technique to be discussed, proposed by Dennard *et al.* [2] in 1974, is known as constant field scaling. The aim of this scaling method is to keep the physical characteristics of the scaled transistors similar to that of the

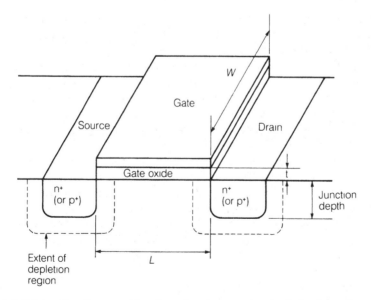

Fig. 2.1 Schematic diagram showing the important physical dimensions and characteristics of an FET.

Table 2.1 Table to show the effect of scaling on device parameters (compiled from Refs [2], [3])

Device parameter	Constant field scaling	Constant voltage scaling	Generalized scaling theory @ 300 K
Dimensions (L, W, t)	$1/\alpha$	$1/\alpha$	$1/\alpha$
Supply voltage	$1/\alpha$	1	$1/\beta$
Doping concentration	α	α^2	α^2/β
Electric field	1	α	α/β
Gate capacitance	$1/\alpha$	$1/\alpha$	$1/\alpha$
Current	$1/\alpha$	α	α/β^2
Power dissipation	$1/\alpha^2$	α	α/β^3
Power-speed product	$1/\alpha^3$	$1/\alpha$	$1/\alpha\beta^2$
Conductor resistance	α	α	α
Current density	α	α^3	α^3/β^2
Power density	1	α^3	α^3/β^3

unscaled devices whose characteristics are well known. The technique involved the scaling down by α of all the horizontal (e.g. L and W) and vertical dimensions (e.g. oxide thickness and junction depth), as well as the supply voltage. Since the supply voltage and gate oxide thickness are scaled by the same factor the electric field in the gate region will remain constant. In order to keep the depletion layer widths in proportion to the new dimensions the doping concentrations were also scaled up by α. The turn-on threshold voltage is decreased by α, mainly due to the decreased gate oxide thickness, therefore the reduced voltages can still operate the new transistor structures. The effects of this type of scaling on device parameters are shown in the constant field column of Table 2.1. The technique was used to produce 1 μm channel length devices ($L = 1\,\mu$m). A drawback to this method, however, is that the current density is increased by α, which can lead to an increased probability of electromigration; however, because of the reduction in area by α^2 it is possible to place more devices on a chip.

The second scaling technique involves keeping the voltage level at its usual value of 5 V in order to allow existing logic families to interface to the new scaled devices. As can be seen from Table 2.1 (constant voltage) not only does the current density problem increase by α^3 but the power dissipation per device is also increased by α, both of these effects are obviously undesirable. This type of scaling could accelerate ageing due to hot electron injection because of the larger electric field across the gate oxide.

Many of these limitations have been surmounted by new scaling theories. In particular the Generalized Scaling Theory (GST) proposed by Baccarani *et al.* [3]. The GST uses two independent scaling parameters, one for physical dimensions (α) and one for voltage (β), in practice $\alpha > \beta$. The independent scaling of both parameters allows the temperature variation of threshold voltage and the non-scalability of built-in potentials to be taken into account. The inability to scale the built-in voltages can lead to shorter than expected channel widths due to the larger than expected depletion regions; this means that doping concentrations cannot be scaled linearly. The GST is intended to scale dimensions to below 0·25 μm and the required scaling factors are quoted at two temperatures, 77 K and 300 K due to the differences in device characteristics at these different temperatures. The scaling parameters are given for 300 K in Table 2.1. However, even using this technique the current density and power dissipation are still likely to increase.

In general therefore as devices are scaled down the current density and power dissipation problems will increase, this will lead to enhanced electromigration hazards, higher junction temperatures and increased electric fields when using constant voltage and GST scaling, with a consequent increase in hot electron effects. As the gate oxide (dielectric) is scaled down manufacturing problems arise in growing such thin layers, this could lead to reliability problems; however, the dielectric thickness may be increased by the use of an ion implant [2] in the gate region. All of the above characteristics lead to difficulties as devices are scaled down. However, scaling down does allow an increase in the number of devices on a chip and an increase in the speed of individual transistors. Although the increase in transistor speed may be offset by the fact that the delays due to signal interconnects remain constant, this can lead to signal skew, which may cause complications when testing and operating a device.

2.3 PERMANENT FAULTS

The types of permanent fault to be discussed are those due to fabrication defects and ageing or wearout mechanisms. Both types of fault need to be considered when devising fault models. However, the faults due to ageing give rise to reliability problems when in the field since they can occur over a period of time, whereas the fabrication faults need to be detected by tests after manufacture. Therefore test strategies need to be applicable to the two different testing aspects: one for post production test and the other for service or maintenance testing.

2.3.1 Fabrication faults

According to Mangir and Avizienis [4] the normal modes of failure of most VLSI chips can be categorized as follows.

(a) Oxide defects

Pinholes in the oxide layer allow two conducting layers not otherwise connected to touch. The photographs in Figs 2.2 and 2.3 show two examples of this kind of fault. The photographs were taken during an investigation into the causes of errors in a 6 μm line width NMOS process. Figure 2.2 shows a pinhole in the oxide layer allowing a ground line to short onto the input of an output pad. Figure 2.3 shows a pinhole in the oxide layer that has allowed a power line to connect via a polysilicon line to another metal signal line. This fault was observable since the output from the affected part of the circuit was always a logic 1. Reliability problems may also occur with the gate oxide layer in scaled-down MOS devices.

(b) Implant layer defects

Improper masking may lead to missing or incorrectly dimensioned implant regions. This can affect the threshold voltages of certain types of devices. In an NMOS process this can alter the operation of the pull-up transistors.

(c) Contamination

This may be due to clean room contamination or bad handling. One particular form of contamination is due to ions, such as sodium ions, trapped in either the thick or thin oxide layer [5,6]. If positive ions are trapped in the thin oxide of an n-channel FET then the threshold voltage will be reduced, making the device very difficult to turn off; similarly the threshold of a p-channel device will be increased, making the transistor very difficult to turn on. However, if the ion contamination occurs under a conductor in the thick oxide region then a parasitic transistor may occur allowing otherwise unconnected structures to conduct together. Figure 2.4 illustrates these two failure mechanisms for sodium ion contamination.

(d) Metal layer defects

These could include surface scratches due to improper handling or bad etching. A bad etch may allow metal conductors to remain in contact when in fact they should be separate; this would produce a bridging fault.

(e) Interconnect defects

These may include shorts or opens in the polysilicon or diffusion layers of a typical MOS process. Figures 2.5 and 2.6 show two examples of this effect. Figure 2.5 shows a break in a polysilicon clock line which has allowed

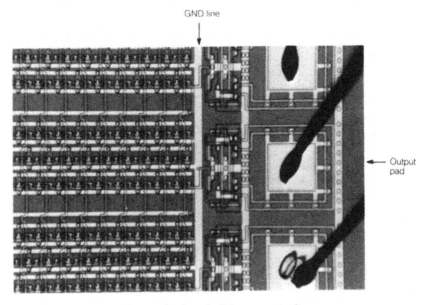

Fig. 2.2 Pinhole in the oxide on the 'input' of the output pad.

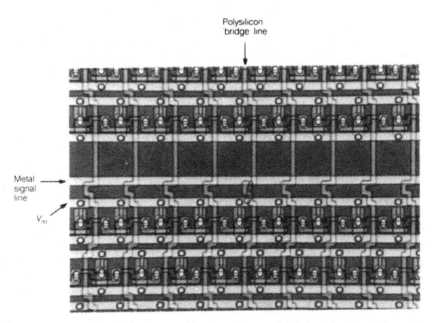

Fig. 2.3 Pinhole in the oxide causing a bridge between V_{dd} and a metal signal line.

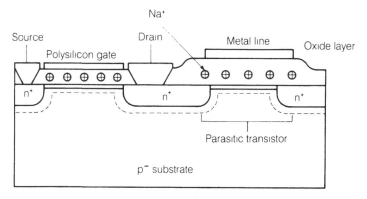

Fig. 2.4 Positive ion contamination of an n-channel transistor (adapted from [5]).

the lower portion of the line to float. The photograph in Fig. 2.6 shows two broken polysilicon lines on an NMOS output pad.

(f) Contact defects

These faults may result in a contact between two layers which is of very high ohmic resistance. A particular form of this fault is due to the fact that silicon is soluble in aluminium, which is used in device manufacture, especially at the temperatures to be found in modern processes. As a consequence the aluminium may penetrate the silicon layer as small 'spikes'. This can degrade the shallow contact junctions found in current VLSI processing forming non-ohmic contacts to the n^+ regions of an n-channel device, as aluminium is a p-type dopant. In severe cases the p–n junction formed by a contact diffusion may be penetrated if narrow contact windows are used, as may be found in scaled-down MOS devices, thus shorting the conductor to the substrate, as shown in Fig. 2.7. A possible solution to this problem involves the use of an extra barrier layer to prevent the aluminium from forming a direct contact to the silicon [7].

(g) Piping [8]

This type of fault is limited to bipolar technologies and is due to the diffusion of impurities along dislocations formed in the structure of the silicon. These impurities can form a low resistance path between the emitter and the collector, via the base region of the transistor, should they occur within the transistor structure. When under test this fault leads to high leakage currents, which can be detected by a change in the expected logic state of the device.

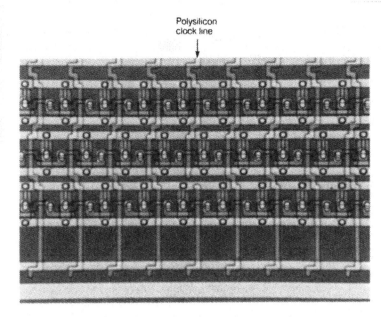

Fig. 2.5 Break in a polysilicon conductor.

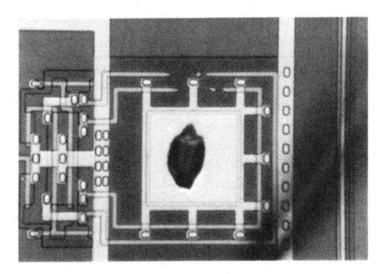

Fig. 2.6 Break in polysilicon lines around an output pad.

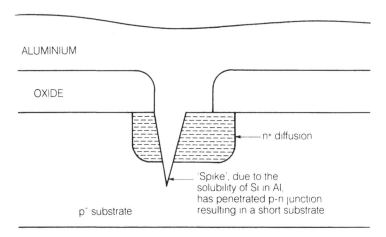

Fig. 2.7 Aluminium 'spiking' in MOS VLSI structures.

(h) Design defects

These may result in device length and widths deviating from those specified for a process. This type of fault is difficult to detect but may lead to reliability problems when the chips are in service. For example a thin metal conductor will be more susceptible to electromigration (covered later in this chapter).

Any of the faults listed above can occur when fabricating a VLSI device although most faults are caused by breaks in conductors or shorts between conductors. It is possible for any of these faults to be randomly distributed across the chip surface due to the effects of incorrect processing [9]. The effects that these faults have on the logical function of the chip will be discussed in Section 2.3.3.

2.3.2 Failure mechanisms due to wearout

Once a chip has been fabricated and initially tested there are still many mechanisms which can cause it to fail. The failure can be caused by hot carrier injection into the gate oxide, oxide breakdown, metallization failures or corrosion. The next few sections will describe some of these ageing or wearout mechanisms:

(a) Hot carrier injection [1], [10]

The MOSFET transistor currently used in VLSI designs has two very important characteristics, threshold voltage and transconductance. Changing just one of these parameters can alter the performance of a circuit. If charge carriers are introduced onto the gate dielectric then the threshold voltage will be changed and the transconductance will be degraded, as has been observed with ion contamination discussed in the previous section. Another mechanism which can bring about the charging of the gate oxide is hot carrier injection. In an NMOS process the carriers in question are usually electrons. These hot electrons are of three different types depending upon where they originate [6]:

(i) CHANNEL HOT ELECTRONS are produced by the electrons traversing from the source to the drain gaining enough energy to penetrate into the gate oxide.

(ii) SUBSTRATE HOT ELECTRONS are thermally generated in the substrate below the inversion layer and are attracted to the gate oxide by the vertical electric field.

(iii) AVALANCHE HOT CARRIERS are formed by impact ionization due to strong lateral electric fields. In this case holes or electrons can be injected into the gate oxide region.

The electrons must have a high energy (hot) in order to surmount the silicon–silicon dioxide energy barrier and penetrate into the gate oxide, placing negative charges within the gate region or producing a leakage current.

It is worth noting that effects (i) and (iii) are due to high electric fields between the source and drain. However, as discussed previously, scaling down the dimensions is likely to increase this field not only due to the reduction in the channel length but also due to the voltages remaining almost constant. Therefore as devices are scaled the probability of hot carrier injection occurring is likely to increase. These effects are also observable in p-channel devices but to a lesser extent because of lower probability of hole injection.

(b) Oxide breakdown [1]

It is thought that one mechanism that could lead to oxide breakdown is the injection of electrons into the gate oxide. If electrons, when entering the gate oxide, have sufficient energy they can create electron–hole pairs by impact ionization. The electrons created by this mechanism are quickly collected by the gate (in an n-channel device). This will leave behind the slower moving holes. A net positive charge is therefore left on the gate, which causes a larger electron current leading to increased electron–hole pair formation, eventually causing oxide breakdown. Currently 50% [1] of all MOS failures are accounted for by these mechanisms, and their importance is increasing due to the difficulties of growing the thinner oxides required by the scaling down process.

(c) Metallization failures

As has already been discussed the scaling procedure for MOS devices increases the current density leading to rising electromigration hazards [1,7,11]. The atoms in a metal are subjected to two different forces; the electrostatic force due to the voltage drop along the conductor and the interaction of the atoms with the electrons driven by the electric field (this is sometimes referred to as the 'electron wind' force). The latter force tends to dominate for the materials used in integrated circuit manufacture. As the electrons collide with the metal atoms they move towards the positive end of the conductor forming hillocks and whiskers while the vacancies left by the atoms move in the opposite direction, forming voids. This has two possible consequences. As the voids begin to grow they cause an increase in the current density which leads to further metal atom migration and hence, eventually, to a complete break in the conductor, causing an open circuit. Alternatively as the material builds up at the positive end of the conductor it may cause a short to occur between itself and an adjacent signal line. These effects are enhanced by any local imperfections such as grain boundaries or scratches on the metal layer due to bad handling.

Obviously as scaling takes place the current density in the conductors will be increased; this will lead to further increases in electromigration hazards. However, a recent discovery [7] may alleviate this problem as VLSI dimensions are scaled down further. If the line width of a conductor is decreased the metal structure becomes composed of single crystal segments which are resistant to electromigration, therefore the mean time to failure of such conductors is actually increased.

Another major effect of electromigration is due to the solubility of silicon in aluminium. The dissolved silicon can also be moved by the electron wind force to the positive end of the conductor. This allows 'spiking' to occur as the silicon is moved away from the contact and the aluminium moves into the voids left by the silicon. This effect cannot be prevented by the presolution of silicon into the aluminium. Therefore special metals such as platinum etc. have to be deposited on the junction before the aluminium [7]; this also prevents aluminium spiking due to the high process temperatures.

(d) Corrosion [1], [7]

Another metallization related problem is that of aluminium corrosion due to galvanic or electrolytic action. Galvanic corrosion may occur due to the bonding of gold wires to aluminium pads producing the so-called 'purple plague'. The intermetallic compounds formed between the gold and aluminium can produce bond embrittlement or voiding, leading to bond breakages and open circuits [12]. This can be controlled to some extent by reducing the length of time that the high temperature bond takes place. Alternatively aluminium bond wires could be used, although they are not as ductile as the gold wire and are therefore more prone to fractures. Another possible solution is to anneal the bonds in a hydrogen atmosphere to reduce the amount of intermetallic compound formed [12].

Electrolytic decay occurs due to the impurities in the overglass used to protect the integrated circuits surface from contaminants. If chlorine is present with a small amount of moisture then the aluminium can be attacked, forming aluminium hydroxide and eventually producing an open circuit, leaving the chlorine free to attack more aluminium. This effect is increased by high electric fields between conductors, a situation that will be found in high density VLSI designs. Phosphorus impurities can also lead to corrosion of aluminium conductors; this element may be present if phosphosilicate glass has been used within the structure, i.e. for passivation.

2.3.3 Test generation models for permanent faults

To determine the quality of a test set it is necessary to have a method of quantifying in a meaningful way how good the fault coverage is, for a particular circuit, from a fault simulator. The two preceding sections have demonstrated how wide ranging and different the set of faults can be for a VLSI process. The requirement, therefore, is for a means of mapping the most serious of these faults into a fault model or models that can be used in a simulator to determine if they can be detected by a set of test patterns. An important point to note is that not all physical faults are equally probable [4].

There are three possible fault models [13] that can be used when generating test patterns. These are the transistor level fault models, gate level fault models and functional level fault models. Of all these modelling techniques the gate level fault models are the most popular and consequently most test pattern generation schemes are based on this model. However, for completeness the transistor and functional models will be briefly described. A more detailed study of gate level models will be made in Chapter 3.

(a) Transistor level fault models [13]

This is the lowest level of abstraction as it is used to represent faults directly at the transistor level. The type of failures represented can vary between the different models depending upon the capabilities of the simulator. Usually shorts and opens are taken into account on most models; however, for analogue work degradations in the performance of the transistors or associated components may also be included. Obviously as more modes of failure are included the complexity of the simulator increases and hence the amount of time required to run the fault simulation also increases. Two transistor level models that have been developed by Schuster and Bryant [14] and Hayes [15] use slightly different techniques to represent the failure modes.

The method proposed by Schuster and Bryant uses extra 'fault transistors' to introduce failure modes into a fault-free circuit. The circuit itself is also made up of similar transistors. The fault transistors are placed between the nodes in the circuit for which faults are to be modelled. The gates of the fault transistors are then used as 'fault inputs' to the circuit (for more detailed explanation refer to Chapter 3). The transistors in the circuit have associated with them a strength based on the conductance of the transistor. The strength value can be different for each transistor depending upon where it is used; this scheme allows for ratioed logic to be simulated; however, certain logic types, such as CMOS, can be modelled with only one transistor strength since they do not rely on transistor ratios [15]. Six types of fault can be modelled using this method; these include nodes directly connected to ground and V_{dd}, transistor stuck open or closed, and shorts and opens between two nodes. To produce the fault condition in the circuit it is simply a matter of turning on or off the fault transistor in question, for short circuit and open faults respectively. Since all the fault inputs are individually controlled it is possible to introduce single or multiple fault conditions. This fault model was used to successfully produce FMOSSIM [16], a concurrent transistor level fault simulator. However, in FMOSSIM, stuck-at nodes and stuck-at transistors are modelled without the need for extra fault transistors. Another simulator called MOSSIM [17], based on a similar level of abstraction and produced by Bryant, allows the bidirectional and charge storage abilities of MOSFET transistors to be used directly in the simulation of the device, allowing a better response to possible faults. This information is not available to the gate level models to be discussed next. Hayes models the faults at the transistor level by representing the circuit as a set of connections, switches and attenuators. The circuit is then built out of these special elements before simulation takes place.

The main drawback with these two models is that they cannot easily be used to model faults within the transistors themselves, such as degraded performance; also they are incapable of representing voltages that are midway between V_{dd} and ground. This is partly due to the models representing the transistors in the circuit as a switch rather than as true devices with transconductance etc.

However, a major problem with all of the transistor models is that as VLSI densities of $>100\,000$ transistors are approached the length of time needed for a simulation to produce reasonable results becomes excessive. Therefore higher level models are required in order to speed up the simulation times; however, the quality of the model used must not suffer as a consequence.

(b) Gate level fault models

The most popular fault model used in gate level simulation is the stuck-at fault. The stuck-at-fault model was originally used as a means of describing faults in early electromagnetic relay computers. However, the model was also found to be applicable to diode transistor logic (DTL); this led to its use in small scale integration (SSI) and medium scale integration (MSI) fault modelling. Thus the model became a standard widely used in the integrated circuit industry [18]. When Roth [19] developed the D-algorithm, in 1966 to automatically generate test sets based on the stuck-at-fault model its continuation was assured. However, as failure modes in modern VLSI circuits are better understood its applicability and usefulness are being challenged [9].

The stuck-at-fault model assumes that a logic gate input or output is permanently set to either a logic 0 or 1 value [18] for the purposes of test generation. Figure 2.8(a) shows a fault-free three-input NMOS NOR gate. Figure 2.8(b) shows a NOR gate with one of its inputs (W) stuck at 0 (s-a-0). The s-a-0 fault on the W input of the NOR gate produces a 1 output. This is the inverse of what would be expected with the pattern shown applied to the NOR gate. Therefore the gate sees an input pattern of 000 instead of 100 due

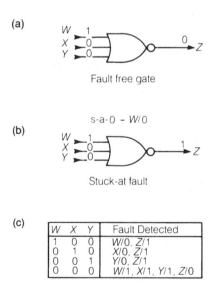

Fig. 2.8 The stuck-at-fault model and equivalencing: (a) fault-free gate; (b) stuck-at fault; (c) table of faults to be detected.

to the stuck-at-fault condition. Consequently the application of the 100 pattern has detected the fault since the output is different for the faulty and fault-free cases. Figure 2.8(c) shows the input patterns necessary to detect particular faults (the stuck-at fault is represented by signal name/X where $X = 0$ or 1 representing s-a-0 or s-a-1 respectively). From this table it can be seen that there are a total of eight faults (i.e. two faults per signal line). However, not all faults are distinguishable, for example Z s-a-0 and W or X or Y s-a-1. These faults are said to be equivalent; this allows the fault simulator using this model to reduce the total number of faults for which it must perform a simulation, since it can combine these equivalent faults into a single fault set. In general most fault simulators only perform simulations for single stuck-at faults since if multiple faults were considered the number of fault simulations required would be of the order of 3^n (where n is the total number of signal nets). The relationship between faults and fault models is obvious for the transistor level model; however, the relationship between the gate level model and the physical failure mechanisms is a little more obscure.

Figure 2.9 shows the transistor connections required to perform the NOR operation in NMOS. X and Y are the two inputs and Z is the output. The figure also shows two faults. Fault (1) represents an open in a diffusion line on the X side. This fault is logically equivalent to the X input s-a-0 since it appears that the gate is incapable of being driven to logic 0. Fault (2) is a short causing the source and drain of the Y transistor to be connected together. This fault is equivalent to either Y s-a-1, i.e. permanently on, or the output Z s-a-0, as shown. Again this shows fault equivalencing, since only one fault needs to be considered from Y s-a-1 and Z s-a-0. However, there is a problem: it is not

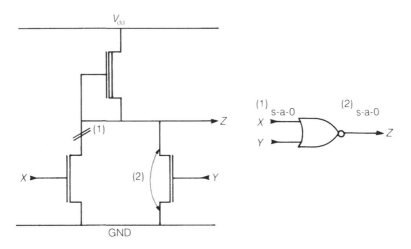

Fig. 2.9 An NMOS NOR gate to illustrate the stuck-at-fault model.

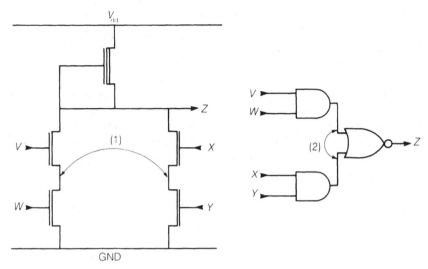

Fig. 2.10 An NMOS circuit containing unrepresentable faults using the stuck-at-fault model.

always possible to represent all probable faults on the physical layout by faults on the logic diagram; the reverse is also true, faults that could be placed on the gate level model are not always likely to occur at the transistor level [20]. Figure 2.10 illustrates this problem. The fault labelled (1), a short, has no logical realization in the gate level circuit. This is due to the fault altering the logical behaviour of the circuit. In this case it would be necessary to test the structure of the gate rather than testing for the gate output stuck at 1/0. Similarly fault (2) in the logic diagram has no physical realization on the layout. There are two possible solutions to this problem. The most obvious is to use the transistor level models already discussed. Alternatively the layout of the transistors on the silicon should be such that it is only possible for stuck-at-type faults to occur [20]. The main drawback with these 'physical design rules' is that in general the size of a particular circuit can be increased by up to 50% in some cases.

Timoc *et al.* [8] also suggest that faults such as ion contamination and hot carrier injection can be modelled as stuck-at-type faults. This is due to carriers present on the gate oxide of a transistor turning the device on or off permanently, hence the input to that device may appear stuck at 1 or 0. A faulty piece of interconnect or degradation in the transconductance of the transistor can affect the speed of operation of the device; this is referred to as a delay fault. Under normal test pattern application this fault may not be easy to detect. However, if the rate at which test patterns are applied is increased then it may be possible for this fault to manifest itself as a stuck-at-type fault.

Bridging faults are also covered by the stuck-at-fault model [8], [13], [21]. Mei has demonstrated that certain types of bridging faults between two inputs of an elementary gate can be detected by a normal stuck-at-fault set. Timoc *et al.* [8] extend this idea to 'asymmetrical' bridging faults on the outputs of logic gates. If the outputs of two gates are shorted then according to Timoc there are four possible results depending upon the resistance ratios of the pull-up and pull-down transistors of the logic gates. The structures can be represented as asymmetrical wired-ANDs or as asymmetrical wired-ORs.

Although the stuck-at-fault model is able to cover almost all the faults in an NMOS design it is only capable of describing some of the faults in CMOS designs. A particular fault not covered by the model can turn a combinational

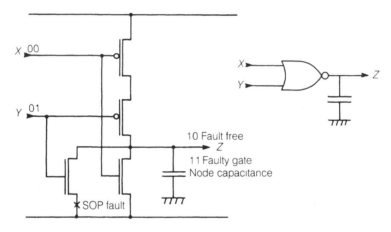

Fig. 2.11 Typical CMOS NOR gate illustrating the stuck-open fault.

CMOS circuit into a sequential circuit; it is called the stuck-open (SOP) fault [22]-[25]. To understand how this type of fault occurs it is necessary to describe the operation of a CMOS gate. Figure 2.11 shows the layout of a simple CMOS NOR gate. The NOR gate consists of two p-channel transistors connected in series and two n-channel devices in parallel. The p-channel devices are turned on by a logic 0 and the n-channel devices are turned on by a logic 1 value. In the case of the NOR gate if both inputs are at a logic 0 then the p-channel network will pull the output up to logic 1 and the node capacitance will be charged up. However, if either of the inputs is logic 1 then the conducting path will be through the n-channel device that is turned on and the node capacitance will be discharged to logic zero. Therefore the logic gate performs the NOR operation as expected. If an open circuit is now introduced at the point marked in Fig. 2.11 this will have the effect of disconnecting the Y-input n-channel device from the ground line. (An SOP fault may occur anywhere in the CMOS gate due to open interconnect layers; this fault has been chosen purely for illustration purposes.) If the node capacitance is now charged up by the inputs X and Y being set to 00 and then input Y is changed to logic 1 the output will remain at its previous state, logic 1, as it will not be discharged by the n-channel device. As the output remained at logic 1 the fault would be detected, the application of the 01 pattern normally producing a 0 output, therefore the faulty and fault-free behaviour are different. However, if the output had been 0 before the application of the 01 pattern then the fault would have gone undetected. Therefore the SOP fault has now given the CMOS gate a memory state since in certain situations the output of the gate depends on the previous inputs. In order to detect this fault it is necessary to apply two patterns in sequence, i.e. first apply 00, the initialization sequence, and then apply 01, the sensitization sequence. This test scheme does not fit well with the stuck-at-fault model proposed earlier. However, Wadsack [25] has proposed a gate level model which will allow a normal stuck-at-fault simulator to generate tests for stuck open faults in CMOS gates. This is achieved by introducing a clocked latch onto the output of the gate being simulated. The clock of the latch is then controlled by six other gates, for an elementary 2-input gate e.g. NOR or NAND gate, which are used to model the stuck-open faults by allowing the simulator to place stuck-at faults on their inputs. If a stuck-open fault is present these control gates configure the latch to retain its previous state. Simulator fault models also exist for a transmission gate, tri-state gate and programmable logic arrays.

Jain and Agrawal [23] have proposed a gate level model for CMOS and NMOS circuits which can also be used to represent complex gate structures. The model retains a memory element ('B Block') at the output; however, this extra element is now controlled by two sets of gates. One set of gates is used to represent the pull-up transistors, p-channel devices in a CMOS process or depletion mode devices in the NMOS process. The other set of gates represent the pull-down transistors, n-channel devices in NMOS and CMOS processes. Consequently using this single gate level model it is possible to generate tests for both CMOS and NMOS circuits. The model has been specifically designed to operate successfully with the D-algorithm [19] (see Chapter 5). However, it is not possible to use the model directly since the D-algorithm must be extended to include the singular covers and the d-cubes for the new memory element. Figure 2.12 shows the gate level model for a CMOS NAND gate using this technique. Stuck-at faults can be introduced onto the gates to represent the stuck-open conditions, i.e. SOP on a p-channel device corresponds to a stuck-at-1 fault on the corresponding gate network driving Sp. The inverse condition applies to the n-channel devices. This model is also used to generate tests for stuck-on faults; that is when a device appears to be permanently switched on.

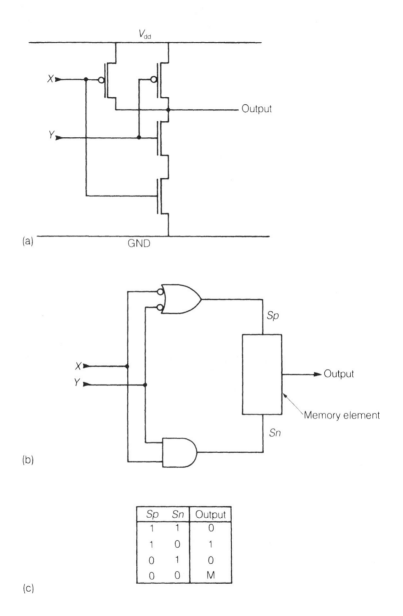

	(a)		V_{dd}			

Sp	Sn	Output
1	1	0
1	0	1
0	1	0
0	0	M

Fig. 2.12 Gate level model (Jain and Agrawal [23]): (a) CMOS NAND gate structure; (b) gate level model; (c) truth table for memory element.

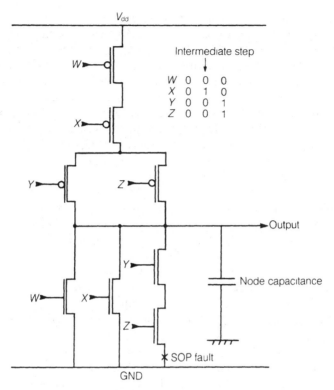

Fig. 2.13 An illustration of test sequence invalidation due to an input hazard condition.

The main problem when generating CMOS test sets is that two input patterns must be applied in sequence. However, the models described above do not take into account any skew that may take place between these inputs. It must be remembered that these patterns may possibly be applied through other gates before they reach the gate of interest. Consequently extra delays could be introduced, producing an invalid test sequence. Figure 2.13 illustrates this mechanism. If it is required to test the SOP fault marked by an X in the diagram then the first test sequence must set the output to logic 1 so that it can be discharged via the Y and Z transistors in series and thus detect the fault. If the initial pattern 0000 is used this will set the output to logic 1. Now the pattern has to be set to 0011 to test for the fault. However, if whilst switching to this new state, W or X should go to logic 1 due to some hazard in the preceding logic then the node will be discharged to 0 before the real test pattern is applied. This hypothetical example illustrates a further complication to test generation for SOP faults. However Reddy *et al.* [26] propose a solution to this problem using the model originally defined by Jain and Agrawal. The intention is to generate 'robust' tests that do not depend on delays through the circuit. This is achieved by choosing the initial test pattern so as to avoid static 0 hazards on the Sp and Sn lines of the faulty gate output latch. Reddy and Reddy [27] have also discussed this problem with regard to arbitrary circuit delays and charge sharing on CMOS nodes. The overall conclusion is that it is possible to design a full CMOS complex gate such that the delays are not significant. This can be achieved by the use of multiple test pattern sequences when required, or the addition of extra control inputs, by means of extra transistors, to the gate. The charge sharing problem is not thought to be too great since internal node capacitances are usually very much less than the output node capacitance, therefore charge redistribution will not greatly affect the output logic value.

A probable solution to the problem of SOP faults in CMOS is to design the circuit so that these types of faults cannot occur. One possible method is to use CMOS domino logic [28] which because of its method of operation is not so difficult to test with regard to SOP faults. Another solution along similar lines is proposed by Murray and Denyer [29]. Their scheme involves the redesign of the CMOS gate so that the CMOS memory problem of a stuck-open fault no longer causes difficulties, since the structure proposed allows at least one input pattern to detect the open fault.

Another solution to the problem of testing for some CMOS faults that are not easily covered by the stuck-at-fault test generation algorithms has been proposed by Zasio [30]. This technique relies on the monitoring of power supply currents to detect the CMOS faults. A CMOS circuit draws very little current when it is not being switched. However, the current required when a circuit switches is usually very large in comparison. Consequently faults such as open input lines, shorted output lines or even signal lines connected to either V_{dd} and ground, which cause abnormal currents to flow when particular inputs are present, can be easily detected by monitoring the supply current after a circuit has been switched to these inputs. The main problem with this method is that when the circuits are switched large current transients are generated, which must settle before any measurements can be taken; this settling time can be several milliseconds.

Although the stuck-at-fault model is widely accepted as an industry standard against which to measure test coverage, it is now becoming inadequate as new and different fault mechanisms are being discovered for VLSI circuits – in particular, faults which cause the output of a gate to take on an intermediate voltage level due to the ratio of the pull-up and pull-down transistors involved.

(c) Functional level fault models [13]

The transistor level and the gate level fault models give reasonably accurate representations of the faults which can manifest themselves in integrated circuits. However, as the density of transistors on a VLSI device increases the ability of these models to cope with the increasing complexity is in doubt. A possible solution to this dilemma is to perform test generation at the functional block level. That is all of the faults modelled at the transistor level would be represented at this higher level. Since the number of functional blocks in a design is usually very much less than the total number of transistors present this test generation scheme may have great promise in the future. A complete functional test could be performed by exhaustively applying all input patterns to a block. This is obviously impractical for blocks with a large number of inputs; however, if a partitioning algorithm [31] is used then the number of inputs may be reduced to a reasonable level. This type of test scheme would eliminate the need for any test pattern generation stage; however, the amount of test data involved would be enormous. Alternatively if the device is already partitioned, i.e. into smaller replicated functional blocks, then it may be possible to generate test patterns for one functional block which could then be used for all the other functional blocks. A major problem with functional testing is the loss of some information in going from the transistor level to the functional model, since it is possible to carry out the same function by numerous different transistor interconnections, thus a direct relationship would not exist between the two levels.

Functional level testing is already implemented to some extent in memories because of the regularity of their structure. Similarly microprocessor [32] and programmable logic arrays [33] can be treated at the functional level rather than the transistor or gate level. Chapter 10 will discuss in more detail the functional testing of programmable logic arrays and memory structures.

2.4 TEMPORARY FAULTS

Several sources of temporary fault behaviour will now be discussed. The various sources include electrical noise, electromagnetic interference and soft errors due to ionized particles striking the silicon. Temporary faults cover two classes of faults – these are transient faults and intermittent faults. The transient faults tend to be random faults which cause no damage to the circuit in which they occur. Such errors may be caused by ionizing radiation or electromagnetic interference. Intermittent faults are usually recurring faults which occur randomly in time and always appear in the same part of a circuit. These types of faults may be caused by bad design, i.e. a race condition which only manifests itself when certain inputs are present, or an overstressed component may be prone to noise. Burgess et al. [9] suggest that intermittent fault behaviour could be linked to either incorrect fabrication, making a particular part of a circuit sensitive to noise or a floating node which is coupled capacitively to another line.

(a) Electrical noise [34]

Electrical noise, in the form of thermal and flicker noise, is a common problem in electronic circuits, therefore it can be a potential source of temporary errors in VLSI circuits. However, electrical noise may not be a serious problem since fundamental limits [11] to scaling will be reached before currents and voltage are small enough to be affected by noise. Several other factors may totally rule out noise problems.

As devices are scaled down certain noise margins must be kept so that variations in the processing do not adversely affect the operation of the device. In such cases extra noise margins may be introduced to cope with the problem of electrical noise. Again reducing dimensions with constant field means that the voltages are reduced, thus dynamic circuits could become impractical. Consequently static devices will have to be used, as they are less prone to electrical noise. However, if devices are scaled with constant voltage instead, the electrical noise effects will not be a significant problem. From this discussion it appears that careful design of a circuit could eliminate the problem of electrical noise altogether.

(b) Electromagnetic interference

Electromagnetic interference has two potential sources, either external to the chip from electromagnetic radiation or internally from power supply fluctuations or capacitive coupling between adjacent signal lines [34], [35].

External transient sources are usually of long duration when compared to the cycle time of a typical VLSI chip. Most of the long duration events can be dealt with by the use of suitable shielding around the device. However, high power RF sources, such as microwave radiation, are very difficult to shield against since they may induce pulses on the wires of a printed circuit board which can then reduce noise margins inside the chip. It is also well known that bipolar and MOSFET transistors can rectify RF signals, due to non-linearities in their characteristics [36], [37]. If this occurs inside a chip it can add small offset voltages onto valid logic signals producing apparent stuck-at faults or introducing intermittent behaviour.

Two major sources of temporary errors that can be generated internally are transients injected from the power supplies or capacitive coupling between conductors. Large injected current transients are produced when the clock is activated in a synchronous circuit and a large number of transistors switch simultaneously. This injected transient can then cause a large number of errors. This problem is so serious that appropriate steps must be taken during the design of a new device to reduce the effect of this problem. Capacitive coupling between signal lines may allow a change in the voltage of one line to produce an effect on the adjacent line, leading to intermittent fault behaviour. The scaling down of device dimensions may only exacerbate the electromagnetic interference problem. This is mainly due to the reduction in switching speeds as devices are scaled down. Therefore the chances of a fault propagating to a point where it may cause a temporary error are greatly enhanced. However, at higher frequencies the capacitive coupling effect is not so great. In general as devices are scaled down electromagnetic interference is likely to become a major problem.

(c) Soft errors [38], [39], [34].

Alpha-particle-induced temporary errors, or soft errors, were first noted by May [38], [39] when the effect was observed in 4K and 16K Dynamic Random Access Memories (DRAMs) and 16K and 64K Charge Coupled Devices (CCDs). However since DRAMs are currently the densest devices available it is likely that designs of VLSI complexity will also suffer from ionization induced soft errors as their density increases. Recently some investigations carried out by Carter and Wilkins [40] have revealed that 64K Static RAMs (SRAMs) are likely to have soft error rates approaching those of similar sizes of DRAMs.

The main source of alpha particle radiation is from the natural decay of uranium and thorium [34], [38], [41] which are found as contaminants in most IC materials, in particular the packaging. Alpha particles produced by natural decay must come from within the packaging material or chip die since they only have a very short range. The decay of these elements also leads to the production of beta particles. However since beta particles have a longer mean free path before ionization collisions occur they are not expected to be a significant problem until device dimensions are scaled down substantially. A second possible source of ionizing particles is due to high energy cosmic rays. However, the effects of these particles may not become a problem until devices are scaled down further [38]. The effect of these particles may cause concern when space applications are involved since there is no atmospheric shield to reduce their effect. It is possible to shield the chip by the use of a special radiation absorbing material; however, this will not stop the alpha particles generated from within the chip die from causing problems [41].

Alpha and beta particles interact with matter by ionizing the atoms in their propagation paths, producing electron–hole pairs. As the particle slows down the energy shedding rate increases producing the maximum number of electron–hole pairs at the end of the propagation path. The electron–hole pairs produced by such ionizing events generate two currents [39], [42], [43]. If the electron–hole pairs are generated in a depletion region, i.e. a region of high electric field, then the particles will be separated quickly; in the case of an n-channel device electrons will be collected by the n-wells and the holes will be taken into the substrate. This current, termed the drift current, occurs over a very short period of time consequently the generated carriers are collected before they have a chance to recombine, therefore the collection efficiency is very nearly 100%. If the end of an alpha particle track occurs within the depletion region then the drift mechanism will cause the most serious problems. The second current occurs when the electron–hole pairs are produced outside a depletion region. In this case the carriers form a diffusion current due to the gradient in carrier concentration. The carriers will diffuse until they reach a depletion region when the drift mechanism will take over. This effect is slower moving than the drift current; hence the diffusion current component can cause the most problems because it is capable of moving large distances producing many correlated errors. Another important consideration is the angle at which the ionizing particle strikes the material. At low angles of incidence the particle is capable of affecting more than one node. Since the radiation is originating from the packaging material low angle strikes are highly probable, therefore large numbers of correlated errors may result [41],[42].

If an alpha particle strikes an NMOS memory node containing a logic 0 it will have no effect [38]. However if a node containing a logic 1 is struck and the amount of charge produced is greater than Q_{crit}, the charge necessary to maintain the logic value, then the value of the bit will change to logic 0, hence a soft error will have occurred. As device scaling takes place Q_{crit} is likely to be reduced, consequently logic nodes will become more susceptible to ionizing radiation strikes; CMOS structures will suffer from similar problems. A further complication in CMOS structures is the effect of ionizing particles on latch-up.

The increase in switching speed of MOS devices, brought about by reduced dimensions, will inevitably increase the probability of an observable soft error since the transient generated by the ionizing particle is more likely to propagate to a latch possibly producing an incorrect output. This effect has already been observed with electromagnetic interference problems. Also if the scaling of the devices is performed at constant field then the transistor saturation current will be reduced; however, injected currents due to ionizing particle strikes will not be reduced, which will make it easier for an ionising particle to produce a soft error. Obviously if scaling is preformed at constant voltage then the saturation current will actually increase, consequently ionizing particles will have less effect on the devices operation. Therefore it seems that as scaling down continues and Q_{crit} is reduced, soft error rates will increase. Even switching to static modes of design may only marginally improve the situation since, as has already been discussed, SRAMs have been shown to be just as susceptible to ionizing radiation.

The main problem with all of the temporary faults described is one of test pattern generation [44]–[46]. The difficulty arises due to these faults occurring randomly in time. Several possible approaches to the problem of test generation for intermittent faults have been proposed, these include the use of two-state Markov chains and a modified D-algorithm [44] to produce test sets. Alternatively the same pattern may be applied many hundreds of times to the device under test in an attempt to detect the fault [45]. This method uses statistical techniques to determine how many times the pattern should be applied to give a bound on the confidence of detecting the fault. As it is unlikely that the contaminants in IC packaging materials will be totally eliminated a structural solution to the problem of soft errors is necessary to produce devices which are not susceptible to soft errors. Sai-Halasz *et al.* [41] have proposed three techniques to achieve this objective:

1. use of a buried p-layer in the substrate to distribute the charge
2. use of the Hi-C technique to enhance the charge storage capability of the storage node [47]
3. use of an n-type grid to collect the excess charge introduced by an alpha strike.

Of course error correcting codes could also be used without affecting the physical structure of the devices. A combination of physical construction and redundancy techniques may well be a solution to this ever increasing problem of soft errors.

In mainframe computers it has been estimated that 80–90% of all faults are temporary in nature [48]. Since these types of faults are extremely difficult to diagnose they account for the largest proportion of servicing costs of mainframe computers. In a recent study of minicomputer reliability it was found that in TTL logic the number of temporary faults was about 87–94% of the total number of faults; this value increased to 97% when LSI logic was involved [35]. Therefore in VLSI, if this trend continues, a large majority of faults are likely to be temporary in nature.

2.5 SUMMARY

In this chapter various types of faults that occur in VLSI devices have been discussed. These faults can range from fabrication faults which produce observable surface defects and are permanent, to temporary faults which may only occur once. Different fault models have been described which allow simulators to produce test patterns to detect these faults within a circuit. The inadequacies present in these fault models have also been demonstrated. However, physical design rules have been suggested to alter the layout of a design so that the use of the stuck-at-fault models can be continued. The stuck-open fault was also described along with possible fault models to allow test generation to take place.

Various types of temporary fault have been discussed, together with their impact on VLSI devices as the dimensions are scaled. This has led to the conclusion that new design techniques are necessary to alleviate problems caused by these faults and that new test strategies are required to detect these faults in future VLSI devices.

2.6 REFERENCES

1. Fantini, F. (1984) Reliability problems with VLSI. *Microelectronics and Reliability*, **24**(2), 275–96.
2. Dennard, R. H., Gaensslen, F. H., Yu, H. *et al.* (1974) Design of ion implanted MOSFET's with very small physical dimensions. *IEEE J. Solid State Circuits*, **SC-9**(5), 256–67.
3. Baccarani, G., Wordeman, M. R. and Dennard, R. H. (1984) Generalised scaling theory and its application to a 1/4 micrometer MOSFET design. *IEEE Trans. Electronic Devices*, **ED-31**(4), 452–61.
4. Mangir, T. E. and Avizienis, A. (1980) Failure modes for VLSI and their effect on chip design. *Proc. 1st IEEE Conf. on Circuits and Components*, 658–88.
5. Fantini, F. and Morandi, C. (1985) Failure modes and mechanisms for VLSI ICs – a review. *Proc. IEE*, Pt. G, **132**(3), 74–81.
6. Kearney, M. A. (1985) MOSFET instability mechanisms. *VLSI Design*, March, 88–91.
7. Sze, S. M. (1983) *VLSI Technology*. McGraw-Hill.
8. Timoc, C., Buehler, M., Grisword, T. *et al.* (1983) Logical models of physical failures. *1983 International Test Conference*, 546–53.
9. Burgess, N., Damper, R. I., Shaw, S. J. and Wilkins, D. R. J. (1985) Faults and fault effects in NMOS Circuits – impact on DFT. *Proc. IEE*, Pt. G, **132**(3), 82–9.
10. Eitan, B. and Frohman-Bentchkowsky, D. (1981) Hot electron injection into the oxide in n-channel MOS devices. *IEEE Trans. Electron Devices*, **28**, 328–40.
11. Mead, C. and Conway, L. (1980) *Introduction to VLSI Systems*. Addison-Wesley.
12. Shih, D.-Y. and Ficalora, P. J. (1979) The reduction of Au–Al intermetallic formation and electromigration in hydrogen environments. *IEEE Trans. Electron Devices*, **ED-26**(1), 27–34.
13. Abraham, J. A. and Fuchs, W. K. (1986) Faults and error models for VLSI. *Proc. IEEE*, **74**(5), 639–54.
14. Schuster, M. D. and Bryant, R. E. (1984) Concurrent fault simulation of MOS digital circuits. *Proc. Conf. on Advanced Research in VLSI*, 1–10.
15. Hayes, J. P. (1984) Fault modelling for digital integrated circuits. *IEEE Trans. Computer Aided Design*, **CAD-3**, 202–8.
16. Bryant, R. E. and Schuster, M. D. (1983) Fault simulation of MOS digital circuits. *VLSI Design*, 24–30.
17. Bryant, R. E. (1981) MOSSIM: A switch level simulator for MOS LSI, *Proc. 18th Design Automation Conference*, 786–90.
18. Williams, T. W. and Parker, K. P. (1983) Design for testability – A survey. *Proc. IEEE*, **71**(1), 98–113.
19. Roth, J. P. (1966) Diagnosis of automata failures: a calculus and a method. *IBM J. Res. and Dev.*, **10**, 278–81.
20. Galiay, J., Crouzet, Y. and Verniault, M. (1979) Physical versus logical fault models in MOS LSI Circuits, impact on their testability. *Proc. 9th Fault Tolerant Computing Symposium*, 195–202.

21. Mei, C. Y. (1973) Bridging and stuck-at faults. *Proc. 3rd Fault Tolerant Computing Symposium*, 786–90.
22. Chandramouli, R. (1983) On testing stuck open faults, *Proc. 13th Fault Tolerant Computing Symposium*, 258–65.
23. Jain, S. K. and Agrawal, V. D. (1985) Modelling test generation algorithms for MOS circuits, *IEEE Trans. Computers*, **C-34**(5), 426–33.
24. Gallace, L. J., Pujol, H. L. and Schable, G. L. (1977) CMOS reliability. *27th Electronic Components Conference*, 496–512.
25. Wadsack, S. L. (1978) Fault modelling and logic simulation of CMOS and MOS integrated circuits, *Bell. Tech. J.*, **75**, 1449–74.
26. Reddy, S. M., Reddy, M. K. and Agrawal, V. D. (1984) Robust tests for stuck open faults in CMOS combinational logic circuits. *14th Fault Tolerant Computing Symposium*, 44–9.
27. Reddy, S. M. and Reddy, M. K. (1986) Testable realizations for FET stuck-open faults in CMOS combinational logic circuits. *IEEE Trans. Computers*, **C-35**(8), 742–54.
28. Okldrzija, V. G. and Kovijanic, P. G. (1986) On testability of CMOS – Domino logic. *IEEE Trans. Computers*, **C-35**(8), 50–5.
29. Murray, A. F. and Denyer, P. B. (1985) Testability and self test in NMOS and CMOS VLSI signal processors. *Proc. IEE*, Pt. G, **132**(3), 93–4.
30. Zasio, J. J. (1985) Non stuck fault testing of CMOS VLSI. *Proc. COMPCON*, **CH2135-2/85**, 338–41.
31. Roberts, M. W. and Lala, P. K. (1984) An algorithm for the partitioning of logic circuits. *Proc. IEE*, Pt. E, **131**(4), 113–18.
32. Brahme, D. S. and Abraham, J. A. (1984) Functional testing of microprocessors. *IEEE Trans. Computers*, **C-33**, 475–85.
33. Hong, S. J. and Ostakpo, D. L. (1980) FITPLA: A programmable logic array for function independent testing. *Proc. 10th Fault Tolerant Computing Symposium*, 131–6.
34. Savaria, Y., Rumin, N. C., Hayes, J. F. and Agrawal, V. K. (1984) *Characterization of Soft Error Sources*. Report No. 84-11, McGill University, VLSI design laboratory.
35. Castillo, X., McConnell, S. R. and Siewiorck, D. P. (1982) Derivation and calibration of a transient error reliability model. *IEEE Trans. Computers*, **C-31**, 58–67.
36. Richardson, R. E. (1979) Modeling a low level rectification RFI in bipolar circuitry. *IEEE Trans. Electromagnetic Compatability*, **EMC-21**(4), 307–11.
37. Forcier, M. L. and Richardson, R. E. (1979) Microwave rectification RFI response in field effect transistors. *IEEE Trans. Electromagnetic Compatability*, **EMC-21**(4), 312–13.
38. May, T. C. (1979) Soft errors in VLSI – present and future. *Proc 29th IEEE Conf. on Electronic Components*, 247–56.
39. May, T. C. and Woods, M. H. (1979) Alpha-particle induced soft errors in dynamic memories. *IEEE Trans. Electron Devices*, **ED-26**(1), 2–9.
40. Carter, P. M. and Wilkins, B. R. (1986) Soft errors in static NMOS RAMs. *ESSCIRC 1986*, Delft, 13–15.
41. Sai-Halasz, G. A., Wordeman, M. R. and Dennard, R. H. (1982) Alpha-particle induced SER in VLSI circuits. *IEEE Trans. Electronic Devices*, **ED-29**(4), 725–31.
42. Yaney, D. S., Nelson, J. T. and Vanskike, L. L. (1979) Alpha-particle tracks in silicon and their effect on dynamic MOS RAM reliability. *IEEE Trans. Electron Devices*, **ED-26**(1), 10–16.

43. Hsieh, C. M., Marley, P. C. and O'Brien, R. R. (1981) Dynamics of charge collection from alpha particle tracks in integrated circuits. *Proc. International Reliability Physics Symposium*, 38–42.
44. Breuer, M. A. (1973) Testing for intermittent faults in digital circuits. *IEEE Trans. Computers*, **C-22**(3), 241–6.
45. Kamal, S. and Page, C. V. (1974) Intermittent faults: A model and a detection procedure. *IEEE Trans. Computers*, **C-23**(7), 713–19.
46. Kamal, S. (1975) An approach to the diagnosis of intermittent faults. *IEEE Trans. Computers*, **C-24**(5), 461–7.
47. Tasch, A. F., Chatterjee, P. K., Fu, H-S and Holloway, T. C. (1978) The Hi-C RAM cell concept. *IEEE Trans. Electron Devices*, **ED-25**(1), 33–41.
48. Tasar, O. and Tasar, V. (1977) A study of intermittent faults in digital computers. *Proc. AFIPS*, **46**, 807–11.

3
SIMULATION TECHNIQUES FOR EMPIRICAL DESIGN VERIFICATION

3.1 INTRODUCTION

In the design of complex digital systems simulation is used extensively both for design verification and fault simulation (Chapter 4), which is a necessary adjunct to test pattern generation. The simulation process comprises creating a model of a circuit under faulty or fault-free conditions, exercising the model with a set of input stimuli and monitoring the output response predicted by the model to the input stimuli.

During the design of a digital system, the simulation process is carried out at several levels of abstraction, each level entails a different trade off between simulation accuracy and computing requirements. The higher levels of simulation are used to check the behavioural/functional aspects of the design, whereas the lower levels relate to design aspects associated with performance and physical implementation. As one descends through the levels of abstraction signal representations change from character representations to discrete logic levels and then to quasi-continuous voltage levels; in a similar way events, i.e. changes in signal value, are described as a number of clock cycles at higher levels and as sub-multiples of seconds at the lower levels.

The levels of simulation [1]–[3] which may be used in the design process are

1. behavioural level
2. register transfer level
3. functional level
4. gate level
5. switch level
6. timing level
7. circuit level.

Switch level is the lowest level of abstraction considered for the purposes of functional and logical design verification and fault simulation; as this text is concerned with these aspects of the design process, the lower levels of abstraction which are concerned with the 'circuit' aspects of a design will not be discussed further.

3.2 APPLICATIONS OF SIMULATION

In the design process simulation has a wide range of applications; it may be used for the following:

(a) Function verification

In this mode the designer uses the simulator to verify that a system or part of a system performs the desired function; for example, if an up–down counter is part of a system, then a gate level model of a counter would be described to the simulator and the basic functions of the counter, i.e. incrementing, decrementing, clearing to zero, etc. would be verified by the simulator.

(b) Fault simulation

A necessary adjunct to test pattern generation is fault simulation which is used to determine the number of faults (fault coverage) detected by the test patterns.

(c) Spike and hazard analysis

An important application of simulation is the detection of switching anomalies which may occur anywhere in the circuit and cause the circuit to malfunction. This simulation function is normally performed at gate level and the designer usually has the ability to specify the minimum acceptable width of a pulse on the output of a gate: if it is less than this value it is deemed to be an unwanted 'spike' or 'glitch'.

(d) Verification of the integrity of decompositions to lower levels of abstraction

The design of complex systems is, in general, implemented using the principle of top-down design. However, as the design is refined through the various levels of abstraction the designer must verify the integrity of the circuit function as it is decomposed from one level of abstraction to another. At present the only way to perform this verification procedure is to use 'lock-step' simulation.

(e) Evaluation of design alternatives

During the initial phase of conceptual design of a system, simulation permits the designer to examine, readily, design alternatives which may, for example, improve the performance of the system or enhance the testability of the system.

3.3 SIMULATOR DESIGN CRITERIA [4]

It is widely accepted that the design of a complex integrated circuit is expensive and that an accurate specification of the function of the circuit is required at the outset; to ensure that the physical realization of the function is correct 'first time', a vast arsenal of CAD tools has also been developed to detect any errors in the design. However, it is not fully appreciated that the development of CAD tools is also an expensive and time consuming process which is not fully understood [5] and that there is an even greater need for an accurate specification of the requirements of a CAD program at the outset in order to get it right 'first time'.

In developing a simulation package a number of modelling and implementation questions must be answered; the answers will reflect the potential capabilities of the simulator in terms of accuracy, efficiency and 'ease of use' which are the three main criteria considered in the design of a simulator.

The main modelling and implementation questions asked relate to the following points.

(a)   The level of simulation

The answer to this question will determine the data structures and algorithms to be used in the simulator together with the level of detail and accuracy of the simulation results. Present day simulators usually have the capability of spanning at least two levels of abstraction, for example gate and functional level. The ability to mix levels of abstraction is considered an advantage with respect to efficiency and 'ease of use'.

(b)   The function of the simulator

Early gate level simulators were used primarily for logic verification, that is gate delays were not considered during simulation. A simulator for logic verification is relatively easy to implement but is of limited use to the designer. Simulators used for design verification, although more complex, have the capability of simulating gate delays, performing spike and hazard analysis etc. In general, design verification simulators have a mode control switch which permits the designer to choose the level of detail used in modelling the gates in the simulator; consequently a simulator developed for design verification can also be used for logic verification. Fault simulation is also used extensively in the design process; ideally, the same simulator should be capable of performing both functions. The decision to include an efficient fault simulation capability in a normal simulator would dictate, at the outset, the data structures and simulation algorithms which can be implemented.

(c) The level of input description

This is an important question with respect to the design criterion of 'ease of use', it also reflects on the techniques used to model the logic functions inside the simulator. The input description of a circuit to a gate level simulator is usually in terms of a special purpose language which describes the interconnection of gates, flip-flops etc., in the circuit; the simulator has a set of built-in subroutines which simulate the function of the gates and flip-flops in the circuit description. However, when simulating large circuits, in the interest of 'ease of use' and efficiency, it is impractical to describe and simulate a circuit at gate level. To overcome this problem hardware or functional description languages have been developed which permit functions to be described without any reference to their gate/flip-flop representation; these functions are, in general, simulated by a 'command interpreter' routine within the simulator. Functional description languages can also be used, either to describe variations in some of the basic switching elements used or to extend the library of switching functions in a gate level simulator, hence making it more flexible and easier to use.

(d) The types of circuit to be simulated

Early simulators were capable only of simulating combinational or synchronous sequential circuits. The algorithms for these simulators are relatively easy to implement and can be made very efficient because of the constrained design environment in which they are used. However, simulation algorithms for asynchronous sequential circuits are more complex but are more applicable to a wider range of circuits including the simulation of synchronous circuits. For design verification purposes it is considered good practice to simulate synchronous circuits asynchronously to ensure that the circuit has settled in a given clock period.

(e) The model detail

This question reflects the accuracy of the simulation results and is related to the delays considered in the switching elements and the number of logic levels used to model logic transitions on the outputs of the switching elements. However, the more detail included in a model adversely affects the simulation efficiency, since more CPU time is required to evaluate a complex model.

(f) The data structures used to model the circuit in the simulator

The data structure used to model the circuit inside the simulator influences the types of circuits which can be simulated, the simulation efficiency, the amount of detail represented in the switching element models and the applications for which the simulator may be used. The data structures used in gate level simulators will be discussed in Section 3.4.1.

(g) Time flow mechanism used

The time flow mechanisms used in simulators, in general, are fixed increment and asynchronous next event. The fixed increment technique is easy to implement and assumes that the circuit is activated by a clock mechanism which only permits events at given multiples of some fixed time increment. Asynchronous, next event time flow mechanism permits signal value changes to occur at any time in the circuit and is used extensively in design verification simulators to detect switching anomalies in a circuit. These time flow mechanisms will be discussed in more detail in Section 3.4.2.2.

(h) Speed-up techniques adopted

Various techniques can be introduced to make the simulation process more efficient, however, they also make the simulation algorithms more complex. Improvements in simulation efficiency can be obtained by exploiting circuit latency and using techniques of selective trace or stimulus bypassing; these will be discussed in Section 3.4.4.

In summary, if a simulator is to be applicable to a wide range of circuit designs and is to be used in a design verification environment, the simulation algorithm will be complex and the development time long and expensive. However, if the simulator is limited to logic verification in a restricted design environment, then a very efficient simulator can be developed in a relatively short time period and at less cost. Hence, the necessity of strictly defining the function of the simulator at the outset of its development. Furthermore, when devleoping a simulator, care must be taken not to implement the simulation routines and algorithms in a restricted way such that the simulator only produces the 'expected' output results for a circuit. The simulation algorithms and routines should predict, as far as possible, the actual behaviour of the circuit; whether or not it is 'expected' by the designer.

3.4 GATE LEVEL SIMULATION [4],[6],[7]

To date gate level simulation has been the 'workhorse' of the design verification process. Many of the algorithms and techniques developed at gate level have been adopted in simulators used at other levels of abstraction, in fault simulation and in hardware accelerators for simulation. Details of the modelling techniques, data structures and algorithms used for gate level simulation are well documented [8], hence only the salient features of the techniques and algorithms used will be summarized.

A gate level simulator, in general, comprises four modules; first a 'pre-processor' module which checks the syntax of the language used to describe the circuit to be simulated to the computer; second, a 'model compiler' which transforms the input circuit description into some internal representation upon which the simulator executive operates; third, the 'simulator executive', which essentially controls the running of the simulator, services simulation control commands and evaluates and schedules signal changes in the circuit; finally, a 'display processor' module which formats and displays the simulation results to the requirements of the user.

3.4.1 Modelling a digital circuit for simulation

The initial task to be performed when simulating a circuit is to describe the circuit topology to the computer in terms of the available simulation primitives and their interconnections; this is done using either a special purpose language or using a graphics input system. The preprocessor module subsequently checks the input description for errors, the actual checks implemented and the subsequent processing performed by the preprocessor module depends upon the format of the input description. It should be noted that the preprocessor module can only detect syntactical errors in an input description; it cannot detect errors where, for example, a given gate has been defined as a NAND gate instead of a NOR gate or that the output of gate A has been inadvertently connected to the input of gate C instead of gate B; the presence of these types of errors can only be detected after the simulator has been run and the output waveforms are found to be incorrect.

A model of the circuit is subsequently generated from the syntactically correct circuit description by the 'model compiler'. The earlier simulators and some present day hardware accelerators use the *compiled code model* of a circuit, in which the circuit description is compiled into a sequence of computer instructions, which will predict the circuit response to a given set of inputs, when executed by the computer. In the compiled code model the instructions, in general, model the gate as a two-valued, zero delay element; consequently, the order in which the gates are compiled and subsequently evaluated is important, if the correct output response to a given set of inputs is to be produced in a single pass. The process of ordering gates so that a gate output is evaluated and updated before it is used as an input to a successor gate is called *levelling*.

An alternative representation to the compiled code technique, for

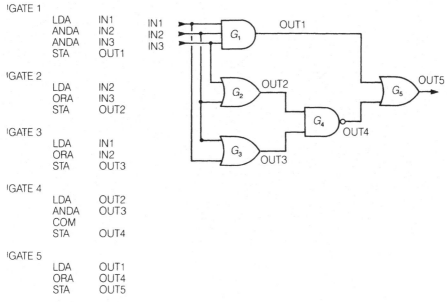

```
ᴵGATE 1
            LDA      IN1
            ANDA     IN2
            ANDA     IN3
            STA      OUT1

ᴵGATE 2
            LDA      IN2
            ORA      IN3
            STA      OUT2

ᴵGATE 3
            LDA      IN1
            ORA      IN2
            STA      OUT3

ᴵGATE 4
            LDA      OUT2
            ANDA     OUT3
            COM
            STA      OUT4

ᴵGATE 5
            LDA      OUT1
            ORA      OUT4
            STA      OUT5
```

Fig. 3.1 Compiled code model of a circuit.

modelling a circuit in a computer for simulation purposes, is called the *table driven model.* In this instance the circuit is translated into a set of interlinked data tables which are operated upon by a separate simulation program called the *simulator executive.*

An example of the compiled code and table driven models of a circuit are shown in Figs 3.1 and 3.2 respectively; in the case of the table driven model only the main tables and associated pointers are shown; some of the tables which have been omitted are the table of switching characteristics for the gates, containing rise and fall times and propagation delays, and the table containing a description of the input waveforms to the circuit. The gate description table in Fig. 3.2 comprises the following entries: gate type and number of inputs, gate output and input identifiers. Within the simulator these identifier names will be replaced by pointers directed to the appropriate column in the signal value table. The function of the signal value table is to store a description of the current state of the source of a signal associated with a given gate input/output identifier. The signal value table also contains a pointer to the fanout table so that, when a particular gate output changes logic value, the gates which may be affected by this change in state can be identified, permitting the technique of 'selective trace' simulation (Section 3.4.4) to be implemented. It is seen with the table driven model of a circuit, that the pointers between various tables, in effect, realize the interconnections between the gates in the circuit.

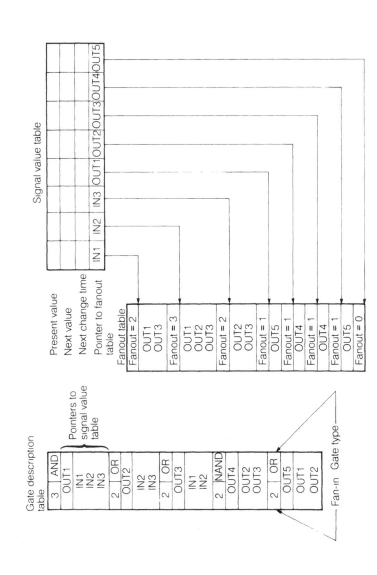

Fig. 3.2 Table driven data structure for the circuit in Fig. 3.1.

The main advantage the compiled code simulator has over the table driven simulator is that it is much faster. However, it has several disadvantages: first, the circuit must be levelled before simulation starts; second, the circuit must be synchronous; third, modifications to the circuit require recompilation of the simulation program; fourth, the techniques of exploiting circuit latency to improve simulator efficiency cannot be employed in the compiled code model. The table driven simulator, however, does not have these disadvantages and is much more versatile in the types of circuits which can be modelled and also on the complexity of the individual gate models, in terms of the number of delays and the number of values used to represent signal transitions on the output of a gate. Consequently the main application of table driven simulators is in design verification, whereas the compiled code simulator is used to model existing hardware systems which are too expensive to duplicate, in order, say, to study their behaviour under fault conditions.

3.4.2 Basic simulation algorithm

The process of simulating the function of a digital circuit comprises

1. evaluating the effects of signal changes in a circuit
2. scheduling events (signal changes) in the correct time sequence.

In this way the output response of a circuit to a given set of input changes can be predicted; this is the main function performed by the simulator executive module in the simulator.

3.4.2.1 Evaluating the effects of signal changes in the circuit

In a logic circuit, in general, several gate outputs will change value concurrently; however, general purpose computers cannot process simultaneous events. To contend with this situation, time is momentarily 'frozen' inside the simulator and changes in logic value are updated one at a time; thereafter, for each signal in turn which changed value, the subsequent effect on the gates in its fanout list is evaluated to determine the effects of the logic changes upon the circuit. Since all signal value updates are performed before individual gate evaluation procedures are implemented, then any gate evaluation performed will include the effect of all signal changes on the inputs to a given gate at that time.

The procedure [9] for evaluating the effect of the changes in logic value, which have occurred at some instant in time in a circuit, is outlined below.

FOR {Each gate G which has changed value at time T} *DO*
FOR {Each gate G_n in the fanout list of G} *DO*
BEGIN

Obtain the input values to G_n;
Compute the new output value of G_n;
IF {Present output value $\neq$ new output value}
    THEN {Schedule the gate output change in the event queue at some time
        $T + dt$}
END

The above procedure does not perform any spike analysis on the gates. This can be incorporated, readily, in the basic algorithm by keeping a record in the signal value table of times when a gate output changed last or when it is about to change, together with the appropriate signal values. In computing the new output value of the gate a table look-up technique [10] may be used or a procedure may be called which simulates the behaviour of the particular type of gate.

3.4.2.2 Event scheduling during simulation

Simulation is the calculation of logic values as a function of time, hence an important aspect of the simulation process is the scheduling of events (gate changes) in the correct time sequence. The techniques normally used for scheduling events are the 'next event list processing' technique or the 'time mapping algorithm' [11].

(a) Next event list processing

This technique is relatively easy to implement and comprises setting up a simple list structure or 'event queue' in which events are ordered with increasing time. Events can be scheduled asynchronously in the list, that is, there are no constraints imposed upon the times when events can occur and the advancement of time in the simulator is accomplished by jumping from one scheduled event to the next, hence avoiding periods of inactivity in the circuit. The main disadvantage of this technique is that the list must be continually processed to ensure that events are placed in the list in the correct time sequence, otherwise mis-scheduling of events will occur. If the list is long, implying that there is much activity in the circuit, time is wasted in processing the list.

(b) Time mapping algorithm

The technique is algorithmic and comprises, as shown in Fig. 3.3, a circular loop (dT-loop) which is used to represent time in increments of dT. The time increment and the number of sectors in the loop are determined prior to the start of simulation from the circuit delays. During the simulation each sector will contain either a signal name or a pointer to a list of signal names which will change at that time. Future events are scheduled into the appropriate time slot by adding the gate delay, expressed in terms of dT, onto the current pointer position in the loop. Consequently there is no need to scan the event list in order to schedule future events. In this technique time is incremented by stepping around sectors in the loop. This process can be inefficient in circuits where there is a large difference between the smallest and largest delays in the circuit and the activity in the circuit is sparse; this results in a loop with a large number of small sectors, many of which will not contain a scheduled event. Time, however, can only be advanced a sector at a time, thus time is wasted in scanning empty sectors. In this situation, an attempt to use a time loop whose

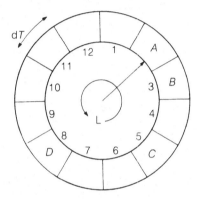

Fig. 3.3 Time mapping event scheduling technique.

length is less than the greatest delay in the circuit will result in events being mis-scheduled. Efficient time scheduling can be obtained in these situations if two loops are used, as shown in Fig. 3.4, where the increment of the first loop is dT and its length represents some delay which is smaller than the maximum delay in the circuit, whilst the increment in the second loop (L-loop) is the length of the first loop. After each revolution of the dT-loop, the pointer to the L-loop is incremented by one and the events found in the appropriate sector (i.e. occurring at $nL + m.dT$, where $m.dT < L$) are placed in the dT-loop; events in a given sector in the L-loop need not be ordered since this will be done automatically when they are placed in the dT-loop. The dT-loop is then scanned as before with new events being scheduled either in the dT-loop or the L-loop as appropriate.

The basic algorithm used to simulate a logic circuit can now be outlined as shown below. It is assumed that the simulator is table driven and that the

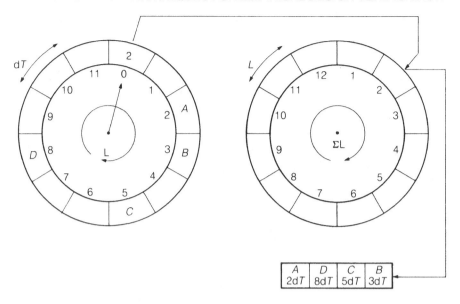

Fig. 3.4 Two loop technique for long delays.

simulation primitives have delays associated with them and gate transitions are modelled using three values, namely, zero, one and unknown.

1. At time zero, initialize all gate outputs to the unknown value;
2. Apply the input stimuli (driving waveforms) to the circuit;
3. Determine the fanout lists of the primary inputs;
4. Evaluate the effect of these signal changes on the gates in the fanout lists;
5. Schedule in an event queue the times when changes in logic value will occur on the outputs of the gates which have been evaluated, or in the driving waveforms;
6. Advance the simulation clock to the next event in sequence in the time queue, or if the duration of the simulation has expired terminate the simulation run;
7. Update the signal changes in the circuit which have been scheduled to occur at this time;
8. Determine the fanout lists for the signals which have changed value – go to (4).

The simulation algorithm outlined above is very efficient when applied to large logic circuits since it incorporates techniques which will be discussed later to exploit circuit latency.

3.4.3 Accuracy of the simulation results [8]

The accuracy of the results produced by a simulator is determined by the accuracy of the gate models and depends upon the delays [12] included in the model and the number of values [13] used to describe the transitions occurring at the gate output.

When gate level simulators were first developed, the only values which could be assigned to the signal lines were logic 1 and 0, since the simulators used the compiled code model, and thus the function of the simulator was limited to logic verification. However, the introduction of a third value, an X or unknown state, greatly increased the usefulness of the simulator. First the presence of switching anomalies [14] could be detected in the circuit, since if a gate switched from a logic 1, to an X-state and back to a logic 1, it would indicate that a glitch had occurred in the circuit. Second, at the start of a simulation run, if all the gates are initialized to an X-state, then the propagation delay time through the circuit can be determined. Third, by setting the outputs of registers to an X-state, the ability to set/reset the registers from any initial value can be checked in only one simulation pass. The number of values which could be assigned to signal lines was increased, subsequently to four, since some designers wished to be more specific about the significance of the X-state, wanting to differentiate between an X-state assigned to a gate output at the beginning of a simulation run and an X-state produced at a gate output as a result of a switching anomaly in the circuit. Other four-valued simulators, rather than considering two different X-states, introduced a high-impedance value, which is necessary when simulating circuits containing transmission gates. To date, the number of values used to describe conditions existing on a gate output ranges from three through to fifteen; however, as the number of values assigned to a signal line increases, the decision as to which value to assign to a gate output becomes more complex and requires some order of precedence to be assigned to the gate input values. The usual convention is that the most pessimistic value dominates unless an input value exists which can categorically define the output value on the gate. Furthermore, as the gate models become more complex, the CPU time required to evaluate gate changes increases and the simulation becomes less efficient although the results will be more accurate.

Since simulation is the calculation of logic values as a function of time, delays which exist in the actual circuit must be included in the gate models. In general, these are

(a) Transport delay

Each gate or a connection that a signal passes through must introduce a time delay, which is referred to as the transport delay and is usually represented as a pure delay block on the output of a gate.

(b) Ambiguity delay

Since the exact delays through a gate are not known, a pair of delays can be assigned to the gate defining a time window when it is uncertain whether a gate transition has occurred or not.

(c) Rise and fall delays

When a gate changes logic value it does not affect the change immediately due to capacitances in the circuit. Consequently gate transitions have a rise and a fall time. In general, since these two delays are not equal the width of the output pulse is distorted.

(d) Inertial delay

If the width of a dominant input pulse to a gate is very short, it will not force the gate to switch. The minimum duration that a logic change must remain on an input before the gate responds is called the inertial delay time of the gate. If the pulse width is greater than the inertial delay time, the effect is similar to the transport delay; otherwise the input pulse is suppressed.

Gate models can have either zero, unit, assignable or precise delays. When using assignable delays groups of gates, either of the same type or with the same number of inputs, are assigned a certain delay value, whilst other groups of gates are assigned different values. The assignment of precise delays to gates usually occurs when a post layout simulation is to be performed and data has been extracted from the layout relevant to individual gates which will affect their delay characteristics. In general the more delay parameters associated with a gate model, the greater the number of values required to model a gate transition.

3.4.4 Exploitation of latency

As the size of the circuit to be simulated increases there is a corresponding increase in CPU time; consequently techniques have had to be developed to improve simulation efficiency. The first technique to be developed was the event driven simulator, where signal evaluation in the circuit is initiated by the occurrence of an event as indicated by the event scheduling algorithm using the next event list processing technique; hence gate evaluations during inactive times is avoided. The efficiency of the simulators was further increased when it was noted that when signals changed logic value at some instant in time; these changes would only affect a small percentage of the gates in the circuit, namely, the gates in the fanout lists of those signals which changed value. Consequently, in order to determine the effect of the signal changes on the circuit at a given time it is only necessary to evaluate the changes in logic state on those gates which are in the fanout list of the signal lines which changed state. This simulation technique is called *selective trace* and can be applied only to circuits where gates exhibit a unilateral switching characteristic. In circuits which have bilateral switching devices [15], it is necessary to partition the circuit into modules, where the interconnections between modules are unidirectional, thereafter the technique of selective trace simulation can be applied.

Further improvements in efficiency can be obtained by using parallel value simulation; however, there is a restriction on the delay models which can be used, or by using special purpose computers which exploit circuit concurrency, that is having the ability to simulate simultaneously all gate changes which are going to occur at a given instant in time.

3.5 FUNCTIONAL LEVEL SIMULATION [16]–[19]

The requirement for higher levels of simulation resulted from the need of digital circuit designers to have a more abstract description of the circuit being developed. Gate level and higher function block level primitives, i.e. flip-flops, counters etc. were too low a level of representation for complex systems comprising in excess of 100 000 gates, requiring vast amounts of storage for the description of the circuit and long simulation times which generated excessive amounts of output data, whose detail essentially masked the macroscopic operation of the circuit, making it difficult to identify functional design errors. Furthermore, with low levels of abstraction the designer has the initial onerous task of entering a detailed circuit description into the simulator. Also, many design houses use standard LSI and MSI parts, the gate level models of which are usually proprietary, and the time taken to generate a gate level model would almost be as long as the time taken to design the part.

Many of these problems have been overcome by the introduction of *functional level simulation*, which attempts to simulate the input/output behaviour of a system/function without specific reference to its internal structure, that is, functional level simulation is concerned with what a system does and not how it does it.

The basic concepts underlying functional level simulation are not new and found their origins in the computer manufacturing industry where either a complete central processor unit or parts of it were described at a high level using Register Transfer Languages (RTL) [20],[21] which provided, in general, algorithmic descriptions of synchronous systems without particular reference to their hardware realization. However, in general, register transfer languages inherently assume that the system will be implemented as a set of registers, interconnected by busses, with a parallel transfer of data occurring between the registers. In addition to providing an input description of a system to a high level simulator, register transfer language description can be used as a form of documentation for a design and also as input to automatic synthesis programs which, directly, generate Boolean design equations from the RTL description. Initially, RTLs could only describe synchronous machines; however, many RTLs have subsequently been evolved which not only can describe asynchronous machines but also include subroutine facilities or special operators which permit subfunctions, for example, counters, adders etc. to be included in an RTL description.

The functional level of abstraction evolved from the register transfer level for a number of reasons [22].

1. The continuum of simulation resolution which exists at the lower levels of abstraction, that is between circuit, timing and gate level, does not exist between gate and register transfer level; hence there was a need for a level of simulation which had the benefits of simulating at the higher levels of abstraction – namely a reduction in storage requirements and simulation time but without too great a decrease in the simulation resolution.
2. In order to offset the costs of developing a more efficient simulator a higher level of simulation had to be developed which could interface readily with existing gate level simulators.
3. Register transfer level descriptions require a knowledge of the register transfer characteristics of the system/device being modelled. On many occasions, when a system is to be constructed out of standard MSI/LSI parts, these characteristics are unavailable or not well defined [23].

The algorithms used in functional level simulation are very similar to those used at gate level, namely selective trace and asynchronous next event simulation. The functional models, as far as the simulator is concerned are treated in a similar way to gates, that is the evaluation of the output value of a function is activated by a change in either the data or control inputs to the function and any subsequent change in the outputs are scheduled in an event queue. However, functional models, unlike gate models, may have internal memory states, and if time delays are associated with internal state changes these will also be scheduled in the event queue. The method of evaluating the new output states resulting from a change in inputs depends upon the technique used to model a given function. Functional models can be produced by

1. synthesis of existing gate level descriptions
2. functional table description
3. subroutine description
4. functional modelling language description
5. use of physical models.

Each of these methods will now be described.

(a) Synthesis from an existing gate level description [24]

This technique assumes the existence of a gate level description of a circuit/ function. The techniques are restricted to single output circuits; if multiple outputs exist a functional model for each output must be generated. The model generated is essentially a truth table in which the input/output timing delays are identical to the gate level model. The preservation of the timing relationships between the two representations is achieved by considering that the gate delays, instead of being associated with a gate output, are associated

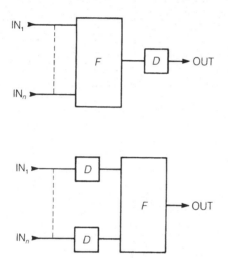

Fig. 3.5 Gate delays referred to inputs.

with the inputs as shown in Fig. 3.5; the transposition process is continued back to the inputs of the circuit, and the functional model generated subsequently is shown in Fig. 3.6. The technique of generating the model inherently excludes reconvergent fanouts within a model. A special scheme has also been devised for handling circuits with feedback loops.

   This technique cannot be used in a top-down design style since it requires the existence of a gate level description, hence its main application is in improving the efficiency of gate level simulation without an appreciable loss in timing accuracy.

(b) Functional table description [25]

This technique is used to model standard MSI parts, in which a simple subroutine is written comprising the function table of the device. During the simulation process, whenever an input to the model is activated the

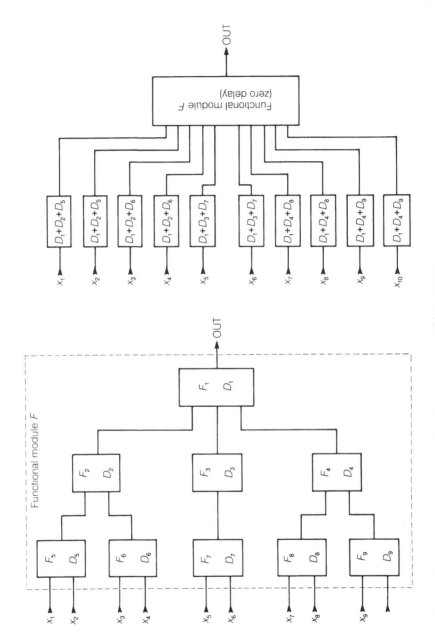

Fig. 3.6 Simulation model generated by the delay transposition process.

subroutine is called and the appropriate output values for the given input condition are derived from the function table and returned as altered output values in the subroutine call, to be used later by the simulator after the appropriate delays associated with the output changes have expired.

(c) Subroutine description

General purpose programming languages, for example, Fortran [26], APL [27], Pascal [28], Modula2 [29], can be used to write subroutines to model device operation. The subroutines, however, are much more complex than those used in the functional table approach, and in general require a good knowledge of programming to write efficient model routines. This technique is used, mainly, to generate simulation models of functions which cannot be produced in any other way, for example, a sign magnitude adder. The main disadvantages of this technique are the time taken to write and debug the model routine and the routines must be changed if the function is to be remodelled.

(d) Functional modelling languages [30]

Functional modelling languages (FML) are special purpose languages which enable the designer and not a programmer to produce a functional model of a circuit or subfunction. Functional modelling languages are usually keyword driven and have an English syntax which makes the functional descriptions very readable. The specification of functional models by means of an FML eliminates the necessity of updating function libraries since only a single interpretive program is necessary to simulate all models.

Since FMLs are a recent development in functional modelling techniques this approach will be discussed in more detail.

(e) Physical models [31], [32]

This technique uses a physical sample of the actual device as a model, which permits performance of the logic simulation of devices whose complexity excludes the generation of a software model from the published documentation.

During a simulation run when inputs to the function change, instead of the output response being evaluated by an appropriate subprogram, the subprogram call will activate the code which controls the physical modelling system. When the physical model has responded to the input changes, the output response is transmitted back to the simulator, if the response of the model is different from its current state an event is scheduled by the simulator for future evaluation. Both static and dynamic parts can be processed by the physical modelling systems; however, the current output state of a static part is maintained continuously in the sample of the device and hence each instance of a given block requires a separate model. When a dynamic part is scheduled to be evaluated in order to ensure the correct output response all previous input patterns, stored in RAM, are first applied to the device, followed by the application of the given input pattern; the repeatability of the logical response of a physical model to the vast number of repeated input patterns is a major requirement of device samples used in a physical modelling system.

A limitation in the use of the physical modelling approach is its inability to represent worst case timing conditions; however, this can be overcome by using the physical model to produce only the logical response and performing the delay modelling in software. There is also the uncertainty that if the simulation results are incorrect the cause may be due to device failure rather than a design fault.

3.5.1 Modelling digital circuit functions using FMLs

The use of FMLs is emerging as the panacea for readily modelling not only high level functions in digital circuits but also permitting analogue functions to be described and simulated in a digital environment.

To date many functional modelling languages have evolved; some of which, in essence, are specification languages which permit the function of a system to be defined at the outset of a design, against which the lower levels of representation can be compared.

Although the detailed format of these languages used to describe the function of a system at a high level may differ, each statement comprises the following components [30]:

1. A cause – *WHEN*;
2. An effect – *MAKE/DO*, followed by an action list which may comprise
   (a) register transfers or signal value assignments;
   (b) conditional statements, i.e. *IF . . . THEN . . . ELSE* statements which may be nested;
3. A temporal constraint – *WITHIN/AFTER*;
4. A nullifying action – *UNLESS*.

A simple example in the use of an FML statement is shown below.

WHEN RESET = 1 *DO* REGA(0:3) = 0000 *WITHIN* 20

This states that when the signal RESET is a logic 1 a four-bit register REGA is set to zero within 20 time units.

Another example is

WHEN CLOCK = 1 *IF* SELECT = 1 *THEN* A = B *WITHIN* 5
ELSE A = C *WITHIN* 5

stating that when the CLOCK is a logic 1, if SELECT is also a logic 1, A is set to B within 5 time units; however, if SELECT is not a logic 1, then A is set to C within 5 time units.

Analogue or quasi-analogue functions, for example Schmitt Triggers and A/D converters, can also be described using FMLs and subsequently simulated in a digital environment. In this instance the analogue signals are represented as quantized values using a binary number on a bus.

The Schmitt Trigger (see for example [33]) would be described as

IF INPUT_BUS > UPPER_THRESHOLD *THEN* OUTPUT = 1
IF INPUT_BUS ≤ LOWER_THRESHOLD *THEN* OUTPUT = 0

Two examples of functional/hardware description languages used are FML [30] and HILO [34]. In practice, however, there are many hardware description languages and no one language is widely used. An attempt has been made, by the VHSIC program, to produce a standard language called VHDL [35].

3.5.2 Anomalies in functional level modelling

In the design of digital systems functional level simulation is also used in a mixed mode of simulation where some functions are modelled at functional level and others at gate level. This mixed mode of simulation is used when designers want to verify that the proposed interfaces between a function block and the rest of the circuit are still maintained when the block is realized at a lower level of abstraction as a gate level implementation without having to simulate the complete system at gate level. Mixed mode simulation may also be used as a mechanism for applying test patterns to an internal function realized at gate level, in order to determine the testability of the gate level representation. The intermediate blocks between the internal function being tested and the primary inputs effectively transform test patterns applied at the primary inputs into suitable test patterns to test the internal function, thus relieving the designer of the onerous task of determining what logic values will be applied to the inputs of the function under test when a test pattern is applied to the primary inputs. Using this approach the fault coverage of a module may be ascertained without waiting until the complete circuit is designed at gate level.

The ability to perform mixed mode simulation implies that functional models can process multivalued logic, in particular unknown values. These unknown values may arise also in a normal functional level simulation due to improper initialization, bus contentions, etc. Several techniques have been devised to process correctly unknown states on the inputs to functional models. The first technique is the exhaustive method, whereby the function is evaluated with the unknown values replaced by the combinations of ones and zeros; if the model output differs for any two evaluations the outputs are set to unknowns. This approach can be time consuming if a large number of inputs are at an unknown value. The problem, however, is more complex if a control input to a functional block is unknown; again the block is evaluated under all conditions and subsequent events are flagged as 'potential'. Thereafter, if the 'potential' event [36] can change the value of a block it is considered as an unknown value; otherwise it has no effect upon the block. An alternative approach is to write the capability of handling unknown values into the functional model. The first of these techniques uses a library of simpler three-valued functions [37] from which more complex functions can be generated, algorithms have been developed for handling effectively three-valued logic in these simpler functions, which include arithmetic and logic functions, count and shift functions, read/write functions in memory etc. The second technique is to write the action to be taken when an input has an unknown value into the model specification; for example, in the functional model of a pass transistor if the condition of an unknown value appearing on the gate terminal must be processed; this can be achieved easily by the following statements [33]:

IF GATE = X *THEN*
   IF (SOURCE_STRENGTH > DRAIN_STRENGTH)
   AND (SOURCE ≠ DRAIN) *THEN*
   MAKE DRAIN = UNKNOWN

A similar procedure can also be used for handling high impedance states in circuits or notifying the designer of the presence of non-functional inputs to the inputs of a block.

3.6 SIMULATING MOS SWITCHING CIRCUITS

The extensive use of MOS devices as autonomous switching elements inside logic circuits has proved to be troublesome when conventional gate level simulators have been used to evaluate the logic function performed by these circuits. The features of MOS circuits which make them difficult to simulate using conventional gate level simulators are the bilateral switching characteristics of the devices, the inherent data storage capability of isolated nodes in the circuit and the ability to group MOS transistors together to perform complex logical functions which cannot be readily realized using basic gate structures, thus rendering the artificial translation of a function into logic gates, for the purpose of simulation, impractical.

The problems of simulating digital circuits containing MOS device configurations, which could not be simulated using conventional gate level simulators, have been solved to a large extent by modelling MOS devices at the 'logical transistor' gate level, and more generally at 'Switch' level.

3.6.1 'Logical transistor' gate level modelling [38]

The 'logical transistor' model for an MOS transistor permits the capabilities of gate level simulators to be extended so that they can model, efficiently and with some degree of accuracy, the unusual characteristics of MOS devices. In addition to the normal values of 0, 1 and X (unknown) used in logic simulation, a fourth, high impedance or Z state is introduced to describe, for example, the output of a pass transistor or a tristate buffer when the clock or control input is inactive (i.e. logic 0). The concept of a 'consensus' node is also introduced which facilitates the evaluation of the logic state on the output of a wired–OR gate, fed by several pass transistors as shown in Fig. 3.7 or on a tristate bus as shown in Fig. 3.8; in the case of the tristate bus the different delays experienced between the source and sinks on the bus can be described

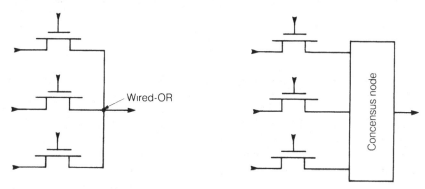

Fig. 3.7 Use of a consensus node to model a wired-OR gate.

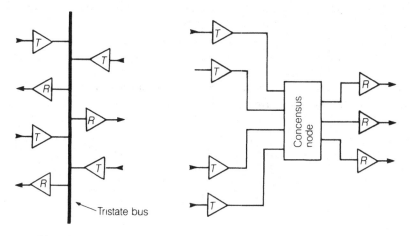

Fig. 3.8 Use of a consensus node to model a tristate bus.

using a 'delay matrix' which defines the delay between each source and each sink.

The behaviour of the MOS device at the logical transistor level is simulated by a simple subroutine which, in the simplest form, implements the truth table shown in Table 3.1; the decay of stored charge in MOS devices in the undriven or high impedance state can be simulated by introducing a 'timeout' parameter in the model, which ensures that if the gate input to the MOS device has not been activated for a given time the output will go into the undefined condition.

Table 3.1 Logical transistor truth table [38]

Source (data)	Gate (enable)	Drain (output)
$0, 1, X$	0	Z
$0, 1, X$	1	$0, 1, X$
$0, 1, X$	X	X

The introduction of the 'high impedance' state as an input to a conventional gate initially proved to be troublesome, as shown in Fig. 3.9, where it is assumed that node A is at a 'high impedance' value; what value is assigned to the output of the inverter? Initially the 'high impedance' state was considered to be an 'unknown' value; this resulted in very pessimistic simulation results. To overcome this difficulty the 'high impedance' state was more strictly defined [39]–[41] as either a high-impedance-one ($Z1$), a high-impedance-zero ($Z0$) or a high-impedance-unknown (ZX), which were considered as 0, 1 and X when applied to conventional gate inputs.

The logical transistor model of an MOS device together with the concept of a consensus node can be used to model one of the most troublesome characteristics of MOS devices with respect to gate level simulation, namely its bilateral switching characteristic. The model of a bilateral switching device which can be used in a normal gate level simulator is shown in Fig. 3.10; the

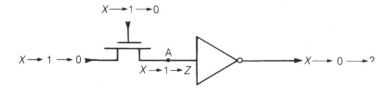

Fig. 3.9 Potential problem in processing high impedance states.

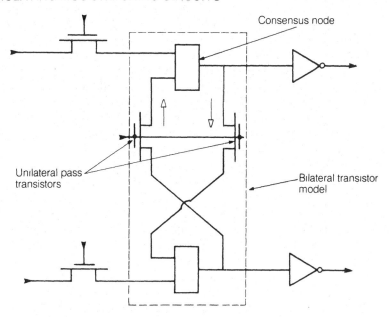

Fig. 3.10 Bilateral transistor model [38].

model comprises two consensus nodes and two unilateral pass transistors. Many other digital MOS circuit configurations can be modelled in a similar way.

3.6.2 Switch level modelling [42]

An MOS circuit can be described as comprising bidirectional FET switches interconnected by wires whose intrinsic capacitance is sufficient to store, dynamically, logic values in the circuit. Consequently the nature of MOS designs is not accurately reflected in gate level simulators which are based on unidirectional logic gates interconnected by memoryless wires. However, as discussed previously, some *ad hoc* extensions can be made to gate level simulators to make them capable of simulating some MOS circuit configurations. To overcome the general problems of simulating MOS circuits the concept of the *switch level simulator* was evolved, in which the basic switching primitive is an MOS device modelled as a voltage controlled switch having, in general, no preferred direction for signal transmission; this is in contrast to gate level primitives which inherently exhibit a unilateral switching characteristic. At switch level the circuit is considered to comprise a series of nodes connected by transistors: it is emphasized that these transistors are regarded as ideal switches and not as analogue devices as used in circuit simulation programs such as SPICE; consequently circuits comprising thousands of devices can be simulated readily without incurring excessive simulation times. Furthermore, since the simulation primitive corresponds to the basic active elements used in the layout, the circuit description for the simulator can be extracted directly from the layout without any reference to the intended function of the circuit, thus enabling designers to verify that the layout does realize the intended function, without the translation from the layout to some suitable simulation configuration being biased by an *a priori* knowledge of the circuit function.

The basic circuit model used in switch level simulators comprises a network of nodes interconnected by transistors. The nodes are categorized as input and storage. The input nodes are considered as power and ground, clock and data input connections to the circuit; storage nodes are any other internal nodes which are capable of storing a logic value dynamically. During the evaluation of logic changes in a circuit a three-valued simulation algorithm is used; the high impedance state is not used in this instance since its effect is considered automatically by the circuit model. In deciding what logic value is to be assigned to a node, the charge handling capabilities of a node and the 'strengths' of the transistors driving the node are examined. In situations where storage nodes must share charge, the state of the node with the largest charge handling capacity dominates; alternatively if a storage node is connected to a set of input nodes by conducting transistors the storage nodes assume the logic value of the input node connected to it via the transistor path of greatest strength, i.e. greatest conductance; transistor strengths are only used in evaluating ratioed logic.

In the switch level simulator the behaviour of the circuit is characterized in terms of its steady state response, that is the dynamic behaviour of the circuit in terms of rates at which nodes either charge or discharge is not considered. Furthermore it is assumed in these simulators that the system is activated by two non-overlapping clock phases which are far enough apart that the circuit can settle before the next clock is activated, but sufficiently close so that charge decay on the nodes does not occur. The effect of a change in inputs is evaluated by performing repeated series of unit delay simulations on the transistors until a stable state is reached or the maximum number of iterations for a given clock phase has been reached.

It would be extremely inefficient when evaluating the effect of a change at the input nodes of the circuit to consider all nodes in the circuit; consequently, several techniques similar to the selective trace technique used at gate level have been developed at switch level. The first technique, prior to the start of a simulation run, assigns a number of transistors to a group such that interconnections between groups are unidirectional. However, within groups bi-directional connections between nodes may exist; consequently each group can be viewed as an autonomous logic block and evaluation of a group occurs only if an input to a group changes state. Many of the transistor groups form well defined logic functions such as NAND and NOR gates and can be simulated at gate level since the input/output relationships are well defined.

A second technique which is used to improve simulation efficiency is called 'perturbation' [43] simulation. In this technique local changes or perturbations in logic value are evaluated incrementally; the computations are localized by identifying nodes in the vicinity of the perturbation and hence require evaluation. After the evaluation of these nodes has occurred an updated set of nodes to be evaluated next is generated; this procedure is repeated until a stable set of node values is obtained.

As an alternative to evaluating the logic state at the output node of a group of MOS transistors by considering ideal switches being opened/closed and storage nodes being charged or discharged, it has been proposed that for small groups of transistors a decision diagram [44] could be used to evaluate the output of the group. The decision diagrams are stored as programs; the size of the diagram can be reduced by only considering observable nodes. The decision diagram for a group of transistors is called only when an input to that group of transistors changes. This principle is demonstrated with the three-input majority voting gate shown in Fig. 3.11.

A major limitation in the use of switch level simulators is their inability to

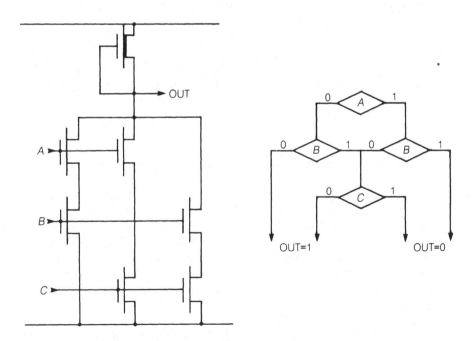

Fig. 3.11 Decision diagram representation of a three-input majority voting gate [44].

handle undefined values efficiently or detect race conditions. The problem with an undefined value at switch level is that there is no way of curbing its sphere of influence. A first attempt at trying to resolve undefined logic values comprised exhaustive simulations in which the value of the unknown was considered to be a logic 1 and then a logic 0; if the same stable condition was reached in both cases the effect of the unknown could be resolved, otherwise the output was set to an unknown; this technique is obviously limited.

The generation and the effect of unknown values in switch level simulators has been solved to some extent by the application of 'ternary' [45] logic. In this instance input transitions are modelled as changes from say a logic 1 to an unknown value, then to a logic 0 or vice versa. The simulation algorithm is implemented in two phases, namely a 'transition phase' where only state changes from either 1 or 0 to X can occur, followed by a 'stabilization' phase when only state changes from X to either 1 or 0 can occur, which would represent a stable state on that node.

Another solution to the problem of handling unknowns and races in switch level simulators proposes a set of criteria to be imposed upon the design which should inhibit the generation of unknown values; some of the design criteria are checked prior to simulation whilst others are checked during simulation. If an unknown state is generated the simulation is stopped since the circuit is deemed not to be a 'clean' design. If required a detailed analysis of the cause of the unknown value, which is probably due to a race condition, can be analysed using a 'switched' graph technique [46].

The basic concepts of switch level simulation strikes an acceptable balance between a detailed electrical model of a circuit and an abstract gate level model; and the simplicity of the basic primitives also permits large circuits to be evaluated efficiently. Switch level simulation has also been implemented for analysing bipolar digital circuits [47].

3.7 HARDWARE ACCELERATORS FOR SIMULATION [48]

The most significant developments in accelerators have been in the area of simulation. The focus, initially, has been directed at gate level simulation because this is the major mechanism used to verify that circuit designs are correct prior to fabrication. The growing complexity of VLSI circuits has increased the amount of simulation to be performed on designs, demanding a greater efficiency from these tools if the simulation process is to be carried out in a reasonable time. Improvements in efficiency can be obtained by increasing the computational power of the host computer, since the simulation process, although not complex, is CPU intensive. However, improvements gained in this way are limited unless very powerful and costly computers are used, since the simulation algorithms do not map efficiently onto general purpose computer architectures which were developed primarily to solve differential equations.

In order to overcome the efficiency barrier special purpose computers or accelerators have been developed. Improvement in the efficiency is achieved in these machines by

1. removing the operating system
2. customizing individual processors to perform particular tasks in an algorithm
3. exploiting algorithmic concurrency by either paralleling or pipelining processors
4. exploiting circuit concurrency, that is the simultaneous propagation of logic changes in the circuit, by using a number of evaluation processors in parallel.

The processing capabilities of hardware accelerators are quoted as ranging from 5×10^5 to 2×10^9 evaluations per second; it should be noted that these speed-up figures usually refer to the processor operating in isolation. A major overhead, however, associated with hardware accelerators is the amount of time (compile time) taken to translate a circuit description into a suitable format for the accelerator. When the translation time is taken into account, particularly for short simulation runs, the hardware accelerator is slower than the software simulator; however, for long simulation runs the hardware accelerator has a distinct advantage over the software simulator.

Several examples of hardware accelerators for simulation will now be briefly described:

(a) A computer architecture for digital logic simulation [49]

The architecture shown in Fig. 3.12 will support a table-driven time based activity directed logic simulator. The host computer is used only for the pre- and postprocessing of data and does not take part in the simulation process.

The system comprises three processors and five memory blocks. The function of the UPDATE processor is that of signal propagation and activity flag setting. The EVAL processor searches for gates which have to be evaluated at each stage and initiates any changes in value which may occur. The event queue processor performs simple manipulations on the event queue memory (see Section 3.4.2.2). The function of the memory blocks FIM and FOM is to store the fanin and fanout data of the gates respectively. The AFM memory block contains the flags of the active gates in the circuit and is used in conjunction with FOM for event directed simulation. The SDM memory block contains data about each gate in the circuit; for example, gate type, number of fanins, number of fanouts, etc. The performance figures for this architecture are quoted as 197 000 gate evaluations per second.

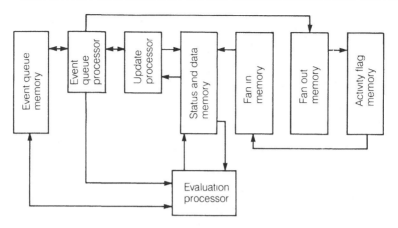

Fig. 3.12 Architecture for a hardware logic simulator [49].

The Tegas accelerator is adapted from this architecture and can simulate approximately one million gate evaluations per second which, depending upon the percentage of simulation activity occurring, is a speed improvement of approximately 200–700 times on a software simulator running on a VAX 11/750.

(b) Simulation accelerators at IBM [50]–[52]

Three simulation accelerators have been developed at IBM, namely the Logic Simulation Machine (LSM), the Yorktown Simulation Engine (YSE) and the Engineering Verification Engine (EVE). The three accelerators are based upon the same architectural concepts and hence the YSE will be the only accelerator discussed in detail.

The YSE was developed to speed up the process of verifying the logic function of a new range of machines. The amount of gate simulation time used in the design process by IBM engineers to verify one quarter of the logic of a medium range machine was approximately 1800 hours of CPU time on an IBM 370/168; it has also been quoted that in order to verify the logic of a 'large' machine, it required four 370/168s running two shifts per day for several months.

The block diagram of a YSE is shown in Fig. 3.13. It is a special purpose highly parallel programmable architecture for performing either rank ordered or unit delay simulation. Timing accuracy is not a serious limitation of the machine since it is used in a restricted Level Sensitive Scan Design (LSSD) environment (see Chapter 6).

The major components in the YSE are the logic processors which are used to simulate combinational logic, array processors for simulating ROM and

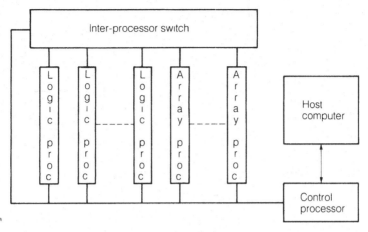

Fig. 3.13 Yorktown simulation engine (YSE) [51].

RAM, an interprocessor switch which permits communications between a maximum of 256 logic/array processors. The control processor and the host computer also run local self-test procedures.

The YSE has the capability of evaluating circuits comprising 2 million gates at the rate of 3 billion gate evaluations per second, implying that it can simulate some commercially available eight-bit microprocessors faster than they can execute their native instruction set.

In the design process the YSE is used for a range of functions: for example, system verification, diagnostic verification, partial hardware 'bring-up', fault simulation, test generation, etc.

The YSE has a compiled code model and does not use stimulus bypassing techniques; the underlying philosophy is that the processing capabilities in the YSE are such that instead of wasting time deciding what should be evaluated at a given instant, together with the associated interprocessor communication problems, simply evaluate all gates at each instant in time. This approach does imply that a large number of gates are re-evaluated which are inactive; consequently the performance advantage of the YSE is not as large in comparison to other accelerators when the throughput is measured in terms of active gate evaluations per second.

(c) Logic simulation machine[53]

Logic activity in a digital circuit is a concurrent process; however, software simulators, by the nature of the architecture of the host machine, do not exploit this characteristic. A special purpose simulator, shown in Fig. 3.14, has been designed by Bell Laboratories, and incorporates a distributed

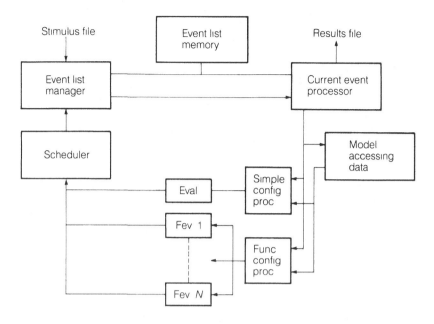

Fig. 3.14 Logic simulation machine [53].

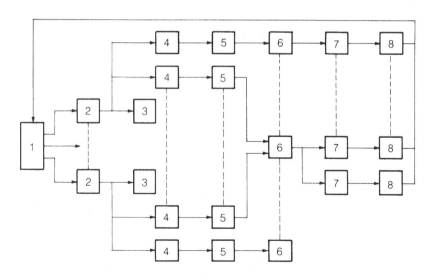

Fig. 3.15 Potential concurrency in logic simulation [53]. 1. Advance time. 2. Retrieve current event. 3. Update config. of source. 4. Determine fanout. 5. Update config. of fanout. 6. Evaluate. 7. Schedule. 8. Insert in event list.

processor architecture to exploit the concurrency characteristics in simulation algorithms, permitting large digital circuits to be simulated at the rate of one million gate evaluations per second, which is approximately fifty times faster than the capability of a software simulator.

The precedence diagram, shown in Fig. 3.15, indicates the activities in logic simulation which, ideally, could be done concurrently, a separate processor being dedicated to each activity. However, true parallelism would require multiple copies of fanout lists, event queues, etc., which would require a vast amount of storage for large circuits. On the other hand, if each processor accesses the same memories then the concept of parallelism is lost. In the logic simulation machine, concurrency is maintained to some degree by using a pipe line data flow algorithm.

The logic simulation machine uses an event-driven simulation technique. When a given event (logic change) occurs in the circuit the configuration processors are activated to update the signal values on the inputs of the elements whose outputs may be affected by the event. If the elements are basic gates or flip-flops the updating of the logic values on the inputs is carried out by the simple configuration processor and the results of the input changes determined by the EVAL processor. If the output of a gate is to change, the time of change is determined, by the block SCHED, by adding on the appropriate delay times for the gate to the current simulation time; thereafter the EVENT LIST MANAGER will schedule it into the appropriate time slot in the EVENT LIST MEMORY. The Functional Evaluation (FEV) blocks are used to process functional elements which are more complex than basic gates and flip-flops. Since the functional elements will be evaluated in parallel it may be considered that a large number of evaluation blocks will be required; however, since an activity-directed algorithm is used the number of FEVs required at any time can be quite small. In general the functional elements are dynamically assigned to FEVs during the simulation process; however, if the functional block is large an FEV is permanently assigned to that function. In some situations two or more large functions may have to be assigned to the same FEV and algorithms have been developed to minimize the probability of two elements assigned to the same FEV being activated simultaneously.

The performance of the logic simulation machine may be enhanced by incorporating a cache memory in conjunction with a 'data look-ahead' technique to further exploit the characteristics of event directed simulation. In event directed simulation, it is known beforehand which sections of the circuit will be evaluated next; consequently the data relevant to this part of the circuit can be brought into the cache memory, whilst the simulator is evaluating the gate changes for the current time, ready to be passed to the evaluation modules when required. The efficiency of the event scheduling processes may also be improved by implementing it as a hardware priority queue in such a way that the processing time of the queue is independent of its length.

The concept of using accelerators for gate level simulation has been extended to perform the simulation process at other levels of abstraction, namely block level [54] and switch level [55]. Accelerators have also been implemented to improve the efficiency of other CAD tools: for example, dimension rule checkers and routers. The use of arrays of general purpose processors has been proposed as an alternative means of obtaining the high performance throughput achieved by hardware accelerators for gate level simulation. The arrays of general purpose processors [56] offer several advantages over conventional accelerators. First, since the processors are programmable the need to perform the tedious process of translating a circuit description into a format suitable for the accelerator is eliminated. Second, the simulation algorithm can be easily modified to handle simulation primitives described at different levels of abstraction. Third, although the circuit must also be partitioned over several processors, the partitioning algorithm can be implemented on the parallel processor, instead of being run on a separate host computer. Finally, the parallel processor can also be used to analyse and subsequently process the simulation results. The potential of using general purpose multiprocessor architectures to exploit circuit concurrency at switch level has also been investigated [57], [58].

Although hardware simulators, in general, offer a considerable speed advantage over software simulators, it is considered that their use is an interim solution to the problem of verifying VLSI circuits and will be superseded by the more formal verification techniques currently used in computer science for program proving [59],[60].

3.8 REFERENCES

1. Walker, R. A. and Thomas, D. E. (1985) A model of design representation and synthesis. *22nd Design Automation Conference Proceedings*, June, 453–9.
2. DeMan, H. (1980) *Computer Aided Design Techniques for VLSI.* NATO Advanced Study Institute Course Notes, Belgium, 1980.
3. Russell, G. (ed.) (1985) *CAD Tools for VLSI Design*, Chapter 6, Van Nostrand Reinhold (UK).
4. Szygenda, S. A. and Thompson, E. W. (1975) Digital logic simulation in a time based table driven environment: Part I Design verification. *Computer*, March, **8**, 24–36.
5. Brooks, F. P. (1978) *The Mythical Man Month*, Addison-Wesley.
6. Russell, G. (1984) Gaelic logic simulator: An interactive CAD tool. *Computer Aided Design*, **12**(4), 195–8.
7. Staff (VLSI System Design) (1986) 1986 survey of logic simulators. *VLSI System Design*, February, 32–40.

8. Breuer, M. A. and Friedman, A. D. (1977) *Diagnosis and Reliable Design of Digital Systems*, Chapter 4, Pitman.

9. Newton, A. R. (1980) *Timing Logic and Mixed Mode Simulation for Large MOS Integrated Circuits*. NATO Advanced Study Institute Course Notes, Urbino, 1980.

10. Ulrich, E. (1980) Table look-up techniques for fast and flexible digital logic simulation. *17th Design Automation Conference Proceedings*, June, 560–3.

11. Ulrich, R. G. (1969) Exclusive simulation of activity in digital networks. *Comm. ACM*, **12**, 102–10.

12. Bening, L. (1979) Developments in computer simulation of gate level physical models. *16th Design Automation Conference Proceedings*, June, 561–7.

13. Hayes, J. P. (1986) Digital simulation of multiple logic values. *IEEE Trans. Computer Aided Design*, **CAD-5**(2), 274–83.

14. Eichelberger, E. B. (1965) Hazard detection in combinational and sequential logic circuits. *IBM J. Research and Development*, **9**, 90–9.

15. Bryant, R. E. (1980) An algorithm for MOS logic simulation. *Lambda*, **1**, 46–53.

16. Szygenda, S. A. and Hemming, C. (1972) Functional simulation – A basis for a systems approach to digital simulation and fault diagnosis. *Summer Computer Simulation Conference Proceedings*, June, 270–81.

17. Szygenda, S. A. and Lekkos, A. A. (1973) Integrated techniques for functional and gate level logic simulation. *10th Design Automation Conference Proceedings*, June, 159–72.

18. Chappel, S. G., Menon, P. R., Pellegrin, J. E. and Schowe, A. M. (1972) Functional simulation in the LAMP System. *J. Design Automation and Fault Tolerant Computing*, May, 203–15.

19. Wilcox, P. (1979) Digital logic simulation at gate and functional level. *16th Design Automation Conference Proceedings*, June, 242–8.

20. Van Cleemput, W. M. (1979) Computer hardware description languages and their applications. *16th Design Automation Conference Proceedings*, June, 554–60.

21. Shiva, S. G. (1979) Computer hardware description languages – A Tutorial, *Proc. IEEE*, **67**(12), 1605–15.

22. Thompson, E. W., Karger, P., Read (Jr), W. R. *et al.* (1980) The incorporation of functional level element routines in an existing digital simulation system. *17th Design Automation Conference Proceedings*, June, 394–401.

23. Fong, J. Y. O. (1981) Microprocessor modelling for logic simulation. *1981 IEEE Test Conference Proceedings*, 458–60.

24. Malek, M., Bose, A. K. (1978) Functional simulation and fault diagnosis. *15th Design Automation Conference Proceedings*, June, 340–6.

25. Kjelkerud, E. and Thessen, O. (1979) Methods of modelling digital devices for logic simulation. *16th Design Automation Conference Proceedings*, June, 235–41.

26. Raeth, P. G., Acken, J. M., Lamont, G. B. and Borky, J. M. (1981) Functional modelling for logic simulation. *18th Design Automation Conference Proceedings*, June, 791–5.

27. Hill, F. J., Swanson, R. E., Masud, M. and Navabi, Z. (1981) Structure specification with a procedural hardware description language, *IEEE Trans. Computers*, **C-30**(2), 157–61.

28. Mokkarala, V. R., Fan, A. and Apte, R. (1985) Unified approach to simulation and timing verification at functional level. *22nd Design Automation Conference Proceedings*, June, 757–61.

29. Robinson, P. and Dion, J. (1983) Programming languages for hardware description. *20th Design Automation Conference Proceedings*, June, 12–16.

30. Noon, W. A. (1977) Design verification and logic validation systems. *14th Design Automation Conference Proceedings*, June, 362–8.
31. Staff (VLSI System Design) (1984) Physical models for logic simulation. *VLSI Design*, June, 62–7.
32. Stoll, P. A. (1985) PMX: A hardware solution to the VLSI model availability problem. *22nd Design Automation Conference Proceedings*, June, 719–23.
33. Robson, G. (1984) Logic design using behavioural models. *VLSI Design*, January, 36–44.
34. Rappaport, A. (1983) Digital logic simulation software supports behavioural models. *EDN*. **28**(10), 95–8.
35. VHDL (1986) VHDL: The VHSIC Hardware Description Language. *IEEE Design and Test of Computers*, **13**(2), Complete Issue.
36. DesMarias, P. J., Shew, E. S. Y. and Wilcox, P. S. (1982) A functional level modelling language for digital simulation. *19th Design Automation Conference Proceedings*, June, 315–20.
37. Alia, G., Ciempi, P. and Martineui, E. (1978) LSI component modelling in a three valued functional simulation. *15th Design Automation Conference Proceedings*, June, 428–38.
38. Sherwood, W. (1981) A MOS modelling technique for 4-state true value hierarchical logic simulation. *18th Design Automation Conference Proceedings*, June, 775–85.
39. Holt, I. D. and Hutchings, D. (1981) A MOS/LSI orientated logic simulator. *18th Design Automation Conference Proceedings*, June, 280–7.
40. Watanabe, J., Miura, J., Kurachi, T. and Suetsuga, I. (1980) Seven value simulation for MOS LSI circuits. *Proc. ICCC80*, October, 941–4.
41. Flake, P. L., Moorby, P. R. and Musgrave, G. (1983) An algebra for logic strength simulation. *20th Design Automation Conference Proceedings*, June, 615–18.
42. Bryant, R. E. (1981) MOSSIM: A switch level simulator for MOS LSI. *18th Design Automation Conference Proceedings*, June, 786–90.
43. Schuster, M. D. and Bryant, R. E. (1984) Concurrent fault simulation of MOS digital circuits. *Proc. 1984 Conf. Advanced Research in VLSI*, MIT, January, 109–38.
44. Roy, C., Demers, L.-P., Cerny, E. and Gecsei, J. (1985) An object oriented switch level simulator. *20th Design Automation Conference Proceedings*, June, 623–9.
45. Bryant, R. E. (1983) Race detection in MOS circuits by ternary simulation. VLSI '83, Elsevier Science Publishers (North-Holland), pp. 85–95.
46. Ramachandran, V. (1983) An improved switch level simulator for MOS circuits. *20th Design Automation Conference Proceedings*, June, 293–9.
47. Hajj, I. N. and Saab, D. (1987) Switch-level simulation of digital bipolar circuits. *IEEE Trans. Computer Aided Design*, **CAD-6**(2), 251–8.
48. Blank, T. (1984) A survey of hardware accelerators used in computer aided design. *IEEE Design and Test of Computers*, **1**(3), 21–39.
49. Barto, R. and Szygenda, S. A. (1980) A computer architecture for digital logic simulation. *Electronic Engineering*, September, 35–66.
50. Howard, J. K., Malm, R. L. and Warren, L. M. (1983) Introduction to the IBM Los Gatos logic simulation machine. *Proc. IEEE Int. Conf. Computer Aided Design: VLSI in Computers*, October, 580–83.
51. Pfister, G. F. (1982) The Yorktown simulation engine: Introduction. *19th Design Automation Conference Proceedings*, June, 51–4.
52. Dunn, L. N. (1984) IBM's engineering design system support for VLSI design and verification. *IEEE Design and Test of Computers*, **1**(1), 30–40.

53. Abramovici, M., Levendel, Y. H. and Menon, P. R. (1983) A logic simulation machine. *IEEE Trans. Computer Aided Design*, **CAD-2**(2), 82–93.
54. Takasaki, S., Sasaki, T., Nomizu, N. *et al.* (1987) Block level hardware logic simulation machine. *IEEE Trans. Computer Aided Design*, **CAD-6**(1), 46–54.
55. Spillinger, I. and Silberman, G. (1986) Improving the performance of a switch level simulator targeted for a logic simulation machine. *IEEE Trans. Computer Aided Design*, **CAD-5**(3), 396–404.
56. Hancock, J. M. and Das Gupta, S. (1986) Tutorial on parallel processing for design automation applications. *23rd Design Automation Conference Proceedings*, June, 69–77.
57. Ashok, A., Costello, R. and Sadayappan, P. (1985) Modelling switch level simulation using data flow. *22nd Design Automation Conference Proceedings*, June, 637–44.
58. Frank, E. H. (1986) Exploiting parallelism in a switch level simulation machine. *23rd Design Automation Conference Proceedings*, June, 20–6.
59. Hantler, L. S. and King, J. C. (1976) An introduction to proving the correctness of programs. *Computing Surveys*, **8**, 331–53.
60. Barrow, H. G. (1984) Proving the correctness of digital hardware designs. *VLSI Design*, July, 64–77.

4
FAULT SIMULATION

4.1 INTRODUCTION

The objectives of fault simulation are to determine how postulated faults, arising through defects in the manufacturing process or due to ageing, affect the operation of the circuit and also how much testing is required to give a certain fault coverage.

In general the types of faults modelled are stuck-at-1/0; although these do not cover all possible faults the justifications for its use are:

1. A set of test patterns which detect all single stuck-at faults will also detect multiple faults; and
2. A set of test patterns which detect all single stuck-at faults will also detect bridging faults.

These justifications have been used by the computer manufacturing industry for many years without any adverse effects, although with changes in technology attempts have been made to model a wider range of faults.

The original fault simulators were true value simulators in which gate outputs could be held at fixed logic values, so that stuck-at-fault conditions could be injected into the circuit. The faults were simulated individually, each test sequence in turn being applied to the circuit. This simple approach is very inefficient when applied to large circuits. In order to improve the efficiency of fault simulation a number of techniques have evolved; although these techniques improved fault simulation efficiency, present-day circuit complexities, increase in demand for circuit reliability and the use of built-in self-test techniques, which use large sequences of test vectors, have demanded an even greater efficiency from the fault simulation techniques; consequently improvements in the process have been sought by performing fault simulation at higher levels of abstraction. Furthermore, alternative approaches, which are deemed to be more efficient than fault simulation in determining the fault coverage of a set of test patterns, have been developed.

4.2 BASIC TECHNIQUES FOR FAULT SIMULATION

To overcome the inefficiencies in using a modified true value simulator for fault simulation several techniques have been developed, namely:

1. Parallel fault simulation;
2. Deductive fault simulation;
3. Concurrent fault simulation;

and, more recently,

1. Parallel Valued Lists (PVL);
2. Parallel Pattern Single Fault Propagation (PPSFP).

Each of these techniques will be discussed in some detail.

4.2.1 Parallel fault simulation [1]

In parallel fault simulation N copies of a circuit are simulated in parallel, one fault-free circuit (reference) and $(N-1)$ faulty circuits. The number of faulty circuits which can be simulated in parallel depends upon the length of the computer word, each bit position in the word being used to identify a particular fault in the circuit. Consequently if there are M faults to be simulated, the number of simulation runs per input test pattern is the smallest integer greater than $M/(N-1)$. The number of simulation passes can be reduced further if independent faults are simulated together, that is faults which do not affect a common part of the circuit. Thus if the circuit has P outputs, then the number of simulation passes required in this instance is the smallest integer greater than $M/((N-1).P)$. This technique, however, can only be used to advantage provided the circuit fanout is low and groups of independent faults are readily identified. The number of faults simulated in parallel may be increased to, say, 2048 if long bit-string machine instructions are used; this, however, makes the program machine dependent. The technique of parallel fault simulation exploits the fact that logical operations are performed on computer words rather than individual bits. To illustrate the process consider the circuit shown in Fig. 4.1. Each bit position represents a fault in the circuit. The method of performing parallel fault simulation is as follows:

1. Set up the input values for each version of the circuit; the underlined values in Fig. 4.1 represent the fault conditions;
2. Create new values for the internal nodes in the circuit by performing the logic function indicated by the gates, inserting the fault conditions where appropriate;
3. Continue (2) until the output of the circuit is evaluated. The list of faults detected by a given test pattern is obtained by comparing the logic values in

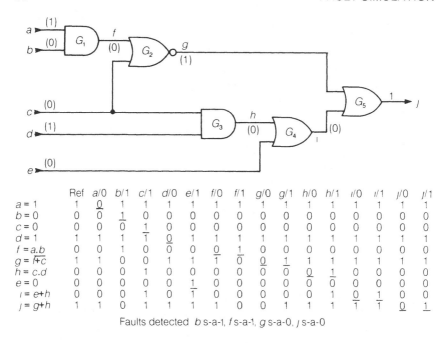

	Ref	a/0	b/1	c/1	d/0	e/1	f/0	f/1	g/0	g/1	h/0	h/1	i/0	i/1	j/0	j/1
a = 1	1	0	1	1	1	1	1	1	1	1	1	1	1	1	1	1
b = 0	0	0	1	0	0	0	0	0	0	0	0	0	0	0	0	0
c = 0	0	0	0	1	0	0	0	0	0	0	0	0	0	0	0	0
d = 1	1	1	1	1	0	1	1	1	1	1	1	1	1	1	1	1
f = a.b	0	0	1	0	0	0	0	1	0	0	0	0	0	0	0	0
g = $\overline{f+c}$	1	1	0	0	1	1	1	0	0	1	1	1	1	1	1	1
h = c.d	0	0	0	1	0	0	0	0	0	0	0	1	0	0	0	0
e = 0	0	0	0	0	0	0	1	0	0	0	0	0	0	0	0	0
i = e+h	0	0	0	1	0	1	0	0	0	0	0	1	0	1	0	0
j = g+h	1	1	0	1	1	1	1	0	0	1	1	1	1	1	0	1

Faults detected b s-a-1, f s-a-1, g s-a-0, j s-a-0

Fig. 4.1 An example of parallel fault simulation.

each bit position with the value in the reference position; those values which differ indicate the faults detected by the test pattern.

It should be noted that in parallel fault simulation the faulty circuit is still simulated although the faulty and fault-free circuits produce the same response; this does not occur in other techniques for fault simulation.

It is considered that this technique can only be used effectively on systems, combinational or sequential, comprising less than 1000 gates when the faults selected to be simulated in parallel have a high probability of remaining active throughout the simulation pass, otherwise a vast amount of CPU time is wasted in processing inactive faults. Furthermore it is not possible to consider time delays efficiently during fault simulation, particularly if the gate rise and fall times differ. The complexity of the simulation algorithm also increases if more than a two-valued simulation is used and, in general, the circuit must be described in terms of the basic Boolean functions of AND, OR and NOT. The main advantage of parallel fault simulation is that it requires less memory for the storage of faults in comparison to dynamic fault list propagation techniques, for example deductive and concurrent fault simulation, and furthermore its storage requirements are predictable, which is not the case for either the deductive or concurrent fault simulation techniques.

4.2.2 Deductive fault simulation [2]

In the technique of deductive fault simulation only the fault-free logic is simulated and from the current state of the fault-free logic it is deduced which faults can be detected at any internal node or primary output of the circuit.

Since all detectable faults are determined at the same time it is only necessary to perform one simulation run per test pattern; this is in contrast to the parallel fault simulation technique where $W/(N-1)$ passes per test vector were required.

In general, the time taken for a single simulation run of the deductive fault simulator is greater than that for a single pass of the parallel fault simulator; however, it is far less than the total time taken to perform the $W/(N-1)$ passes required by the parallel fault simulator for each test vector.

The main step in deductive fault simulation is that of generating fault lists which comprise

1. faults detected at predecessor gates whose effect is propagated to the input of a given gate
2. faults which originate at a given gate and produce an incorrect output at that gate for a given set of input conditions.

The fault lists generated on a gate output are determined from the gate type, the fault-free inputs to the gate, the input fault lists (i.e. fault effects to be propagated) and the postulated faults on the gate. The process of fault list propagation is implemented using a fault list algebra which uses set union and intersection operators.

The rules for deriving the fault lists are summarized below [3]

Consider an arbitrary Boolean function $F(X_1, X_2, X_3, \ldots, X_n)$ realized as either a sum of products or a product of sums expressions:

1. If the value of X_i in the fault-free circuit is zero, replace all occurrences of X_i and $\overline{X}_i$ by L_{X_i} and $\overline{L}_{X_i}$ respectively, where L_{X_i} is the fault list associated with input X_i;
2. If the value of X_i in the fault-free circuit is 1, replace all occurrences of X_i and $\overline{X}_i$ by $\overline{L}_{X_i}$ and L_{X_i} respectively;
3. Replace AND and OR operators in the function by the set operators $\cap$ and $\cup$ respectively;
4. Simplify the fault list expression using the set operations defined in (3);
5. Append the appropriate stuck-at fault on the output of the function to the simplified fault list.

The time per pass in the deductive fault simulation technique is a function of the length of the fault lists, which is determined by the number of faults uncovered by a given test pattern. The length of the fault lists are continually changing as they are propagated through the circuit, not only by more faults being added to the list, but also by faults being removed from the list, that is faults which produce the same response as the fault-free circuit. Thus, unlike parallel fault simulation, deductive fault simulation processes only active faults. A major disadvantage of 'dynamic fault list' propagation techniques is the inability to predict the amount of storage required during the process, consequently a deductive fault simulator requires a vast amount of storage which must be allocated dynamically to accommodate the varying size of fault lists. The number of faults processed can be reduced to some extent by using fault collapsing and fault dropping techniques; fault collapsing reduces the number of faults by using a single fault to represent a number of faults which produce the same effect upon the circuit and are, hence, indistinguishable for the applied test pattern; fault dropping implies removing a fault from further processing after it has been detected by, say, two test patterns; these techniques not only improve the efficiency of the process but also reduce memory requirements. If the amount of storage available at any time is too small several passes of the simulator may be made with different fault sets considered; in this way storage capacity is traded against time.

Since faults are essentially considered individually during deductive fault simulation, circuit delays can be considered during the process, although the algorithms become more involved particularly when a gate has dissimilar rise and fall times and the faulty circuit reacts faster than the fault-free circuit [4].

Deductive fault simulation can be used with three-valued logic although the complexity of the process increases. The exact treatment for handling 'unknown' conditions, employs an exhaustive technique, where fault lists are computed for all possible combinations of determinate states, this technique is obviously very time consuming. However, an inexact treatment can also be used, and has been shown to work in most cases. In the inexact treatment several rules have been developed for processing unknown values.

1. If the unknown values on the inputs to a gate produce an unknown value on the output, a single entry fault list called an 'indeterminate' list is assigned to the output;
2. If the output value can be defined although some input values are undefined, the output fault list is generated from the fault list on the defined inputs only and these are flagged as 'star' faults [3] since they may or may not be detected.

When deductive fault simulation is applied to memory devices such as flip-flops the computation of the fault list is also more involved since the computation must include not only faults currently on the inputs but also faults on the output list from the previous state; any fault list derived from conditions which may produce a 'race' are subsequently flagged as 'star' faults.

To illustrate the technique of a deductive fault simulation consider the simple example shown in Fig. 4.2. The propagated fault list will be first derived heuristically and thereafter using the set of rules previously defined for deriving fault lists.

With reference to Fig. 4.2, the fault list to inputs A, B, C and D are $L_a = \{a,b,c,e\}$, $L_b = \{a,b,d\}$, $L_c = \{b,c\}$ and $L_d = \{d,f\}$ respectively. The fault-free inputs to the gate are $A = B = 1$, $C = D = 0$; under these conditions the fault-free output is a logic 0. If a fault is to be detectable on the output of the gate it must be capable of making the output a logic 1 in this instance, that is the fault must make inputs A and B have the value logic 0 but must not affect inputs C and D. On searching the fault lists on the gate inputs, the only fault which can satisfy the given requirements is fault a.

The fault conditions on the inputs which will propagate to the output can also be determined systematically using the rules for fault list derivation discussed previously.

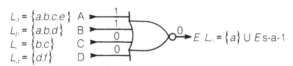

Fig. 4.2 A simple example on deductive fault simulation [2].

Since inputs A and B are at a logic 1, these inputs are replaced with $\bar{L}_a$ and $\bar{L}_b$ (Rule 2).

Inputs C and D are both logic 0; these inputs are replaced with L_c and L_d (Rule 1).

The OR operator is replaced with set operator $\cup$ (Rule 3). The output fault list

$$L_e = \overline{\bar{L}_a \cup \bar{L}_b \cup L_c \cup L_d} = L_a \cap L_b \cap \bar{L}_c \cap \bar{L}_d$$

i.e. if a fault is to be propagated to the output it must be in fault lists L_a and L_b but not in L_c and L_d, resulting in fault a as the only fault to satisfy these requirements. The complete fault list on the gate output is obtained by appending the appropriate stuck-at fault on the output detected by the given input conditions. In general the literals in the fault lists have a 0 or 1 suffix to indicate a stuck-at-0 and a stuck-at-1 fault respectively.

It has been shown by experiment that deductive fault simulation is more cost effective than parallel fault simulation when large circuits have to be simulated with a large number of faults [5].

4.2.3 Concurrent fault simulation [6]

This is another dynamic fault list technique, it combines both the features of parallel and deductive fault simulation. However, in concurrent fault

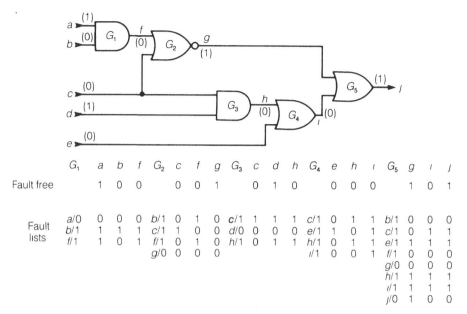

G_1	a	b	f	G_2	c	f	g	G_3	c	d	h	G_4	e	h	i	G_5	g	i	j
Fault free	1	0	0		0	0	1		0	1	0		0	0	0		1	0	1
a/0	0	0	0	b/1	0	1	0	c/1	1	1	1	c/1	0	1	1	b/1	0	0	0
b/1	1	1	1	c/1	1	0	0	d/0	0	0	0	e/1	1	0	1	c/1	0	1	1
f/1	1	0	1	f/1	0	1	0	h/1	0	1	1	h/1	0	1	1	e/1	1	1	1
				g/0	0	0	0					i/1	0	0	1	f/1	0	0	0
																g/0	0	0	0
																h/1	1	1	1
																i/1	1	1	1
																j/0	1	0	0

Faults detected b s-a-1, f s-a-1, g s-a-0, j s-a-0

Fig. 4.3 An example of concurrent fault simulation.

simulation the faulty gates are only simulated when their output value differs from that of the fault-free circuit. The major disadvantage with this technique is again the unpredictability of the amount of storage required; the storage requirements are much greater than that required by deductive fault simulation since all the differences between the faulty and fault-free circuits are stored in the fault lists, whereas in the deductive technique only the differences in the gate output list are stored. Concurrent fault simulation is faster than deductive fault simulation since the simulation process used to propagate fault effects is much faster than the deductive process. Furthermore since faults are processed individually and simulation is used the technique can be used with multivalued logic systems, also circuit delays can be processed readily. The technique of fault collapsing, fault dropping and multiple passes can also be used to reduce storage requirements. Concurrent fault simulation is probably the most widely used fault simulation technique at present.

As an example of this technique consider the circuit shown in Fig. 4.3.

As in the previous example on parallel fault simulation the fault free inputs are $a = d = 1$ and $b = c = e = 0$.

The postulated faults on the first gate are input a stuck-at-0 and input b and output f stuck-at-1, i.e. $a/0$, $b/1$ and $f/1$. Since $a/0$ produces the same effect as the fault-free circuit it is removed from the fault list. It should be noted that the fault conditions, although evaluated concurrently, are all considered as single fault conditions.

The postulated faults on the second gate are $b/1$, $c/1$, $f/1$ and $g/0$. All of these conditions produce an output response which differs from the fault-free response.

The faults postulated on the third gate are $c/1$, $d/0$ and $h/1$; since $d/0$ produces the same effect as the fault-free circuit it is removed from the fault list.

On the fourth gate the postulated faults are $c/1$, $e/1$, $h/1$ and $i/1$; again all of these faults produce an output response which differs from the fault-free response.

On the final gate the faults postulated are $b/1$, $c/1$, $e/1$, $f/1$, $g/0$, $h/1$, $i/1$ and $j/0$. Faults $c/1$, $e/1$, $h/1$ and $i/1$ produce the same effect as the fault-free circuit and thus cannot be detected by the test pattern. Hence, as before, the faults detected are $b/1$, $f/1$, $g/0$ and $j/0$. It should be noted that the fault condition $c/1$ is propagated to the output gate along two paths.

4.2.4 Parallel valued lists (PVL) [7]

The parallel valued list technique is a hybrid fault simulation method which monopolizes the best features of parallel and concurrent fault simulation without incurring their disadvantages to any great extent.

The PVL technique stores faults in densely packed groups (words) as in parallel fault simulation, but only propagates active fault groups as in concurrent fault simulation. This technique was implemented so that it could be used in circuits where high impedance states could be produced on circuit nodes, consequently two-bits are assigned to each logic value. Although two adjacent bits in a computer word could be used to represent the state on a node it was found to be more advantageous to use two separate words, i.e. $A(1)/B(1)$, $A(2)/B(2)$ etc. Since faults are grouped together each fault is identified by its group number and bit position pair. In the simulation process all faults are processed in a single pass, fault groups being created and deleted dynamically; a fault group is deleted when all faults in the group become inactive, hence time is not wasted in continually processing inactive faults as in conventional parallel fault simulation. Furthermore, as the values are stored in parallel much less storage is required per node for the fault lists in comparison to the amount required in concurrent fault simulation. If the faults are grouped together judiciously, then the activity produced by the fault groups will generate shorter lists which will contain a higher number of active faults per group. If necessary faults from sparsely populated groups can be compressed into a smaller number of groups.

The evaluation and propagation of parallel valued lists are carried out using set intersection and union operators, as in deductive fault simulation, when the fault-free inputs to a gate change (i.e. a gate event); however, if the same fault group appears on more than one input Boolean equation processing must be performed on the entries stored in parallel in the fault groups. Alternatively, if the gate output does not change but the input fault lists do (i.e. a fault list event) a technique similar to concurrent fault simulation is used, in this instance; however, only the active fault lists are considered; these have been flagged previously in the simulation procedure.

This technique has been applied successfully to a range of circuit designs which have included MOS transistor elements, for example bidirectional transistors, wired-ORs etc. and has also been implemented to work with levels of abstraction higher than gate level.

4.2.5 Parallel pattern single fault propagation (PPSFP) [8]

This method was developed in order to have an efficient technique to determine the fault coverage of built-in test systems whose input patterns were derived from pseudo-random binary sequence (PRBS) generators (see Chapter 6), since some circuit designs are resistant to random pattern testing.

The standard techniques of fault simulation were considered too inefficient in this instance, since the PRB sequences used for testing are much longer than deterministic test pattern sequences, although they are much easier to generate. Consequently several analytical approaches were developed (see Section 4.5) which were more efficient than fault simulation but required lengthy computations of detection probabilities and were inexact in their analysis since some faults, although detectable, would be classified as undetectable, particularly if the circuit analysed contained a reconvergent fanout or the fault required multiple path propagation for its detection; both situations would be correctly analysed using fault simulation. Consequently there is a need for an efficient method of fault simulation which can be used to determine the fault coverage of PRB sequences since this technique has been proposed as a method of reducing testing costs; the *parallel patterns single fault propagation* technique was developed to achieve this objective.

The PPSFP technique assumes that unknown logic values cannot be generated as a stable state in a circuit and also any high impedance states can be controlled to a logic one or zero; these assumptions permit a two-valued, zero delay simulation algorithm to be used. The technique of PPSFP comprises:

1. A good circuit simulation, which in a single pass will evaluate the effect of applying 256 input patterns to the circuit. All circuit functions are evaluated in turn using the Boolean operations AND, OR and complementation;

2. For each stuck-at fault considered, the faulty circuit response to the 256 input patterns is propagated forward, beginning at the site of the fault and continuing until either the faulty and fault-free circuit responses coincide or the fault effect reaches an observable output, in which case no further processing of the fault condition is carried out. This step is repeated until all faults have been detected or the number of test patterns has been exhausted.

This technique, in which only one fault is simulated with N patterns applied, is in contrast to conventional parallel fault simulation in which one pattern at a time is applied, whilst simulating M single faults in parallel, which is inefficient since the effect of each fault is evaluated on all gates even though there is little possibility of the fault affecting that gate. The reduction in run time achieved by PPSFP in comparison to parallel fault simulation is of the order of 400 times. Furthermore PPSFP can start in the middle of a circuit at the site of a fault, unlike parallel fault simulation.

Benchmarks run on a range of circuits using PPSFP produce most encouraging results: for example, a 4000 gate circuit with 7500 faults was simulated with 512 000 patterns, giving 97% fault coverage in 87.3 seconds; from the test results derived from ten circuits of varying complexities it was estimated that for a 100 000 gate circuit, fault simulated with 512 000 patterns, 30 minutes of CPU time would be required, thus demonstrating the feasibility of using fault simulation to determine the fault coverage of PRB sequences.

4.3 FUNCTIONAL LEVEL FAULT SIMULATION

As circuit complexities increase, the cost of fault simulation rises rapidly (cost is proportional to either the square or the cube of the number of gates). Furthermore the amount of storage required by the fault simulator also increases with circuit complexity, particularly with the methods which use dynamic fault lists; in this instance the problem is aggravated since the amount of storage required by these techniques is unpredictable.

Attempts have been made to reduce the amount of storage by introducing techniques such as fault collapsing, where a single fault is used to represent an equivalent class of faults which produce the same effect upon the circuit, or by performing multiple passes, in this instance a trade-off is made between storage requirement and CPU time.

An alternative approach to reducing the costs of fault simulation is to perform this process at a higher level of abstraction, namely functional level. The ability to perform fault simulation at functional level also overcomes the problem of performing fault simulation on systems comprising MSI and LSI packages, where gate level descriptions of these packages are not available and the cost of generating a gate level description is high. Although functional level fault simulation offers a solution to some of the problems of fault-simulating complex circuits at gate level, some difficulties do arise in its use. First, it is difficult to simulate faults internal to a function block; second, problems can arise when non-functional inputs have to be processed through a block; third, there is the general difficulty of processing multivalued logic at functional level; however, at the higher levels of abstraction the number of logic levels used should be only two, with the addition of an unknown state for initialization purposes. Finally, the most difficult problem of all is that of proving to designers that a test set which gives a 'good' fault coverage of functional level faults will also give a 'good' fault coverage of gate level faults; however, an untested functional fault will highlight an untestable gate level region. At present there is insufficient data available to make any conclusive statement on how close functional level fault coverage matches gate level fault coverage. It should be noted, however, that the stuck-at-fault model used at gate level has never been 'proven' to be effective; experience in its use has 'shown' it to be effective [9].

Functional fault level simulation has been attempted using the concurrent and deductive fault simulation techniques; however, parallel fault simulation, in general, is not considered suitable since it relies heavily on the logical operations on words, requiring the word to be unpacked before signal values could be processed through some functional blocks, for example a counter, and then repacked. However, the technique of parallel fault simulation could be applied to the Boolean representation of some functional blocks, this approach is used in the application of deductive fault simulation at functional level.

4.3.1 Concurrent fault simulation at functional level [10]

The technique of concurrent fault simulation can readily be applied to functional level blocks, since the fault simulation process is based on the individual evaluation of blocks. Normal functional level simulators can readily be adapted to perform fault simulation; the major additional data structure required is the assignment of a Concurrent Fault List (CFL) to each function block, which identifies the 'faulty' and 'error' blocks associated with a 'good' block. The CFL stores the configuration of a block, that is, an n-tuple which defines the inputs, outputs and internal states of a function, under given conditions. If a fault is local to a block the CFL for that block will store the configuration for the fault until the fault is detected. However, if a block is propagating the effect of a fault somewhere else in the circuit, the block is referred to as an 'error' block and the configuration of the error block is only stored in the CFL for the block provided the 'error' configuration differs from the normal block configuration. The amount of storage required by the CFLs must be allocated dynamically since configurations will be continually added to, and deleted from them as the simulation process continues.

4.3.2 Deductive fault simulation at functional level [11]

In the technique of deductive fault simulation the ability to propagate fault lists through functions, in general, depends upon the function being specified for all input sequences it may receive in both the fault and fault-free circuits; under these conditions, provided the function itself is fault-free, the output sequences resulting from the incorrect inputs can be deduced from the true input values and the input fault lists.

The process of fault list propagation at gate level was implemented using a fault list algebra which used the set union and intersection operators; this technique, unfortunately, cannot be applied directly to a high level description of a function. It is noted, however, that the output of a function and the output fault lists are independent of the circuit realization of a function; consequently, any computational convenient realization of the function may be used to determine the output response of the function for a given set of inputs or fault lists. Since the propagation of fault lists is implemented using set union and intersection operators a computationally convenient realization of a high level description of a function is a set of Boolean equations.

The essence of the transformation of the high level description into a set of Boolean equations is to assign a control variable to each conditional statement in a given section of the description and thereafter express the outcome of an action in a form which includes the effect of control variables. For example consider the conditional statement

$$IF\ a.AND.b\ THEN\ y1\ =\ AND\ (c,d,e)$$
$$ELSE\ y2\ =\ OR\ (f,g)$$

The transformation is obtained as follows.

First, assign a control variable $C1$ to the conditional part of the statement i.e. $C1 = ab$.

Second express the assignments to $y1$ and $y2$ in terms of $C1$ i.e.,

$$y1 = (c.d.e).C1$$
$$y2 = (f+g).\overline{C1}$$

The technique of deductive fault simulation can then be applied to the Boolean equations.

Although this technique is conceptually straightforward, in practice it can be extremely difficult to perform the transformation, particularly if the description contains conditional statements with many levels of nesting and also when the actions themselves are functions other than simple assignments or logical operations: for example, arithmetic or bit manipulation functions.

4.3.3 Modelling of internal faults for functional level fault simulation

When fault simulation is performed at gate level, faults can be systematically assigned to gates, that is inputs and outputs stuck-at 1/0, as the simulation proceeds. A similar procedure can be adopted at functional level for the 'pin' faults on the function blocks. However, there is no systematic method for modelling faults internal to a functional block; in general the internal faults have to be modelled manually before simulation begins; several of the techniques used are outlined below.

If the functional model comprises a look-up table or sequence of micro-operations (micro-operation model), internal fault models can be generated by either changing the contents of the state table such that when the correct entry is decoded the contents moved to the outputs are incorrect, or if the micro-operation model is used the micro-operations may be modified. This approach to modelling internal faults is called *model perturbation* [12].

The technique of modelling internal faults, for deductive fault simulation, which affect the output behaviour of a block is obtained by modifying the method of propagating faults through the functional blocks. In this instance, however, not only must the behaviour of the fault-free block be described but also the behaviour of all the faulty blocks under consideration, which imposes a limit upon the number of internal faults which can be represented. Although the technique has the advantage of modelling the effects of non-classical faults, that is short circuits or timing faults. The technique used to model internal faults is outlined below.

Let $g(x)$ describe the fault-free behaviour of a block and let some internal fault cause the behaviour of the circuit to change to some function $g_1(x)$. Let us introduce a function [11]

$$h(x,f) = \bar{f} g(x) + fg_1(x)$$

where f is the fault variable which can assume a true value of 0 and a fault list value of α. Consequently an internal fault, which transforms the block function into $g_1(x)$, can be considered to be a stuck-at-1 fault on the input f of the function $h(x,f)$. Thus the internal fault α is transformed to a fault on an input variable f and its effect is simulated, by propagating the fault list through on f for the function $h(x,f)$.

The technique adopted in concurrent fault simulation for modelling internal faults depends upon the type of internal fault. If the fault can be represented by one of the internal state variables stuck-at 1/0, that is a 'state fault', the internal fault can be modelled by modifying the appropriate state vector in the configuration of the block. However, if the fault cannot be represented as a state fault, then the functional block is interchanged with its gate level equivalent. The interchanging of functional and gate level models can be done either statically or dynamically [9]. A static interchange is performed before the fault simulation process commences and the internal faults in a block which could not be represented by state faults are continually processed until they have all been detected by some given number of test patterns, whereupon the gate level model is dynamically changed for the functional level model; this change, however, can only take place when all the activity within the gate level model has ceased. This technique of inter-changing model levels is also used when a functional block is incapable of processing a set of non-functional inputs or when some of the inputs have unknown values; in these conditions the interchange is performed dynamically and when the gate level model has resolved the problem the models are interchanged again. However, great care must be taken when switching levels to ensure that any state events in the functional model can be mapped into gate level events and any changes in the model outputs at functional level must be consistent with future events scheduled from the gate level model.

However, the major problem in general with functional level fault modelling is the absence of a generalized fault model as exists at gate level.

4.4 FAULT SIMULATION OF MOS CIRCUITS [13]

In present-day implementations of digital circuits using either NMOS or CMOS technologies numerous fault conditions can occur which cannot be modelled adequately using gate level fault models since many MOS circuit configurations do not readily map into gate level functions. The obvious solution is to model the faults at transistor level. It may be considered that since fault simulation at gate level, using the techniques which are generally available, is an inefficient process, the problems of fault simulation at a lower level abstraction would be enormous, due to the increase in the number of simulation primitives; however, the simplicity of the model used at this level, namely switch level, renders the simulation process quite practical. In many ways the switch level model is a more realistic representation of MOS circuits under faulty and fault-free conditions.

At switch level faults are injected into the circuit by including extra 'fault' transistors at appropriate parts in the circuit. The inputs to these fault transistors are considered as extra inputs to the circuit; in this way multiple faults can be represented. The type of faults which can be simulated are either open or shorted transistors, nodes shorted to power or ground, resistive faults in ratioed logic, etc. These fault conditions are implemented by a series or parallel connection of a 'fault' transistor in a given location in the circuit: for example, an open circuit transistor is simulated by connecting a 'fault' transistor in series with it and maintaining the input to the 'fault' transistor at a logic zero.

The method of concurrent fault simulation is the most commonly used technique at switch level. The presence or absence of a fault condition is determined by the logic value assigned to the inputs of the 'fault' transistors; the input test vectors are extended to include the signals which will activate a number of faults in the circuit. The process of fault simulation proceeds in a manner similar to the normal switch level simulator, except that the logic state of a node is stored as a pair of values called 'a state set' which identifies the test or fault number and associated node state; if the faulty and fault-free values differ the node is said to have 'diverged' from the fault-free value. If a node is classified as 'diverged' any transistor whose gate is connected to the node is also said to be 'diverged'.

During normal switch level simulation the technique of perturbation simulation is used to improve the efficiency of circuit evaluation; this technique is similar to the concept of 'selective trace simulation' used at gate level. Perturbation simulation requires that a list of nodes be kept which, if a given node changes value, will require to be re-evaluated. In fault simulation it is also necessary to store the identifier of the test sequence which will perturb a node if the node is not affected by the reference sequence, thus indicating that this part of the circuit is operating in a different manner under fault conditions. In most cases the faulty and fault-free circuit responses will be identical; hence it is only necessary to evaluate the fault-free circuit response. However, during the evaluation process it is necessary to check whether any transistors in the vicinity of a perturbation have 'diverged', that is, have been affected by a fault condition, in which case it is necessary to recompute the node states for the fault condition since the faulty and fault-free values may differ; this involves a search of the node 'state sets' for the appropriate node values resulting from the fault condition. The simulation process is continued until a primary output is reached; whereby examining the state sets on the output nodes which differ from the fault-free values, the faults detected by the test can be determined.

This fault simulation technique was applied to a 64-bit RAM which contained precharged busses, bidirectional transistors, etc., 428 faults were detected using 407 patterns and required 11 minutes of CPU time; this was approximately thirty times faster than serial fault simulation applied to the same circuit.

4.5 ALTERNATIVES TO FAULT SIMULATION

It is recognized that fault simulation costs increase as the square or the cube of the circuit complexity; consequently to try and reduce these costs several techniques have been proposed as substitutes for fault simulation, namely

1. testability analysis
2. statistical fault analysis
3. critical path analysis.

These techniques will be briefly described.

4.5.1 Testability analysis [14]

Testability analysis attempts to quantify the testability properties, usually of an unstructured design, identifying nodes in a circuit whose logic value is either difficult to control or observe. The designer may then add a few test points to the circuit or modify the design to simplify the testing of these nodes. Testability measurements may be either 'extrinsic' or 'intrinsic'; extrinsic measurement of testability is derived from the fault coverage and requires a knowledge of both circuit structure and associated test sequences. However, the 'intrinsic' measurement of testability is derived directly from the structure of the circuit without any recourse to simulation or test pattern generation.

Testability analysers are topological analysis tools, which without the use of simulation or test pattern generation give the designer a quantitative measurement of the degree of difficulty involved in testing nodes inside a circuit. This information is then used by the designer in deciding which sections of a circuit require to be modified to improve the testability of the circuit.

The testability of a circuit is derived from the controllability/observability attributes of each node in the circuit.

The controllability of a node is a measure of the ease with which a given node in the circuit may be set to a particular logic value by an assignment of logic values to the primary inputs.

The observability of a node is the measure of the ease with which the logic value on a node can be observed at a primary output. The observability value will implicity include the controllability of the nodes instrumental in sensitizing a path from the node to a primary output.

In some instances the testability of a node is defined in terms of its combinational and sequential testability values. The combinational testability values are a measure, in the spatial sense, of the difficulty of obtaining a complete set of node justifications to control/observe a given node in the circuit. In contrast the sequential testability values provide an indication of the number of time frames required to control/observe a given node in a circuit.

Several testability analysis programs [15]–[17] have been developed, and although their objectives are the same the programs differ in the rigour of the analysis that is performed on the circuit. For example CAMELOT [15] does not distinguish between the controllability figures for setting a node to a logic one or logic zero, which is in contrast to SCOAP [16], which not only derives different controllability values for setting a node to a one or a zero, but also distinguishes between combinational and sequential controllability/observability values.

In addition to using controllability/observability values to establish the testability profile of a circuit, the results of testability analysers have been used to

1. improve the efficiency of automatic test generation programs (Chapter 5)
2. select sets of nodes to be connected into scan/set registers (Chapter 6) to improve the testability of a circuit
3. identify gross design errors before simulation is used to verify a design.

The major advantage of testability analysis is that the computational complexity of the algorithms grows linearly with circuit size; this is maintained by making, in some instances, certain simplifying assumptions: for example, that the degree of difficulty in controlling the output node of a gate to a logic 0 and to a logic 1 is the same. Its major disadvantage is that the testability data itself does not convey a great amount of information to the designer; however, a vast amount of information can be extracted from the results of a testability analysis, for example [18], [19]:

1. given the testability value for a node in a circuit, the probability of detecting a fault on the node can be determined
2. the percentage fault coverage for a given length of test sequence can be calculated
3. regardless of circuit size, provided the overall testability index for the circuit can be calculated, it is possible to estimate the test length required to achieve a given fault coverage.

Although the cost of performing a testability analysis grows linearly with circuit size attempts to reduce the cost of analysing large circuits have been made by performing a testability analysis at higher levels of abstraction [20].

4.5.2 Statistical fault analysis [21]

Fault simulation costs grow rapidly with increase in circuit size. Attempts to reduce the costs have been made by performing statistical fault simulation which only processes a random sample of faults in order to estimate the fault coverage. The main disadvantage of this technique is that a fault simulator is still required and there is no information on the faults which were not sampled.

Statistical fault analysis also makes use of the controllability/observability attributes of a connection line in a circuit; however, these are redefined as the probability of controlling or observing lines in the circuit. The data required to perform a statistical fault analysis is obtained from a fault-free simulation of the circuit for a given set of test patterns; subsequently, the detection probability of a fault is calculated from the product of the controllability and observability values. Faults can then be graded in terms of their detection probabilities.

The controllability value of a line in the circuit is obtained by assigning two counters to that line, to determine the number of times the line is set to a logic 1 and a logic 0. If the line has unknown value the counters are not incremented. However, if a high impedance state occurs the counter incremented previously is incremented again. The one and zero controllability is determined by dividing the appropriate count by the number of patterns applied to the circuit.

In determining the observability of a line a 'sensitization' counter is first assigned to each input line to a gate; a counter is only incremented if for a given input pattern the output is made sensitive to that input. The sensitization probability of that line is then determined by dividing the final value of the counter by the number of test patterns applied to the circuit. The calculation of line observabilities starts by assigning the value of unity to the zero and one observability values for each of the primary outputs. The observability values of the inputs to the gates attached to the primary outputs is calculated in terms of the output observability values; the calculation involves the use of the sensitization probabilities and controllability values for the inputs to the gate. The process of calculating the observability of the inputs to a gate in terms of the observability value for the output of the gate is continued until the observability values of all lines in the circuit have been calculated; the calculations performed can be adjusted to account for fanouts and feedback lines in the circuit. Thereafter from the calculation of the detection probability of a fault, which is determined from the product of the line observabilities and controllabilities, the probability of detecting that fault with a given length of test set can be calculated.

When the fault coverage determined by statistical fault analysis was compared with that derived from fault simulation for a range of circuits the average deviation was between 0.25% and 3.65%. The overheads, essentially, incurred using statistical fault analysis are those of updating the counters in the fault-free simulation and performing the calculation for the controllability and observability values.

4.5.3 Critical path analysis [22]

Critical path analysis has been proposed as an alternative to fault simulation as a means of determining the fault coverage of sets of test patterns. In

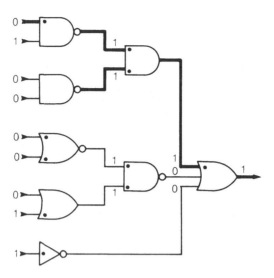

Fig. 4.4 An example of critical path tracing in a fanout-free circuit [22]: ● denotes a critical input; — denotes a critical path.

comparison to fault simulation the technique of critical path analysis does not simulate the faulty circuits to determine which faults are determined by a given test, hence it is not involved with the propagation of fault list and does not require unspecified amounts of storage. Furthermore faults are dealt with implicitly, and there is no need for fault enumeration, fault collapsing, fault insertion, etc.

The technique of critical path analysis comprises simulating the fault-free circuit with a given input test pattern and then, using the computed signal values on the circuit nodes, a critical path is traced backwards from the primary outputs to the primary inputs, along which the faults detected by the test pattern will be located. A line inside a circuit is said to have a 'critical' value, say a logic 1, if the input test pattern will detect the fault condition in which the given line is stuck-at 0. The process of determining the critical path through a circuit comprises identifying a continuous path from the output to inputs consisting of critical lines; the lines which are critical for a given input test pattern identify the faults detected by the test pattern.

To assist the backtracking algorithms to identify the critical lines in a circuit certain gate inputs are marked as 'sensitive' during the fault-free simulation; a gate input is declared to be sensitive if the output value changes when the given input is complemented. For example, if only one input to a gate has been assigned the dominant logic value for the gate, this input is declared sensitive, or if all inputs to a gate have the non-dominant logic values, then all inputs are declared sensitive. A simple example of critical path tracing in a fanout-free circuit is illustrated in Fig. 4.4. When critical path analysis is applied to circuits with reconvergent fanouts some faults which can be detected by the input test patterns may be flagged as undetectable; hence this technique will only give approximate values of fault coverage. The situations in which critical path analysis fails to identify a detectable fault are those which require multiple paths to be sensitized in order to detect the fault. In general, however, many circuits do not have reconvergent fanouts and many test pattern generation algorithms (for example, RAPS, PODEM, SMART, FAST, etc. – See chapter 5) only sensitize single paths through a circuit.

When critical path analysis was compared to concurrent fault simulation incorporating the technique of fault dropping after first detection, critical path analysis was found to be 60% faster on average.

In view of the performance advantages offered by the above techniques it may be considered that fault simulation will be replaced by these techniques. However, the costs of running fault simulations can be reduced by using

1. design for testability techniques to partition the circuit into smaller subfunctions
2. the technique of statistical fault simulation
3. hardware accelerators.

It is also argued that fault simulators are more versatile in the faults that can be processed and that these alternative techniques do not generate any diagnostic data which may be required by field service.

4.6 REFERENCES

1. Szygenda, S. A. and Thompson, E. W. (1975) Digital logic simulation in a time based table driven environment: Part II Parallel fault simulation. *Computer*, **8**, 39–49.
2. Armstrong, D. B. (1972) A deductive method of simulating faults on large circuits. *IEEE Trans. Computers*, **C-21**(5), 469–71.
3. Breuer, M. A. and Friedman, A. D. (1977) Diagnosis and reliable design of digital systems. Chapter 4, Pitman.
4. Giambiasi, N., Miara, M. and Muriach, D. (1980) Methods of generalised deductive fault simulation. *17th Design Automation Conference Proceedings*, June, 386–93.
5. Chang, H. Y. (1974) Comparison of parallel and deductive fault simulation methods. *IEEE Trans. Computers*, **C-23**(11), 1132–8.
6. Ulrich, E. G. and Baker, E. T. (1974) Concurrent simulation of nearly identical digital networks. *Computer*, April, 39–44.
7. Moorby, P. R. (1983) Fault simulation using parallel valued lists. *Proc. IEEE International Conference on Computer Aided Design*, September, 101–2.
8. Waicukauski, J. A., Eichelberger, E. B., Forlenza, D. O. *et al.* (1985) Fault simulation for structured VLSI. *VLSI System Design*, December, 20–32.

9. Davidson, S. (1984) Fault simulation at the architectural level. *Proc. 1984 International Test Conference*, November, 669–79.
10. Abramovici, M., Breuer, M. A. and Kumar, K. (1977) Concurrent fault simulation and functional level modelling. *14th Design Automation Conference Proceedings*, June, 128–37.
11. Premachandran, R. M. and Chappel, S. G. (1978) Deductive fault simulation with functional blocks. *IEEE Trans. Computers*, **C-27**(8), 689–95.
12. Gupta, A. K. and Armstrong, J. R. (1985) Functional fault modelling and simulation for VLSI devices. *22nd Design Automation Conference Proceedings*, June, 720–6.
13. Schuster, M. D. and Bryant, R. E. (1984) Concurrent fault simulation of MOS digital circuits. *Proc. 1984 Conference on Advanced Research in VLSI, MIT*, January, 109–38.
14. Goldstein, L. H. (1979) Controllability/observability analysis of digital circuits. *IEEE Trans. Circuits and Systems*, **CAS-26**(9), 685–93.
15. Bennetts, R. G., Maunder, C. M. and Robinson, G. D. (1981) CAMELOT: A computer aided measure of logic testability. *Proc. IEE*, **128**(E5), 177–89.
16. Goldstein, L. H. and Thigpen, E. L. (1980) SCOAP: Sandia Controllability/Observability Analysis Program. *17th Design Automation Conference Proceedings*, June, 190–6.
17. Grason, J. (1979) TMEAS – A testability measurement program. *16th Design Automation Conference Proceedings*, June, 156–61.
18. Agrawal, V. D. and Mercer, M. R. (1982) Testability measures – what do they tell us?, *Proc. 1982 IEEE Test Conference*, November, 391–6.
19. Singer, D. M. (1984) Testability analysis of MOS VLSI circuits. *Proc. 1984 International Test Conference*, November, 690–6.
20. Fong, J. Y. O. (1982) On functional controllability analysis. *Proc. 1982 IEEE Test Conference*, November, 170–5.
21. Jain, S. K. and Agrawal, V. D. (1985) Statistical fault analysis. *IEEE Design and Test of Computers*, **2**(1), 38–44.
22. Abramovici, M., Menon, P. R. and Miller, D. T. (1984) Critical path tracing: An alternative to fault simulation. *IEEE Design and Test of Computers*, **1**(1), 83–93.

5
AUTOMATIC TEST PATTERN GENERATION TECHNIQUES

5.1 INTRODUCTION

A major barrier to the full exploitation of the capabilities offered by VLSI, is the problem of the increased cost of testing the complex devices immediately after fabrication. The need to test devices results from imperfections in the fabrication process producing a wide range of defects in the devices, for example, pin-holes in the gate oxide, shorted or open interconnect lines (polysilicon, diffusion and metal), contact hole defects, crystalline defects on the wafer, etc. There may also be some design faults, such as a gate output having insufficient drive capability for its output capacitance, which may not be identified by the simulator, unless a post layout simulation is performed; although simulation may have been used extensively many design faults may go undetected since simulation is an incomplete process based on an abstracted model.

To isolate these faulty devices, it is necessary to apply a set of input waveforms which will result in a different response from the faulty and fault-free circuits. In the past these input waveforms were generated manually; however, as circuit complexities increased the manual methods have, to a large extent, been superseded by automatic test generation methods [1],[2]. With present-day circuit complexities the cost of test pattern generation is extremely high and grows to a first approximation as the square of the number of gates; however, if the effect of increase in gate interaction is considered, which also increases with complexity, test generation costs grow as the cube of the number of gates. The consequence of inadequate testing at chip level is increased cost of system testing and field repair; it is generally accepted that the cost of detecting a fault increases by a factor of ten at each level of test from chip, to printed circuit board, to system test, to field test and repair.

Attempts to reduce the costs of testing have been made by developing more sophisticated gate level test generation algorithms and also by performing test generation at higher levels of abstraction; these techniques will be discussed in this chapter. An alternative approach to reducing test generation costs, or maintaining them at an acceptable level, involves the use of design for testability techniques (discussed in Chapter 6) which make circuits more testable by including hardware which, at test time, permit the circuit to be reconfigured into less complex subfunctions or permit the circuit to adopt a self-test mode of operation. It should be noted that the process of test generation comprises not only techniques to generate test patterns but also the techniques of fault simulation and fault modelling; these aspects of the test generation process have been discussed in previous chapters.

5.2 TEST GENERATION FOR COMBINATIONAL CIRCUITS

Test generation techniques can be broadly categorized as either algebraic or structural. The algebraic techniques derive test patterns from the logic equations realized by the circuit; the best known algebraic technique is the Boolean difference method [3]. The structural techniques derive test patterns from a topological gate description of the circuit; at present these are the most widely used techniques in practice, consequently the ensuing discussion on test generation techniques will consider only the structural methods.

Invariably, structural test pattern generation methods use the technique of *path sensitization*, so this technique will be described first, followed by what may be considered as refinements on this basic technique of test pattern generation.

5.2.1 Single path sensitization

The concept underlying the technique of test pattern generation using path sensitization methods comprises tracing a signal path from the site of the fault to an observable output, in which the logic state at any gate output along the path is dependent upon the logic value at the site of the fault, i.e. the path is sensitive to the fault condition.

The method of test pattern generation using path sensitization techniques can be summarized as follows:

1. At the site of the fault, specify the inputs to the faulty gate to make its output response sensitive to the fault condition;
2. Sensitize a path, from the site of the fault to an observable output, by assigning non-dominant logic values to each input of a gate, except the input propagating the fault condition, so that the output of that gate is sensitive to the fault condition. In this way the effect of the fault is propagated to an observable output. This propagation of fault information is the essence of structural methods for test pattern generation;

3. By the process of backward simulation, determine the set of primary input values necessary to make the gate at the site of the fault sensitive to the fault condition and also to sensitize the path to an observable output. The set input values so derived constitute the test for the fault condition.

As an example of the technique consider the circuit shown in Fig. 5.1, where it is considered that the output of gate 1 is stuck at 0.

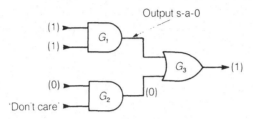

Fig. 5.1 Example on path sensitization.

Method

1. Make the output of gate 1 sensitive to the fault condition by assigning a logic 1 to both inputs, so that in the fault-free circuit the output would be a logic 1, i.e. opposite to that of the fault condition.
2. Propagate information about the fault condition through gate 3 to an observable point by assigning a logic 0 to the second input to gate 3. In this way the logic value on the output of gate 3 depends upon the presence or absence of the fault on gate 1.
3. By a process of backward simulation justify the logic assignments made in making the faulty gate sensitive to the fault condition and also propagating it to an observable output. This process starts from the observable gate output and regresses to the primary inputs to the circuit. Thus the assignment of logic 0 to the output of gate 2 must be justified by assigning a logic 0 to either of the inputs to gate 2. Since no other logic values require justification the test for the fault condition comprises the assignment of a logic 1 to both inputs of gate 1 and a logic 0 assignment to either input of gate 2. The output of gate 3 under fault-free conditions is a logic 1, whilst if the fault is present it is a logic 0.

The path sensitization method described above sensitizes only a single path from the site of the fault to an observable output and fails to generate a test for a fault in a circuit containing a reconvergent fanout structure. This weakness in the method is overcome in the *D*-algorithm, described below, which sensitizes all paths from the site of a fault to an observable output. To demonstrate the problem of sensitizing only a single path of a reconvergent fanout, consider the circuit shown in Fig. 5.2. It is assumed that the output of

G_8 is stuck-at-0, requiring x_1, x_2 and G_6 to be logic 0s. To set up a single sensitization path say through G_9, G_{11} to G_{12}, requires both x_1 and G_{10} to be a logic 0. The requirement for a logic 0 on the output of G_{10} results in the assignment of logic 1 to x_2. However, x_2 has already been assigned a logic 0, resulting in an inconsistent assignment of logic values. The output of G_8 cannot be assigned a logic value since this is a fault-sensitive node. An attempt to propagate a sensitive path through G_{10} and G_{11} to G_{12} will also result in an inconsistent assignment of values. However, if both paths, i.e. G_9, G_{11} and G_{10}, G_{11} are sensitized simultaneously the inconsistency in the assignment of logic values does not occur.

5.2.2 *D*-algorithm [4]

The *D*-algorithm is a more formal specification of the path sensitization method, described above, in which a symbol D is assigned to the fault-sensitive nodes in the circuit, permitting all possible sensitized paths to be readily identified, and systematically processed until a path is traced to an observable output.

The *D*-algorithm is based on the 'calculus of *D*-cubes'; therefore, before the processes within the *D*-algorithm are described some of its associated terminology will be explained:

(a) D-*cube*

The mapping of the minterms of a Boolean function onto a Karnaugh map is well known; alternatively, the minterms may be mapped onto the vertices of an *n*-dimensional cube. Each vertex is identified by an ordered *n*-tuple comprising the inputs or variables in the Boolean function, which normally have the values 0, 1 and X (don't care). In the context of the *D*-algorithm, a variable may also assume the fault-sensitive value D, hence the *n*-tuple associated with some vertex is referred to as a *D*-cube.

(b) *Primitive* D-*cube of failure (PDCF)*

The primitive *D*-cube of failure simply defines the minimal assignment of logic values to the inputs of a gate in order to make the output sensitive to the fault condition. For example:

1. Consider a three-input AND gate whose output is stuck-at-0 (s-a-0). Assuming that the inputs are a, b, c and the output is d, the PDCF is $1^a 1^b 1^c D^d$, implying that a logic 1 is assigned to each input a, b and c, and the output d is a logic 1 in the fault-free gate and a logic 0 in the faulty gate.
2. Consider a three-input NAND gate in which the output is stuck-at-1 (s-a-1),

again assuming the inputs are a, b, c and output d. The PDCF is $1^a\ 1^b\ 1^c\ \bar{D}^d$, implying that each input is assigned a logic 1 and the output of the fault-free gate is a logic 0 and that of the faulty gate a logic 1.

If the output of the NAND gate was stuck-at-0, then three PDCFs would exist, namely $0^a\ X^b\ X^c\ D^d$, $X^a\ 0^b\ X^c\ D^d$ and $X^a\ X^b\ 0^c\ D^d$, where X denotes a don't care value.

The simple rule to generate a PDCF for a fault on a gate input is to assign to the faulty input the logic value opposite to the fault condition and to all other inputs the non-dominant logic value for that particular type of gate. The output of the gate is then assigned a $D\ or\ \bar{D}$ depending upon whether the fault-free response for the assignments is a logic 1 or a logic 0 respectively.

(c) Propagation D-cube (PDC)

The propagation D-cube defines the assignment of logic values to the inputs to a gate, other than those propagating fault information, in order to make the output of the gate sensitive to the incoming fault information. It should be noted that more than one input may be propagating this information (e.g. in a circuit with a reconvergent fanout); in this case the PDC is referred to as a multiple propagation D-cube. In a multiple propagation D-cube, if both a D and $\bar{D}$ occur on the inputs the output of the gate will assume a fixed logic value inhibiting the propagation of fault information beyond this gate. Table 5.1 illustrates the propagation D-cubes for the four standard logic functions. It should be noted that the Ds and $\bar{D}$s in the propagation D-cubes may be interchanged, since the PDC simply defines the conditions to be satisfied to propagate the fault information and indicate whether or not the fault

Table 5.1 Single propagation D-cubes for the four standard logic functions. It is assumed that each gate has three inputs a, b, c and an output d

NAND gate				AND gate			
a	b	c	d	a	b	c	d
D	1	1	$\bar{D}$	D	1	1	D
1	D	1	$\bar{D}$	1	D	1	D
1	1	D	$\bar{D}$	1	1	D	D

NOR gate				OR gate			
a	b	c	d	a	b	c	d
D	0	0	$\bar{D}$	D	0	0	D
0	D	0	$\bar{D}$	0	D	0	D
0	0	D	$\bar{D}$	0	0	D	D

information is inverted in the process. However, at the outset of the D-algorithm process, once the significance of the D and $\bar{D}$ has been defined the definition must not be altered until the process is complete.

(d) Primitive cube (PC)

The primitive cube simply defines the minimal assignment of logic values to the inputs of a gate in order to achieve some defined output value. Table 5.2 illustrates the primitive cubes for the four standard logic functions.

Table 5.2 Primitive cubes for the four standard logic functions. It is assumed that each gate has three inputs a, b, c and output d

NAND gate				AND gate			
a	b	c	d	a	b	c	d
0	X	X	1	0	X	X	0
X	0	X	1	X	0	X	0
X	X	0	1	X	X	0	0
1	1	1	0	1	1	1	1
NOR gate				OR gate			
a	b	c	d	a	b	c	d
1	X	X	0	1	X	X	1
X	1	X	0	X	1	X	1
X	X	1	0	X	X	1	1
0	0	0	1	0	0	0	0

(e) D-intersection process

The D-intersection process is a means of simultaneously matching the logic assignments which already exist in a circuit as a result of the D-algorithm up to a particular point with those required to propagate the fault condition through a successor gate. The D-intersection rules are shown in Table 5.3.

The basic steps in the D-algorithm method of test pattern generation are outlined below:

1. Select a fault.
2. Generate the primitive D-cube of failure.
3. Sensitize at least one path from the site of the fault to an observable output node. This path sensitization procedure is referred to as the D-drive process. The D-drive process starts at the output of the faulty gate; the effect of the fault is then propagated through all the gates in the fanout list of the faulty gate, by performing a D-intersection initially between the PDCF of the faulty gate and the propagation D-cube of the first gate in the

Table 5.3 D-intersection rules

$\cap$	0	1	X	D	$\bar{D}$
0	0	$\emptyset$	0	ψ	ψ
1	$\emptyset$	1	1	ψ	ψ
X	0	1	X	D	$\bar{D}$
D	ψ	ψ	$\bar{D}$	μ	λ
$\bar{D}$	ψ	ψ	D	λ	μ

$\emptyset$ Empty intersection resulting from an inconsistent assignment of logic values to a node in the circuit.

ψ Undefined intersection in which an attempt has been made to match a fault-sensitive value with a fixed logic value.

μ Permissible intersection between fault-sensitive values, i.e. $D \cap D = D$ or $\bar{D} \cap \bar{D} = \bar{D}$.

λ Attempt to intersect fault-sensitive values of different polarity. This inconsistency can be resolved, usually, if in the propagation D-cube all occurrences of D and $\bar{D}$ are changed to $\bar{D}$ and D respectively. However, if during a given intersection process a μ-intersection has already occurred, interchanging Ds and $\bar{D}$s will not resolve the conflict.

fanout list of the faulty gate. The results of the intersection process are then stored in the 'test cube' of the circuit for this stage of the process. This test cube is then D-intersected with the propagation D-cube of the next gate in the fanout list of the faulty gate, forming another test cube, which is then D-intersected with the propagation D-cube of another gate in the fanout list of the faulty gate. The process is repeated until all gates in the fanout list of the faulty gate have been processed. Thereafter, for each gate through which the fault information has been propagated, an attempt is made to propagate this information through the gates in their fanout lists. This procedure is continued until an observable output is reached.

4. The final stage in the D-algorithm is the *consistency operation* where, by means of a backward simulation technique, the assignments of logic ones and zeros used in the formation of the primitive D-cube of failure and the propagation D-cubes are justified. The consistency operation starts at the inputs to the gate where the D-drive process terminated, because the output of this gate was observable, and regresses to the primary inputs. The assignment of logic ones and zeros to the primary inputs is the test for the fault condition.

As an example of the D-algorithm method of test pattern generation, consider the circuit, shown in Fig. 5.2, in which the output of gate 6 is s-a-0 (see Table 5.4).

Comments on Table 5.4

LINE 1.   Defines the primitive D-cube of failure for the fault condition, i.e. x_3 and x_4 are set to a logic 0; under fault-free conditions the output of G_6 is a logic 1, hence it is assigned the value D.

Table 5.4

	x_1	x_2	x_3	x_4	x_5	G_6	G_7	G_8	G_9	G_{10}	G_{11}	G_{12}
(1)			0	0		D						
(2)	0	0				D		$\bar{D}$				
(3)	0	0	0	0		D		$\bar{D}$				
(4)	0							$\bar{D}$	D			
(5)	0	0	0	0		D		$\bar{D}$	D			
(6)		0						$\bar{D}$		D		
(7)	0	0	0	0		D		$\bar{D}$	D	D		
(8)								$\bar{D}$	D	D	D	
(9)	0	0	0	0		D		$\bar{D}$	D	D	D	
(10)							0				D	$\bar{D}$
(11)	0	0	0	0		D	0	$\bar{D}$	D	D	D	$\bar{D}$
(12)				X	1		0					
(13)	0	0	0	0	1	D	0	$\bar{D}$	D	D	D	$\bar{D}$

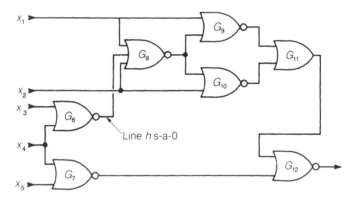

Fig. 5.2 Circuit for D-algorithm example.

LINE 2. Defines the propagation D-cube for G_8. Inputs x_1 and x_2 are assigned non-dominant logic values, so that the output is sensitive to the fault information propagating from G_6. Since G_8 is a NOR gate, the output is inverted hence the assignment of $\bar{D}$ to the output.

LINE 3. Results of the D-intersection between the primitive D-cube of failure and the propagation D-cube (PDC) for G_8. The D-intersection essentially propagates the fault information to the output of G_8 and also ensures that the assignment of logic values to the gate inputs, so far, is consistent.

In the D-algorithm the fault information is propagated along all possible paths simultaneously. Lines 4–7 propagate the fault information from G_8 through G_9 and G_{10} to the inputs of G_{11}.

LINE 4. Defines the PDC for G_9.

LINE 5. Results of the D-intersection between the PDC for G_9 and the current assignment of logic values in the circuit.

LINE 6. Defines the PDC for G_{10}.

LINE 7. Results of the D-intersection between the PDC for G_{10} and the current assignment of logic values in the circuit.

LINE 8. Defines a multiple-propagation D-cube for G_{11}, since the fault information is being propagated via G_9 and G_{10}.

LINE 9. Results of the D-intersection between the PDC for G_{11} and the current assignment of logic values in the circuit.

LINE 10. Defines the PDC for G_{12}.

LINE 11. Results of the D-intersection between the PDC for G_{12} and the current assignment of logic values in the circuit.

Since a path has been sensitized to a primary output, the D-drive process stops and the consistency operation starts. The first logic assignment to be justified by the consistency operation is the assignment of a logic 0 to the output of G_7.

LINE 12. Defines the primitive cube to produce a logic 0 on the output of G_7; in this instance x_5 is assigned a logic 1 and x_4 a 'don't care' value which avoids an inconsistent assignment of logic values since x_4 has been assigned a logic 0.

LINE 13. Results of the intersection between the primitive cube for G_7 and the current assignment of logic values in the circuit.

In this instance all the logic assignments to internal nodes in the circuit have been justified and the consistency operation stops. The assignment of logic values to the primary inputs is the test for the fault condition, i.e. $x_1 = x_2 = x_3 = x_4 = 0$ and $x_5 = 1$; the output of the circuit has the value $\overline{D}$ indicating that the output of G_{12} would be a logic 0 if fault-free and logic 1 in the presence of a fault.

5.2.3 TEST-DETECT [5]

In an attempt to reduce test pattern generation costs, once a test has been generated a fault simulator is generally invoked to determine what other faults can be detected by a given test pattern. However, fault simulation is an expensive process, and several alternatives, as discussed in Chapter 4, have been proposed. An alternative form of fault simulation called TEST-DETECT has been integrated into the D-algorithm, to determine what other faults can be detected by a given test pattern.

TEST-DETECT simply uses the existing logic 1 and 0 assignments in the fault-free circuit generated during the normal D-algorithm procedure. Potentially, any fault which will create a logic value on a given line opposite to its current value will be detected by a given input test pattern; the purpose of TEST-DETECT is to determine which of these faults will be detected by the input pattern. The procedure to determine what other faults are detected by a given test pattern comprises constructing a D-chain from a particular node in the circuit to a primary output; it may be thought that this process requires repeated use of the D-drive process and the consistency operation in the D-algorithm; however, the construction of the D-chain is relatively simple and consists of matching the existing inputs to a gate to determine if they are consistent with the condition necessary to propagate fault information on some input through the gate. For example, if all the inputs to a three-input AND gate are at a logic 1, this gate is capable of either propagating fault information through the gate on any input or acting as a PDCF for a s-a-0 fault on any input. Similarly, if an input was at a logic 0 and the other two inputs at a logic 1, the gate has the potential to propagate fault information through on the input assigned a logic 0; it is also capable of acting as the PDCF for the fault condition in which the input assigned the logic 0 is stuck-at-1. However, if two of the inputs to the AND gate were at a logic 0, fault information could not be propagated on a single path through the gate. If a D-chain cannot be completed from the site of a fault to a primary output, the given test cannot detect that fault condition. The TEST-DETECT process starts at the primary outputs and systematically backtraces through the circuit, and a list of lines where faults can be detected are stored. As TEST-DETECT attempts to create D-chains from nodes internal to the circuit, the list of nodes where faults have been detected is continually examined, since if a D-chain picks up a node whose identifier exists in the list of detected faults for this test pattern, this implies that a D-chain can be formed from this node to a primary output; hence the current D-chain need not be propagated any further. The TEST-DETECT procedure is repeated until the primary inputs are reached.

The major advantage that the D-algorithm has over the single path sensitization technique is its ability to generate tests in circuits with reconvergent fanouts. This advantage, however, is in fact a disadvantage as it makes the test generation process very inefficient, since the algorithm inherently assumes that a circuit has a reconvergent fanout, and attempts are made to sensitize all paths from the site of the fault to the primary outputs; however, many circuits do not have reconvergent fanouts; consequently a vast amount of CPU time is wasted needlessly. In order to improve upon the efficiency of the basic D-algorithm process a nine-valued path sensitization algorithm [6] has been developed. The nine-valued algorithm sensitizes only a single path from the site of a fault to a primary output, but incorporates partially specified values in the test generation process which permit multiple path sensitization to occur if and when it is required.

Although the D-algorithm guarantees to find a test for a fault if such a test exists, it may use a vast amount of CPU time in the process, particularly in trying to resolve inconsistent assignments of logic values. This deficiency in the D-algorithm was highlighted by workers in the computer industry who, in order to improve system reliability, incorporated into their circuits Error Correction and Translation (ECAT)-type functions [7] which comprised trees of EXOR gates. As an example of poor performance of the D-algorithm when used on ECAT circuits, the D-algorithm could only generate tests for six faults out of seven thousand faults after 2 hours of CPU time on an IBM 360/85 computer system. This deficiency in the D-algorithm is a result of the implementation of the algorithm rather than its concept. When a test is generated by the D-algorithm a decision structure is set up in which there is more than one choice at each decision node. Whenever an inconsistency occurs in the test generation process, the D-algorithm systematically processes all possible choices in the decision structure until a solution is found; in the case of ECAT circuits the possible solution space is vast due to the large number of reconvergent fanout paths which exist in these structures.

In order to improve the efficiency of test pattern generation in ECAT circuits several programs have been developed containing heuristics which, in essence, limit the number of possible choices available at a given decision node and also identify possible solutions, at an early stage, which will lead to inconsistencies. The heuristics contained in these programs not only render the programs more efficient than the D-algorithm for ECAT circuits but also for all types of combinational circuits. Several of these programs will now be described.

5.2.4 PODEM-X [8]

PODEM-X is an automatic test generation system comprising three test generation programs, a fault simulator and a test pattern compaction program, which has been incorporated into IBM's design for testability methodology called Level-Sensitive Scan Design (LSSD) (see Chapter 6). It has been reported that this test generation system has been used successfully on logic modules comprising 50 000 gates.

5.2.4.1 Test generation programs

A feature of an LSSD design is that for the purposes of testing, a large circuit can be partitioned into smaller subcircuits, consisting entirely of combinational logic, by blocks of shift register cells which are implicit in the LSSD style of design. These shift register blocks are instrumental in applying the test inputs to the subcircuits and also in observing the output responses from the subcircuits; consequently the registers must be tested initially for possible faults before testing can begin on the combinational subcircuits; this is the function of the first of the three test generation programs called SRTG

(Shift Register Test Generator) [8]. The tests performed by SRTG comprise a 'flush' and a 'shift' test; however, instead of clocking blocks of 1s and 0s or a pattern 001100, which checks all combinations of initial and next states through the register block, which can be concatenated into a single serial shift register, a pattern of either 1111...., 0000...., 1010.... or 0101.... is loaded initially into the registers and the main system clock is pulsed once, producing a one-stage shift in the register, and the resulting pattern is then clocked or scanned out of the register; these patterns exercise each stage of the register through all possible combinations of initial and next states.

The testing strategy incorporated into PODEM-X is to generate initially a set of 'global' tests designed to detect a large number of faults, and thereafter to generate a set of 'clean-up' tests to detect faults not covered by the global tests.

In generating both the global and clean-up test patterns continual use is made of a backward trace procedure as part of the process employed to assign logic values to nodes in the circuit. Consequently before describing individual test generation methods, the backward trace procedure will be summarized:

1. Define an 'objective'; for example setting the output node of a gate to a given logic value;
2. Define the input conditions to the gate to justify this objective;
3. Define a new objective of setting one of the inputs to this gate to the required logic value;
4. Transfer this objective of setting a gate input to a given value to that of setting the output of its predecessor gate to the same value;
5. Define the necessary input conditions to the predecessor gate to achieve the required output value;
6. Define the objective of setting one of the inputs to the predecessor gate to the required value and subsequently transfer this objective to the output of its predecessor gate; and
7. Repeat steps 5 and 6 until a primary input is reached.

The above procedure is shown diagrammatically in Fig. 5.3, where the initial objective is to set the output of gate *B* to logic 1.

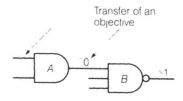

Fig. 5.3 An example of the backward trace procedure.

In PODEM-X global tests are generated by a program called RAPS (RAndom Path Sensitization test generator) [8], whose objective is to derive an input pattern which will sensitize a large number of random paths through a circuit and then use a fault simulator to determine which faults can be detected by the input pattern. The main steps in the RAPS procedure are outlined below:

1. Initialize all circuit nodes to a 'don't care' value;
2. Arbitrarily select a primary output and assign to it the logic value 1 or 0;
3. Perform a backward trace from the output until a primary input is reached, assign to this input the value decided upon from the backward trace procedure and simulate. The main objective in simulating the circuit is to determine, with the current assignment of logic values in the circuit, if the primary output is set to its specified value. A secondary effect of the simulation is the assignment of defined logic values to other nodes in the circuit;
4. If the objective of setting the primary output to its defined value is not achieved, the backward trace procedure is repeated using nodes which have not been assigned a defined logic value; this procedure will result in another primary input being assigned a logic value. The circuit is again simulated to determine if the objective has been achieved, and if not this section of the procedure is repeated until the objective is achieved;
5. Once the objective is achieved, another unassigned primary output is chosen and the above procedure is repeated. The process is continued until all primary outputs have defined values;
6. If any primary inputs have not been assigned a logic value, a search is made for a gate output, internal to the circuit, whose output is assigned but has an unassigned input. This input is then assigned a non-dominant logic value for that type of gate, and the backtracing and simulation procedures are repeated until this objective is achieved. If any primary inputs are still unassigned, this part of the procedure is repeated. However, if all the internal circuit nodes have assigned values, and some primary inputs remain unassigned, then these inputs are arbitrarily assigned logic 1 and logic 0 values.

The primary input pattern so generated is then fault simulated to determine its fault coverage.

It should be noted that in the backtrace procedure no reference is made to the controllability values of the individual gate inputs; this ensures that for each iteration of the overall procedure arbitrary choices are made in the backtrace procedure, resulting in random paths being sensitized by each of the generated test patterns.

Once a set of global tests has been generated by RAPS, a procedure called PODEM (Path Oriented DEcision Making algorithm) [7] is called to generate a set of tests for any remaining fault conditions not covered by the global test set. The essential steps in the PODEM procedure are outlined below.

1. Perform a structural analysis on the circuit to determine the controllability indices [9] for the nodes in the circuit and the 'distance' that each gate output is from a primary output. The controllability indices are used in the backtrace procedure and the 'distance' that a gate is from a primary output is used to determine the shortest path to primary output when propagating fault information.
2. Set all nodes in the circuit to a 'don't care' state.
3. Select a fault condition on a gate.
4. Define the objective of making the output of the gate sensitive to the fault condition. Attempt to satisfy this objective by using the backtrace and simulation procedure as performed in RAPS. In this instance, however, reference is made to the controllability indices during the backtrace procedure to improve the efficiency of the test generation technique. The controllability indices indicate the ease with which a given node can be set to some logic value; thus if, say, a NAND gate has three unassigned inputs and its output value must be a logic 1, this objective is satisfied more readily if the backtrace procedure is performed starting with the input which has the best controllability index. However, if the objective is to set the output of this gate to a logic 0, requiring a logic 1 to be assigned to each input, the backtrace procedure would start on the input with the worst controllability index, since if the conditions on this input cannot be satisfied, CPU time has not been wasted in satisfying the conditions on the other inputs.
5. Once a value has been assigned to a primary input a five-valued simulator $(0, 1, X, D$ and $\bar{D})$ is used to determine the effect of the logic assignment to the primary input. If the initial objective of making the gate output sensitive to the fault condition is achieved, then an attempt is made to propagate the fault-sensitive information to a primary output. If the initial objective is not achieved, the backtrace and simulation procedures are repeated until it is.
6. Propagate the fault information from the site of the fault to a primary output, using the backtrace and simulation procedures to satisfy the necessary input conditions on the gates in setting up a sensitized path to a primary output. In some situations the fault information may be propagated along multiple paths. However, instead of systematically propagating this information along all paths, as in the D-algorithm, the distance of each gate output in the multiple path from a primary output is determined, and the shortest path is pursued.
7. Once the final objective of propagating the fault information to the primary output is achieved, the set of logic values assigned to the primary inputs is

the test for the fault condition. The input pattern is then applied to a fault simulator to determine what other faults may be detected by the test pattern.

On occasions the assignment of logic values to the primary inputs may produce conflicting logic assignments to internal nodes in the circuit. When this situation occurs the 'PI-remake' procedure [7] is called in an attempt to resolve the conflict, if possible. In this process the values of previously assigned primary inputs are successively complemented in an attempt to resolve the conflict. This process can be very time consuming and the PI-remake procedure is repeated only a limited number of times, usually equal to the number of inputs to the circuit.

5.2.4.2 Fault simulator

The fault simulator used in PODEM-X is called FFSIM (Fast Fault SIMulator) [8] and is used to determine the faults detected by the RAPS and PODEM procedures. Since the simulator is used in an LSSD environment it need only have the capability of simulating combinational circuits.

5.2.4.3 Compaction program

Test pattern compaction programs are incorporated into PODEM-X to reduce the number of individual test patterns required to test a system by merging as many test patterns into a single test vector as possible. The compaction process can be performed either statically or dynamically.

In static compaction, the merging process is performed after the test generation phase has been completed. Extensive use is made of any unassigned primary inputs in a given set of test patterns, since tests can be merged only if the assigned values in both sets of test patterns match or if a given input in one test pattern is assigned whilst the same input in another test pattern is unassigned.

In dynamic compaction, tests are merged during the test generation phase. Initially a test for a given fault is generated, and in general not all the primary inputs are assigned logic values. Another fault is then chosen and an attempt to generate a test for this fault is made using the currently assigned primary inputs and some of the remaining unassigned inputs. This process is repeated until almost all the primary inputs have been assigned logic values. Thereafter a fault simulator is used to determine the fault coverage of the compacted test.

5.2.5 FUTURE [10]

FUTURE is a test generation system developed by NEC. In a similar way to PODEM-X, it also has a 'global' test generator and a fault-oriented test generator. The 'global' tests simply comprise a set of pseudo-random binary patterns; the fault coverage is determined using a concurrent fault simulator. The criterion for switching between one test generation technique and the other is very basic: that is, the switch is performed when, for a user-specified number of iterations, the random patterns produced by the generator fail to detect any new fault conditions.

The fault oriented test generation algorithm used in FUTURE is called FAN (FANout oriented test generation algorithm) [11]. This algorithm is much more efficient than PODEM, because of the various heuristics included in the algorithm which attempt to identify situations, early in the test generation process, where conflicting assignments of logic values will occur, and thus reduce the amount of CPU time wasted in driving the test generation algorithm into the conflicting situation and the subsequent time backtracking in trying to resolve the situation.

The essence, in general, of the fault oriented test generation procedure is that of setting up and traversing a decision tree for the circuit. At each node several possible decisions can be taken, the choice, at the outset, being arbitrary. However, when the number of arbitrary choices is large, the possibility of making conflicting decisions is high and a vast amount of time can be wasted subsequently in attempting to resolve these conflicts. The heuristics integrated into FAN are designed to reduce the number of arbitrary choices and hence reduce the backtracking time by computing, as early as possible, the logic values which can be uniquely assigned in the circuit at each iteration of the test generation algorithm.

The first heuristic is to assign all logic values to the inputs of a gate which can be uniquely implied from the fault condition, that is, if the output of an OR gate is stuck-at-1 all inputs are immediately assigned logic 0s; however, if the output is stuck-at-0, all the inputs are left unassigned since the particular value assigned to each input cannot be uniquely determined from the fault condition. A second heuristic called 'complete implication' determines, whenever a logic assignment is made, what other logic values can be uniquely implied, both forwards and backwards throughout the circuit, by this assignment. FAN also incorporates a 'unique sensitization' heuristic; the implementation of this heuristic requires some preprocessing of the circuit. In this heuristic gates are identified in paths throughout the circuit through which fault information must propagate between the site of a fault or a partially sensitized path to a primary output. The gates to which this heuristic is applied are all single fanout gates, and whenever the algorithm propagates fault information onto the start of one of the paths comprising gates which can be uniquely sensitized, non-dominant logic values are assigned immediately to those inputs which are not propagating the fault information, on the gates along the sensitized path; the implications of these assignments are then computed throughout the circuit. The FAN algorithm also uses the concept of *headlines* to improve its efficiency. A headline is a node in the circuit where the structure of the circuit up to the given node comprises gates with single fanouts. Beyond the 'headlines', gates become involved in multiple fanout paths [12]; the identification of headlines again requires the circuit to be preprocessed. The advantage of identifying headlines is that the backtracing procedure, similar to that used in PODEM to justify given objectives, can terminate at a headline, since it is known that the structure of the circuit preceding the headlines comprises an interconnection of gates with single fanouts; hence the justification of a headline node can be performed without introducing conflicts. If it is discovered that the value assigned to a headline node causes an inconsistency, its value is simply inverted without the need to rejustify this value back to the primary inputs, which would happen in PODEM in similar circumstances. Since the justification of headlines can be performed without introducing conflicts, the justification of these nodes is performed after a path has been sensitized to a primary output.

In PODEM a vast amount of CPU time is used in backward tracing in order to justify some objective; this procedure is particularly time consuming in PODEM since it uses a single backtrace algorithm. That is, for example, if the output of a three-input AND gate must be a logic 1, in essence three backtrace procedures would be initiated in an attempt to achieve the given objective. FAN, however, uses a multiple backtrace algorithm, which identifies a set of initial objectives and, using a breadth first approach, works backwards to determine the next objectives and so on until the headlines or a fanout point are reached. At each stage of the backtrace the objective to be achieved at a given line is defined in terms of a triple $(S, Z(S), O(S))$, where S identifies the objective line, $Z(S)$ is the number of lines dependent upon a logic 0 assignment at S, $O(S)$ is the number of lines dependent upon a logic 1 assignment at S, i.e. if the given objective is to set line S to a logic 1 the triple becomes $(S, 0, 1)$ and if the line has to be set to a logic 0 the triple becomes $(S, 1, 0)$. When a fanout point is reached, the values of $Z(S)$ and $O(S)$ depend upon the number of lines at the fanout point. If either $Z(S)$ or $O(S)$ is zero, the fanout node is further processed until a headline is reached. If $Z(S) \geq O(S)$ the fanout point is assigned the value logic 0, if $O(S) > Z(S)$ the fanout point is assigned the value logic 1; further processing on this node is halted since there is the strong possibility of a conflict occurring.

5.2.6 LAMP2 test generator (LTG) [13]

LTG was developed at AT & T Laboratories in order to improve test generation efficiency for faults in large scan-design circuits. Test generation times for a combinational circuit comprising 75 000 gates, using LTG on an IBM 3081K is 1.6 hours of CPU time; previous techniques of test pattern generation when applied to circuits of a smaller size, namely 45 000 gates and

32 000 gates, took 7.2 hours and 14 hours, respectively, on an IBM 3081 K, and 23 hours and 45 hours respectively on an IBM 370/168.

In a similar way to PODEM-X, LTG uses two test pattern generation philosophies: first a set of 'global' tests is generated using a fault independent test generation procedure and then a fault oriented test procedure is used to generate a set of 'clean-up' tests, which will detect those faults not uncovered by the global test set. Both of the test generation procedures rely on a fault simulator to determine the fault coverage of a test; however the fault simulator in this instance uses the technique of 'critical path tracing' [14] rather than simulation to identify the faults detected by a test. LTG also incorporates a dynamic compaction algorithm to reduce the size of test sets generated for a given circuit.

The test generation algorithms used in LTG are more efficient, in general, than those used in PODEM-X; the salient features of these algorithms will now be described briefly.

At the outset of the test generation procedure LTG attempts to identify any undetectable faults in the circuit, so that time is not wasted in attempting to generate tests for these faults. The undetectable faults arise from using predefined function blocks in which not all the input/outputs are used; in general unused inputs are connected to fixed logic values, and unused outputs are left unconnected. Furthermore, in order to reduce the number of cells or predefined blocks in a library, multifunction blocks are sometimes produced. The function of the block is defined by connecting the 'personality' inputs to fixed logic values; any undetectable faults, which subsequently arise from not using certain functions in the block are readily identified.

The global test set in LTG is generated by a program called SMART (Sensitizing Method for Algorithmic Random Testing) [13]. In general, global testing is done using random pattern test sets which are very easy to generate; however, some circuits are resistant to random pattern testing and furthermore random patterns have the characteristic of detecting a large number of faults at the outset and thereafter the fault coverage drops off rapidly; a large amount of CPU time is wasted in attempting to determine the faults covered by these tests. SMART, however, combines some features of random and deterministic test generation which, overall, results in a smaller set of test patterns, comprising tests which have a high fault coverage, being generated with considerably less computational effort than other techniques.

The SMART procedure is very similar to RAPS which attempts to sensitize a large number of paths between the primary inputs and primary outputs of a circuit. However, the RAPS algorithm is inherently inefficient since it may repeatedly sensitize a given primary output to the same logic value, preventing the detection of faults at other primary outputs dependent upon the same primary inputs, which are repeatedly assigned the same logic values required to satisfy the condition on the given primary output. Furthermore, the procedure used in RAPS to assign logic values to unassigned inputs, when all primary outputs have been assigned values, can also inhibit the propagation of faults through other gates in the circuit. These deficiencies in RAPS are overcome in SMART by using information generated automatically by the fault simulator which uses a Critical Path Tracing algorithm (CRIPT).

The technique of critical path tracing is more efficient than conventional fault simulation techniques since the algorithms used do not compute the logic values of faulty circuits and only process detected faults rather than all simulated faults; furthermore the faults to be processed do not require to be enumerated.

During the fault simulation process CRIPT accumulates information which is used, subsequently, in the SMART procedure to maximize the number of faults detected per test. During a simulation run CRIPT identifies 'stop lines' and 'restart gates'. A 'stop line' essentially delimits areas in a circuit where no additional fault coverage can be obtained, since all the faults which would make a given line a logic 0 (0-stop line) or a logic 1 (1-stop line) have been detected by the tests generated to date. A 'restart gate' identifies areas where new faults have a high probability of being detected since the outputs of 'restart gates' are on a critical path, although the inputs are not defined as 'critical'; furthermore, exactly one input to a 'restart' gate must be at the dominant logic value for that gate, and this input line must not be a stop line for the dominant logic value. Under these conditions, if the remaining inputs are assigned non-dominant logic values, the dominant input becomes 'critical'; for example, if the dominant value is a logic 1, this permits a stuck-at-0 fault to be detected on this line together with any other faults which, potentially, make this line a logic 0 under the given input conditions.

The objective of SMART is to sensitize a large number of paths, at random, through the circuit from the primary outputs to the primary inputs. The process of setting up a sensitive path initially comprises randomly assigning a logic value to a randomly selected primary output; an attempt is then made to justify this logic assignment to the given output. The justification process consists of transferring the 'objective' of setting the primary output to a given value to that of setting the unassigned inputs to the primary output gate to the appropriate value. This backward trace process of transferring 'objectives' from gate outputs to gate inputs continues until the primary inputs are reached. So far, the SMART and RAPS algorithms are similar; however, as the SMART procedure continues and the fault simulator CRIPT starts to accumulate data about stop lines and restart gates, the SMART procedure starts to use this data to make the algorithm more efficient. For example, when transferring objectives from a gate output to its inputs, preference is given to those gate inputs which are not stop-lines; in this way sensitive paths are not repeatedly set up through parts of a circuit where all the faults have been detected by previous input assignments, and in this respect the backward trace process used in SMART is said to be 'selective'. In a similar way when randomly assigning logic values to primary outputs, a given output will not be assigned the logic value for which it is a stop-line. When the primary inputs have been determined, which will justify the assignment of a given logic value to a primary output, the CRIPT program is invoked to do a fault simulation on the partially generated input test vector. In this process other stop-lines and restart gates will be identified. SMART then selects the restart gates and attempts to justify the assignment of non-dominant logic values to the un-assigned inputs to these gates. When the inputs to all the current restart gates have been justified, SMART then selects another primary output and the procedure is repeated; any new restart gates identified in the process are subsequently processed before another primary output is selected. Again, in this respect SMART differs from RAPS which would process all of the primary outputs first and then, if any primary inputs are unassigned, arbitrarily attempt to justify non-dominant logic values to unassigned inputs to certain gates in the circuit which satisfied certain conditions. In many cases the assignment of logic values to these inputs was not only a waste of time but also blocked the detection of some faults; this resulted from the inability of RAPS to identify restart gates and critical paths through the circuit. Further-more fault simulation is an integral part of the test generation procedure used in SMART whereas RAPS only uses fault simulation once a test vector has been generated to determine its fault coverage.

When it has been deemed that SMART can no longer generate tests efficiently, the fault oriented test generation procedure called FAST is in-voked; the criteria for switching between test generation procedures will be discussed later. In selecting a target fault, that is a fault which has not been detected by any of the currently generated test patterns, FAST selects one which is closest to the primary inputs; in this way during the test generation procedure long sensitive paths will be generated which will uncover, it is hoped, other faults which have not been detected. In an attempt to reduce the time wasted in generating tests for target faults which cannot be detected, FAST, having decided that a particular fault cannot be detected, labels all other faults equivalent to the given fault as undetectable.

The improvements in efficiency gained by other algorithms, for example PODEM and FAN, over the D-algorithm have been obtained by making intelligent decisions during the backtracing process which attempts to avoid conflicting assignments of logic values early in the test generation process. FAST has adopted the basic concepts used in PODEM and FAN for making intelligent decisions during the backtracing process, and has expanded and improved upon them.

FAST, like PODEM, uses controllability values in order to decide which input should be used, or the order in which inputs should be processed, to achieve a given objective at a gate. It also uses observability values to assess the degree of difficulty encountered in propagating fault information to a primary output. In FAST, however, the controllability/observability values are computed in a different way from the methods used in testability analysis programs. The connotation applied to controllability values used in FAST is that if a controllability value of zero is calculated for a node then this node can be assigned a logic value without introducing any conflicting logic assignments; as the controllability values increase the possibility of creating conflicting assignments increases. The calculation of controllability values starts by assigning 0 values to the primary inputs; the calculations of the controllability values on the internal lines proceed systematically to the primary outputs. When performing the calculation on internal lines several factors are taken into consideration, namely the logic value to be set on the line, the type of gate to which the line is attached and the number of reconvergent fanout paths with which a given line is involved. Any line which has a controllability value of zero is considered as a backtrace-stop line. In FAST backtrace-stop lines can, in general, be either 0-stop lines or 1-stop lines and differ from the 'headlines' used in FAN, which are backtrace-stop lines for both values, which is a result of the constraints placed on the circuit structure up to a node, declared in the FAN algorithm, as a headline.

The calculation of observability values in FAST includes the controllability values; hence it reflects the relative potential of incurring conflict when a given internal line is observed at a primary output.

The procedure used by FAST to generate a test for a fault is similar to that used in PODEM. First a target fault is selected and the initial objective is defined of generating the necessary input conditions to the given gate to activate the fault. In FAST, however, the backtracing procedure used to justify line assignment terminates on backtrace-stop lines, rather than continuing to the primary inputs, because a backtrace-stop line for a given value can be assigned that value without creating any conflicts in the circuit, hence a backtrace-stop line can be considered as a pseudo-input to the circuit for a given logic value. The assignment of the logic value to this pseudo-input is simulated to determine if the initial objective is achieved; if not the backtracing procedure coupled with simulation is repeated until the input conditions required to activate the fault are achieved. Thereafter an attempt to propagate the fault information to a primary output is made. If a choice of paths to a primary output is available, the path with the best observability values is chosen; that is the path least likely to result in conflicting logic assignments. Thereafter, the fault information is systematically propagated through each gate in the path, by justifying the assignment of the appropriate non-dominant logic value to all the inputs to a gate except the input propagating the fault information. The justification process on each gate starts with the input which has the worst controllability value since this assignment is most likely to cause a conflict; if the conflict cannot be resolved time has not been wasted justifying the other input assignments. Again the justification process comprises repeated backtraces to stop lines or primary inputs followed by a simulation run to determine if the given assignment to the stop line or primary input satisfies the given objective. Finally, when the fault information is propagated to the primary output, the logic values assigned to the backtrace-stop lines are fully justified back to the primary inputs.

In test generation systems which have schemes for generating 'global' and 'clean-up' test sets, in order to maximize the effectiveness of each technique, there must be some criterion whereby the system can decide that a particular test generation philosophy is becoming less effective and switch to the other technique. In LTG, when the 'global' test generator SMART is in use, the average number of new faults detected over say the last n tests (approx. 30) is computed; and if this value is less than the number of 'useful' primary outputs, that is primary outputs which are not stop lines, then the fault oriented test procedure FAST is invoked.

When LTG was compared with other state-of-the-art test generation schemes it was observed that the test sets generated by LTG were smaller and provided either the same or a better fault coverage; this improved performance is accredited to the close interaction between the test generation programs and the fault simulator.

5.3 TEST GENERATION IN SEQUENTIAL CIRCUITS

Although adequate test generation methods exist for detecting faults in combinational circuits, the same cannot be said for sequential circuits [15]. The major difficulty in testing sequential circuits is that the output response of the circuit depends not only upon the input test pattern but also upon the internal state of the circuit; these internal states, however, are not directly observable. Several techniques have been developed to drive the circuit into some known state before testing begins; the first technique simply uses a master reset on the circuit, the second technique employs 'homing sequences' which drive the circuit into some known state regardless of its present state; however, 'homing sequences' can be very long and not all circuits have a 'homing sequence'. Both of these techniques must be valid under faulty and fault-free conditions. Therefore in order to detect a fault in a sequential circuit a sequence of test patterns must be applied rather than a single pattern as in combinational circuits.

Due to the problem of testing sequential circuits, several design techniques have been developed which permit sequential circuits to be reconfigured in such a way that, for the purposes of testing, the circuit is essentially combinational; these design techniques will be discussed in Chapter 6. Very few algorithmic techniques exist for testing sequential circuits and these are restricted to small circuits [16]. The techniques generally used to generate tests in sequential circuits can be categorized as

(a) functional
(b) heuristic.

5.3.1 Functional techniques

Sequential circuits can be tested by specifying input sequences which exercise the functional characteristics of a circuit, for example clearing registers, incrementing/decrementing counters. The effectiveness of the input patterns is determined by fault simulation. This technique is used for highly sequential circuits. An alternative technique is to verify that the circuit under test operates in accordance with its state table, however several assumptions must be made, namely:

1. Under fault conditions no more states are created;
2. In the fault-free circuit there is an appropriate input sequence which can cause a transfer from each state to every other state; and
3. The next state of the circuit is uniquely defined by the inputs and the present state of the circuit, i.e. the circuit is fully specified and deterministic.

This technique is, again, limited to highly sequential logic functions; furthermore, for complex circuits the state table can become very large. In many instances, the state tables may not be available, particularly for random sequential circuits, and are very expensive to generate.

5.3.2 Heuristic techniques

Heuristic test generation techniques refer to a class of procedures which, in general, work but do not guarantee to generate a test for a fault even though one does exist. Several heuristic techniques are described below:

(a) Sequential analyser [17]

The sequential analyser was probably the first acceptable technique for generating tests to diagnose faults, primarily, in asynchronous sequential circuits. It was, in essence, a compiled code fault simulator and when used with Seshu's heuristics [18] could automatically generate tests for faults in sequential circuits. The heuristics used were

(i) BEST NEXT OR RETURN TO GOOD This heuristic attempts to generate further test patterns from an existing pattern which has been declared as 'good' by the fault simulator, that is, it may have detected a large number of faults which had not already been detected or, since the test patterns were to be used for fault diagnosis the pattern may distinguish between a number of fault conditions. Additional test patterns are generated from the existing good pattern by computing all the input patterns which differ from this pattern by one bit, since successive input patterns to an asynchronous sequential circuit are allowed to differ by only one bit. The 'goodness' of these candidate test patterns is then evaluated by the fault simulator; those which fall below the given criterion are discarded. The 'best' pattern is then selected for further processing to generate additional patterns, the remaining 'good' patterns are placed on a push-down stack. If all the patterns generated from a 'good' pattern fall below the given criterion for selection, the next pattern on the push-down stack is selected. If the push-down stack becomes exhausted without finding an acceptable test pattern the heuristic is deemed to have failed.

(ii) WANDER Again potential test patterns are generated from an existing pattern by generating all the patterns which differ from it by one bit. If none of the potential patterns is acceptable, the bits are changed cyclically until a pattern is found which does not cause the circuit to oscillate or generate a critical race. The 'wander' procedure is then applied to this pattern. This process is repeated until an acceptable pattern is found or the number of iterations of the procedure exceeds the user-defined limit, whereupon the heuristic is deemed to have failed.

(iii) RESET This heuristic simply uses the 'permissible' reset states of the machine as potential test patterns. The 'permissible reset state' which, for example, detects the largest number of undetected faults, is then selected as a test pattern.

(iv) COMBINATIONAL This is the last of the heuristics and is most effective in circuits with a small number of feedback lines. It is assumed that the feedback lines in the circuit are open; a test is subsequently generated for a fault in the modified combinational circuit, and provided it does not cause any races or oscillations it will be used as a test for the same fault in the sequential circuit.

The sequential analyser has been used, effectively, on circuits having ninety-six inputs, ninety-six outputs and forty-eight feedback loops. The technique was subsequently modified [19] to operate on large sequential circuits comprising 200 inputs, 260 outputs, 700 feedback loops and 3800 logic blocks.

(b) Iterative Test Generation method (ITG) [20], [21]

This heuristic technique was developed to generate tests for faults in asynchronous sequential circuits, without any major restrictions being applied to the feedback loops; although it is best suited to circuits which only have a few feedback loops, for example control logic. The test generation process is carried out in two phases, first a potential test is generated for a given fault and second fault simulation is used to verify that the test will detect the fault in the circuit.

At the outset of the test generation process the sequential circuit is converted into an iterative combinational circuit by opening up all the feedback lines in the circuit as shown in Fig. 5.4; this procedure essentially maps the time domain response of the sequential circuit, onto the spatial domain response of the iterative circuit. The single fault which was assumed to exist in the sequential circuit is now considered as a multiple fault in the iterative

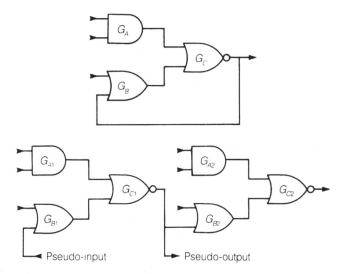

Fig. 5.4 Iterative circuit.

circuit; the broken feedback lines are considered as pseudo-inputs and -outputs of the circuit. The test generation process, which is based on the D-algorithm, produces a sequence of tests; the maximum number of tests in the sequence is equal to the number of iterations in the model used for test generation. The process of generating a test starts by selecting the appropriate copy of the circuit in the iterative model, and hence the time frame in which an attempt to sensitize a path to a primary output will start. The primitive D-cube of failure is then generated for the fault condition in this iteration of the circuit and the D-drive process is initiated and continued through one or more iterations of the circuit until a path is sensitized to a primary output, whereupon the consistency operation is invoked to justify all the logic assignments made in the circuit.

The version of the D-algorithm used in the ITG method differs from the normal D-algorithm process, in that

1. Only a single path is sensitized to a primary output.
2. Pseudo-inputs can only be assigned 'don't care' values.
3. A sensitized path cannot terminate at a pseudo-output.
4. It must be remembered that the fault condition exists in each iteration of the circuit and that the faulty node can only be assigned logic values consistent with the fault condition that is if the node is stuck-at-0, the node can only be assigned a logic 0 or a D-value.

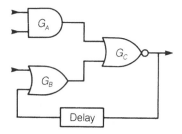

Fig. 5.5 Reconstructed circuit with delay.

The test sequence generated for the multiple fault in the iterative circuit is subsequently converted, by a process called 'completion' to a potential test sequence for the single fault condition in a sequential circuit which differs from the original circuit by the insertion of delays in the feedback lines as shown in Fig. 5.5. The duration of these delays is equal to the time delay between the application of the test patterns in the test sequence applied to the iterative circuit. The potential test generated by the 'completion' process is then applied to a fault simulator to determine if the fault condition can be detected by this test pattern sequence in the actual circuit.

The 'completion' process used to derive the potential test sequence for a single fault in the sequential circuit from the test sequence derived from the iterative circuit eliminates the 'don't care' states from the input sequences in such a way as to minimize the number of input signals which change from one input test pattern to the next, in order to reduce the possibility of creating race conditions when the test patterns are applied to the actual circuit.

(c) Macro Test Generator (MTG) [21], [22]

If the circuit is highly sequential, for example, a shift register or counter, the iterative test generation technique does not produce acceptable tests, since the delays which have been introduced into the model used for test generation purposes produce large differences in the operation between the model and the actual circuit.

In the macro test generation technique, flip-flops are considered as macro blocks or logical primitives in the same way as one considers NAND and NOR gates, and only feedback lines external to the macro blocks are broken when deriving the iterative circuit.

In the MTG procedure failures are not allowed to exist inside a macro block; it is assumed, in general, that many internal failures can be detected by tests generated for faults on the input/outputs of the macro blocks. During the test generation procedure, which is based on the D-algorithm, the propagation of fault-sensitive values through the macro blocks is performed by using D-tables [23] which comprise sets of sequential propagation D-cubes (D-sequences); these may extend over several time frames. The consistency operation on macro blocks is implemented using C-tables [23] which comprise sets of sequential singular cubes (C-sequences) which may also extend over several time frames. The C- and D-tables may contain a mix of 'reset' or 'non-reset' sequences. In the 'reset' sequences the present output is defined uniquely by the inputs; in 'non-reset' sequences the present output is a function of not only the present inputs but also the past outputs.

In order to demonstrate the advantage of MTG over ITG a highly sequential circuit was used as a test vehicle; the fault coverage obtained from the MTG procedure was approximately 90%, whereas that obtained from the ITG procedure was only 17%; furthermore, since test generation is essentially performed at a higher level of abstraction in the MTG method, namely macro blocks instead of basic gates, it is much more efficient than the ITG procedure.

(d) Simulator Oriented Fault Test Generator (SOFTG) [24]

The objective in developing SOFTG was to provide a practical automatic test generator for complex sequential circuits, capable of generating long sequences and also of modelling sequential circuits accurately, since the available methods are inadequate. For example an extension of the D-algorithm for sequential circuits called the iterative test generation method has two disadvantages: first, for the purposes of test generation a circuit larger in size than the original circuit must be produced; second, in the final stages of the procedure the circuit model is not an accurate representation of the actual circuit.

SOFTG depends heavily on the use of a fault simulator to

1. provide nodal logic values for the faulty and fault-free circuits
2. store internal states for one pattern so that the test generator has the ability to discard a pattern if the simulation results are not suitable
3. determine the other faults that a possible test pattern can detect.

Test sequences are generated incrementally, that is each pattern differs from its predecessor in only one bit, in this way race conditions are avoided in the circuit. Furthermore SOFTG does not work its way systematically through a fault list, but dynamically selects the next fault condition based on the presence of test values (i.e. Ds or $\bar{D}$s) at nodes in the circuit. In this way the lengths of test sequences are reduced since the end of one test is merged into the beginning of the next.

Before the test generation phase commences a structural analysis of the circuit is performed in order to determine the controllability values for the nodes in the circuit and also to determine their distance from the output.

At the start of the test generation procedure all the primary inputs are assigned the value of logic 0. The fault simulator is then invoked to determine what faults if any are detected by the pattern; all the inputs are now at a known value. A target fault is then selected and the objective, of sensitizing the gate output at the site of the fault, is defined. A backward trace procedure is subsequently performed until a primary input is reached. The primary input is then set to the specific value as dictated by the backward trace procedure and the circuit is simulated. An analysis phase, on the outcome of the simulation results, is performed to decide whether or not to proceed with this new pattern or to discard it and restore the node values in the circuit to the settings defined by the previous input pattern. A pattern is retained if it does not produce a conflict in node values or if it causes any faults to be detected, otherwise it is discarded. A check is then made to determine if the objective has been achieved: in the first instance this is to make a gate output sensitive to the fault condition and in the second to propagate fault information to a primary output. If the objective is not achieved the process of backward tracing and simulation is continued. If at any time the backward tracing procedure cannot reach a primary input the target fault is flagged as undetected and the process is abandoned for this fault condition. An attempt to generate a test for this fault condition may be made later in the process if the conditions are right: that is, for example, if the logic values in the circuit resulting from the generation of a test for another fault also propagate fault information about this fault closer to a primary output.

When SOFTG was compared with the iterative test generation method on a range of circuits whose complexity was of the order of 600 gates with approximately 80 feedback loops, SOFTG produced a fault coverage which was approximately 10% better than that achieved by the iterative test generation method, although the run times were somewhat longer.

5.4 TEST GENERATION TECHNIQUES AT FUNCTIONAL LEVEL

It is widely accepted that unless the rising costs of testing complex circuits can be curtailed then full advantage cannot be taken of the potential offered by current and future VLSI fabrication capabilities. Several approaches have been adopted in an attempt to curtail the increase in test generation costs: first, as discussed previously, more efficient test generation algorithms have been developed, including the use of special-purpose engines [25], [26]; second, design for testability techniques have been introduced (Chapter 6); finally, test generation schemes have been developed to operate at higher levels of abstraction. The rationale for attempting to generate tests for higher levels of circuit abstraction is identical to that used in the early days of digital circuit testing when it was realized that circuits were becoming too complex to test at component level and it was necessary, in the interest of efficiency and cost effectiveness, to test groups of related components together at gate level. Present-day circuit complexities are such that gate level test generation is costly; hence groups of gates must be tested as functions. Furthermore, many digital systems comprise LSI functions, for which gate level descriptions are unavailable, necessitating test generation to be performed at a higher level of abstraction. Several of the techniques proposed to perform test generation at functional level will now be described.

5.4.1 Extensions to the *D*-algorithm

Two approaches have been proposed to extend the capabilities of the *D*-algorithm for use on higher level functional blocks. The first technique [27] employs algorithms to generate specific 'solution' sequences to perform the *D*-drive and consistency operations of the *D*-algorithm for particular function blocks such as shift registers, counters, decoders, etc. During the test generation process, when a higher level functional block is to be processed a 4-tuple is sent to the 'problem decoder' defining the element type, its complete state, the function to be performed (that is *D*-drive, consistency operation, etc.) and the input/output line to be processed. The appropriate algorithm is then selected and the required 'solution' is returned in symbolic form which defines a sequence of operations such as left or right shift, increment, decrement, hold or clear to be applied to the block in order to perform the *D*-drive or consistency operation. These symbolic or functional solutions are then converted by a 'translator' into the appropriate logic values required by the given function. One advantage of describing the solutions in symbolic form is that the algorithms can be independent of the hardware implementation of the function; the 'translator' can then generate the necessary signals required for a specific implementation of a function.

The second approach [28], [29] is a more general approach which can be applied to functions described in either procedural or non-procedural hardware description languages. In hardware description languages the behaviour of a function is described as a series of statements comprising one or more conditional or unconditional transfers of information between functional variables. Control of the data transfers is governed by either binary cause–effect statements, i.e. IF . . . THEN . . . ELSE statements, or by case statements, which are non-binary cause–effect statements. In a procedural language the interpretation of the statements must be done in sequence, whereas in non-procedural languages the statements are interpreted in parallel; the differences in interpretation is illustrated in Fig. 5.6 where delays are associated with data transfers in a block. In the function described in a non-procedural language an event occurring at time t will cause several events to be generated as shown; whenever one of these new events happens the function block is re-evaluated since the event may be an internal state change. In the function described in the procedural language individual statements and subsequent actions are performed sequentially.

During the test generation process fault information is propagated through the functional level description of the blocks in a circuit by first generating sets of D-equations [29] which are used to analyse the different modes of propagation through the control or data variables in the functional description, subsequently graph traversal techniques are used to perform the actual propagation.

The concept of the D-equations is simply a concise method of enumerating all possible conditions to realize some output value; for example consider an OR gate with two inputs a and b and output c:

$$
\begin{aligned}
c^0 &= (a^0 + b^0) = a^0\, b^0 \\
c^1 &= (a + b)^1 \;\; = a^1 + b^1 + a^D b^{\bar{D}} + a^{\bar{D}} b^D \\
c^D &= (a + b)^D \;\; = a^D b^0 + a^0 b^D + a^D b^D \\
c^{\bar{D}} &= (a + b)^{\bar{D}} \;\; = a^{\bar{D}} b^0 + a^0 b^{\bar{D}} + a^{\bar{D}} b^{\bar{D}}
\end{aligned}
$$

Similar expressions can be generated for the AND and INVERT functions. Subsequently using these expressions the D-equations can be generated for any sum of products switching expression.

Unfortunately the 'action' part of a cause–effect statement used in a functional description of a block requires the evaluation of functions other than simple switching expressions: for example, shift, add, increment/decrement, encode/decode, etc.

The propagation of fault information through some of these functions is trivial, for example the shift function, others however are much more difficult, for example the addition and incrementing functions. To propagate information through an addition function recourse is made to sets of 'sum' and 'carry' tables generated to process the five values encountered in the D-algorithm; in the case of the increment function, a table is compiled containing six basic counter patterns, comprising 0, 1, D and $\bar{D}$, together with the next state of the counter for each pattern.

The technique adopted to propagate fault information through the control structures in a high level description of a function comprises replacing the

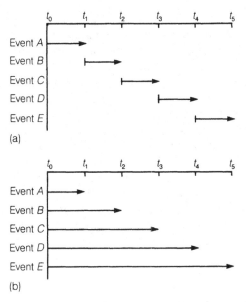

Fig. 5.6 Interpretation of events in procedural and non-procedural HDLs: (a) sequential activation of events; (b) parallel activation of events.

statement with the equivalent algebraic switching expression, as shown below [29], and subsequently generating the corresponding D-expressions: for example the following binary cause-effect statement:

$$IF\ A\ THEN\ Z := B$$
$$ELSE\ Z := C;$$

is rewritten as

$$Z = AB + \bar{A}C$$

the corresponding D-expression, assuming single D-propagation, becomes

$$Z^D = A^1 B^D + A^0 C^D + A^D B^1 C^0 + A^{\bar{D}} B^0 C^1.$$

The D-expression indicates that fault information can be propagated through either the control expression or the data transfer expression. In practice, however, the D-expressions are not used directly to propagate fault information but algorithms are derived from a verbal description of the conditions necessary to propagate fault information via the given terms.

A similar technique is employed, although much more complex, when the 'action' part of the cause–effect statement are functions other than simple assignment statements, or when the binary cause–effect statement is replaced by a case statement. In situations in which the cause–effect statements are 'nested' in a functional description, a 'graph' approach is adopted and branches in the graph are made depending upon the values of the control variables; the problem of propagating fault information is subsequently perceived as a graph traversal problem.

The above approach can also be used in procedural language descriptions of function blocks; however, the D-propagation technique must be modified to reflect the sequential interpretation of events generated within a procedural language description of a function. The sequential interpretation of events is accounted for through the use of 'path' functions [29] which are a shorthand notation to define the combinations of input or internal variables necessary to control the transfer of data through a graph representation of the procedural language description. The D-expression of the procedural description is subsequently formulated in which the path functions are simply considered as literals in the expression and may be assigned the values 0, 1, D or $\bar{D}$. The algorithm for the propagation of fault information through the functional description is subsequently generated from the D-expression.

5.4.2 S-algorithm [30], [31]

The S(symbolic)-algorithm is a systematic technique of generating functional level test patterns when a circuit is described using a register transfer language. The test generation system in which the S-algorithm is embedded comprises three modules:

1. The processor module, which checks the current description for syntax errors, partitions the circuit into subfunctions for the purposes of test generation and computes the order in which the subfunctions will be processed;
2. The test generation module, which generates the tests for the given functional faults;
3. The postprocessor module, which determines what other undetected faults are uncovered by a given test pattern and generates a list of hard-to-test faults which require further processing.

A system is described as fault-free if it operates in accordance with its specification, otherwise it is deemed to be faulty due to an RT (Register Transfer) level fault in an RT-component. Nine RT-level faults can be derived from the syntax of a basic RT-language statement, which comprises the following RT-components, labels, conditional expressions, operators, source and destination registers, etc. The faults which can be considered using these components are label faults, jump faults, condition faults, operation decoding faults, operation execution faults, data transfer and storage faults, etc. In general, however, the nine RT level faults are collapsed into a smaller group of faults which are functionally equivalent. Furthermore, the same

RT-statement may appear several times in a circuit description; hence it is sufficient to test this statement once for the given set of RT-faults, since each instance of the given RT-statement will refer to the same piece of hardware.

The method of test pattern generation using the S-algorithm relies on the technique of symbolic execution, which is used extensively in computer program testing. In symbolic execution symbolic values are processed rather than actual values, which results in the generation of very large expressions which become difficult to manipulate. However, RT level descriptions are relatively simple and symbolic execution of the description can be performed quite efficiently. During the symbolic execution process, every internal register is described in terms of the symbolic values on the external inputs to the function and as the process continues a symbolic execution tree is generated which highlights the paths of all possible symbolic execution flows; at the end of each path the symbolic results of the path together with a set of path constraints will be obtained. Furthermore, since symbolic values are used all possible input combinations are processed in a single symbolic execution run. From a simple comparison between the symbolic results generated by the faulty and fault-free circuits the necessary tests can be derived to detect the given faults. The symbolic execution module for RT-level descriptions comprises a monitor which controls the overall symbolic execution process; a symbolic execution interpreter which interprets the semantics of the RT-statements; a symbolic expression simplifier; and a symbolic inequality solver, which compares the faulty and fault-free results and produces the appropriate test pattern if possible.

The process of test pattern generation using the S-algorithm starts by locating all the covering paths in a given subfunction, that is finding all the symbolic execution paths which cover all distinct RT-components, that is, RT-components which have not been tested previously in other subfunctions. The process of finding all the covering paths is not a very onerous task since the subfunction descriptions are usually quite simple and a single path is sufficient, except in cases where there may be conditional branches; furthermore an attempt is always made to locate the shortest path. Thereafter, a fault-free symbolic execution of the covering paths is performed. For the purpose of test pattern generation, the fault conditions are classified as data transfer faults and others. A set of 'transfer-test-finding' heuristics [31] are employed to detect bridging faults or multiple stuck-at faults in each transfer path in every distinct RT-statement. Thereafter, the non-data transfer faults occurring along a given symbolic execution path are processed; these faults are injected individually into the RT-statements and a fault-injected symbolic execution is performed. Subsequently an attempt is made to generate a test for a given fault by invoking the symbolic inequality solver, which examines the fault-free and fault-injected symbolic execution results. If any inequality in the symbolic execution results is not detected another symbolic execution path is tried; if all paths have been processed and an inequality cannot be generated, the fault is categorized as 'hard to test' and is scheduled for further processing by the postprocessor. On some occasions, since the inequality solver uses heuristics, the solver may request the designer to inspect the inequality expressions in order to determine if an inequality exists.

The additional processing performed on the 'hard-to-test' faults usually comprises an exhaustive search of all symbolic paths, which may lead to a test being generated or using tests which have detected hard-to-test faults in other subfunctions. Alternatively it may be necessary to redesign that part of the circuit.

5.4.3 *P*-algorithm [32]

The *P*-algorithm is a high-level test pattern generation technique which can be applied either to sequential circuits incorporating scan paths (Chapter 6) or combinational circuits. In this technique the circuit described in a Functional Description Language (FDL) is translated into a graph where each operator in an FDL statement represents a node in the graph and each data/control line corresponds to an edge in the graph; primary inputs and scan-in flip-flops are defined as controllable nodes and primary outputs and scan-out flip-flops are defined as observable nodes. During the test generation process explicit modelling of the classical faults is not used, only the differences between faulty and fault-free operations is considered. The test generation process comprises selecting an observable node and assigning the symbolic value *P* to the node. A fanin branch to the node is selected and subsequently sensitized to this symbolic value; this backward propagation is continued until a controllable node is reached. A justification procedure is then implemented to assign a consistent set of values to the fanin branches activated during the backward propagation of the symbolic *P*-value. A test set for the circuit is obtained by repeating this procedure until each branch in the graph has been sensitized at least once.

Although high-level test pattern generation techniques potentially offer a cost reduction over gate level test generation techniques, several obstacles [33] must be overcome before they will gain general acceptance. First, gate level techniques are well tried and tested and a vast amount of information has been accumulated about gate level testing; hence there is a basic reluctance to change. Second, there must be some mechanism which will permit the significance of the fault coverage obtained at functional and gate level to be readily correlated, particularly during the transition phase. However, an alternative solution to reducing gate level test generation costs may be to introduce a 'mixed mode' test pattern generator which permits the function block, where a fault is assumed to exist, to be performed at gate level and the test generation processes outside the block performed at a higher level of abstraction, thus permitting the accuracy of gate level test generation to be merged with the efficiency of high-level test pattern generation.

5.5 TEST GENERATION IN MOS CIRCUITS

The classical fault model used to represent stuck-at-1/0 faults on gate inputs and outputs has been shown to be inadequate for modelling a large number of fault conditions (for example, stuck-open and stuck-short faults in transistors) which can exist in either NMOS or CMOS circuits. Furthermore, certain faults, which occur in pass transistors and bus structures, necessitate the generation of a test sequence rather than a test pattern for their detection, although the circuit is combinational. Some of the approaches used to improve the fault coverage in NMOS and CMOS circuits and also to handle pass transistor and bus structures will now be briefly described.

The first technique [34], which can readily be incorporated into an existing gate level automatic test generation program, comprises modelling the structures of either an NMOS or a CMOS gate as an interconnection of basic switching elements. An additional 'memory' function block is introduced to model the stored charge effect or high impedance state which certain MOS structures exhibit; it is also used to eliminate power and ground connections from the model. The MOS-to-logic transformations used to generate an equivalent gate model for an MOS structure are shown in Table 5.5. An MOS gate structure, in general, is considered to comprise a pull-up and a pull-down network, connected to the 1 and 0 inputs to the 'memory' block in which the 0 input is dominant; if both inputs are at a zero value the output of the block remains in its last state. As an example of modelling an MOS gate as an 'equivalent' gate model consider the example of the three-input AND-OR-INVERT NMOS structure shown in Fig. 5.7. The normal input/output stuck-at faults in the MOS structures are represented as stuck-at faults on the inputs/output of the equivalent gate model; stuck-open and stuck-short transistor faults are represented as stuck-at faults on the input to the gate which is used to replace the transistor.

When the circuit has been transformed into its equivalent gate model the D-algorithm test generation technique can then be used, with some minor modifications, to generate tests to detect faults in the circuit. First, the necessary cubes to perform the D-drive and consistency operations through the 'memory' block must be generated; second, since certain faults in MOS structures result in a gate retaining its last output value, it is necessary in some instances to initialize the faulty gate output into some known state; this is achieved, essentially, by performing the consistency operation from the faulty gate back to the primary inputs, consequently a test sequence of length two will be generated for these types of faults instead of a single pattern. Care

Table 5.5 NMOS to logic transformations [34]

Input to the gate of an NMOS transistor		Input to a logic gate	——
Depletion load		Logic '1'	▷—
Series devices	A—⊣ ⊢ B—⊣ ⊢	AND gate	A, B → AND —D—
Parallel devices	A ⊣ ⊢ B	OR gate	A, B → OR
MOS gate output	V_{dd} / S_1 —○Y / S_0 —○ 0V	$\begin{array}{ccc} S_1 & S_0 & Y \\ 1 & 1 & 0 \\ 1 & 0 & 1 \\ 0 & 1 & 0 \\ 0 & 0 & M \end{array}$	S_1 1 / B / 0 S_0 → Y

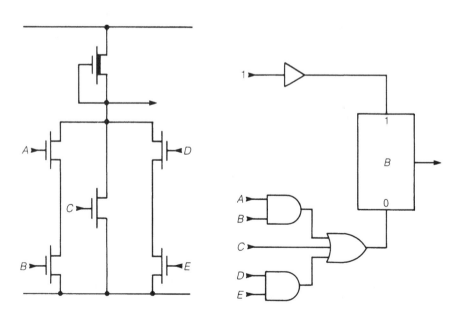

Fig. 5.7 An MOS AND–OR–INVERT gate and its 'equivalent gate' model.

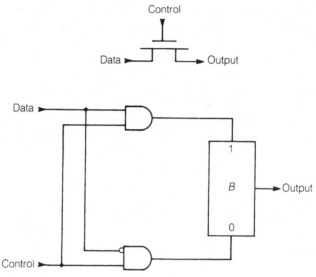

Fig. 5.8 Logic model for a transmission gate [34].

must be taken in the generation of the two sequence patterns in order to avoid the generation of a hazard condition which may corrupt the initialized value when the sequence is applied to the actual circuit.

Pass transistors can also be represented, as shown in Fig. 5.8, by an equivalent gate model. Bus structures can be represented as multi-input wired-AND or wired-OR gates depending upon the technology; when processing a bus structure during test pattern generation all the gate inputs tied to the bus must be considered. The cubes for the bus structure must also reflect the situation that the bus may be in a high impedance state.

A more direct approach has also been proposed by several researchers [35],[36] to model MOS circuits at switch level and perform the test pattern generation at this level using the *D*-algorithm, again a test sequence of length two will be generated when the given fault requires an initialization sequence before it can be detected.

Although the above techniques permit a large number of fault conditions peculiar to MOS structures to be modelled, they have the disadvantage of increasing the complexity of the circuit, in terms of the number of primitives – hence potentially increasing the test generation time. An alternative procedure, which does not increase the size of the circuit for test generation purposes, has been developed for detecting stuck-open faults in CMOS circuits; the technique is called 'matching sequences' [37] and can be incorporated readily into any existing gate level test pattern generation system. Two algorithms are used to generate tests for stuck-open faults; the first algorithm attempts to identify the stuck-open faults which can be detected by normal stuck-at-fault test patterns and also lists those stuck-open faults which are detected. During the process of generating a test for a given stuck-at fault, the algorithm examines the logic state on the individual nodes in the circuit generated by the previous test pattern. If the previous test pattern sets up the initial conditions which would enable a given stuck-open fault on a gate to be detected, and if the present test pattern supplies the necessary input condition to distinguish the faulty and fault-free gate, then, provided the present test pattern can detect the corresponding stuck-at fault on the output of the gate the stuck-open fault will also be detected. This technique of examining the logic values on circuit nodes generated by previous test vectors to determine if the conditions are correct for the present test vector to detect a stuck-open fault is called *matching sequences.* Invariably, the technique of matching sequences will not detect all stuck-open faults; a second procedure is then invoked to generate tests for the remaining faults. This comprises explicitly generating two pattern sequences necessary to detect a given fault condition: the first pattern preconditions the gate and the second pattern distinguishes the faulty and fault-free circuits.

At present the above approach has only been developed for conventional gate structures, i.e. AND, NAND, OR and NOR gates. Further work is required to extend this approach, if possible, to circuits containing pass transistors and busses, which to date can only be processed by using the MOS-to-logic transformation technique or by performing test generation at switch level.

In conclusion the development of more sophisticated test generation algorithms has, to some extent, reduced the overall cost of testing VLSI circuits. However, this approach can only be considered to be an interim solution to the test generation problem since it is directed at the symptoms of the problem of generating tests for faults in VLSI circuits, i.e. complexity, and not at the root cause of the problem, which is the inability to control and observe signal nodes in the circuit. The root cause can only be attacked by adopting design-for-testability techniques, which are discussed in the following chapters.

5.6 REFERENCES

1. Muehldorf, E. I. and Savkar, A. D. (1981) LSI logic testing – an overview. *IEEE Trans. Computers*, **C-30**(1), 1–17.
2. Abadir, M. S. and Reghbati, H. K. (1983) LSI testing techniques. *IEEE Micro*, February, 34–51.
3. Sellers, F. F., Hsiao, M. Y. and Bearnson, L. W. (1968) Analyzing errors with Boolean differences. *IEEE Trans. Computers*, **C17**(7), July, 676–83.
4. Roth, J. P. (1966) Diagnosis of automata failures: A calculus and a method. *IBM J. Research and Development*, **10**, 278–91.

5. Roth, J. P., Bouricius, W. G. and Schneider, P. R. (1967) Programmed algorithms to compute tests to detect and distinguish between failures in logic circuits. *IEEE Trans. Electronic Computers*, **EC-16**(5), 567–80.

6. Cha, C. W. (1978) 9-V algorithm for test pattern generation of combinational digital circuits. *IEEE Trans. Computers*, **C-27**(3), 193–200.

7. Goel, P. (1981) An implicit enumeration algorithm to generate tests for combinational logic circuits. *IEEE Trans. Computers*, **C-30**(3), 215–22.

8. Goel, P. and Rosales, B. C. (1981) PODEM-X: An automatic test generation system for VLSI logic structures. *18th Design Automation Conference Proceedings*, June, 260–8.

9. Patel, S. and Patel, J. (1986) Effectiveness of heuristic measures for automatic test pattern generation. *23rd Design Automation Conference Proceedings*, June, 547–52.

10. Funatsu, S. and Kawai, M. (1985). An automatic test generation system for large digital circuits. *IEEE Design and Test of Computers*, **2**(5), 54–60.

11. Fujiwara, H. and Shimono, T. (1983) On the acceleration of test generation algorithms. *IEEE Trans. Computers*, **C-32**(12), 1137–44.

12. Roberts, M. W. and Lala, P. K. (1987) Algorithms to detect reconvergent fanouts in logic circuits, *IEE Proceedings*, **134**, Pt. E, 105–11.

13. Abramovici, M., Kulikowski, J. J., Menon, P. R. and Miller, D. T. (1986) SMART and FAST: Test generation for VLSI scan design circuits. *IEEE Design and Test of Computers*, **3**(4), 43–54.

14. Abramovici, M., Menon, P. R. and Miller, D. T. (1984) Critical path tracing: An alternative to fault simulation. *IEEE Design and Test of Computers*, **1**(1), 83–93.

15. Miczo, A. (1983) A sequential ATPG: A theoretical limit. *Digest of Papers, 1983 International Test Conference*, October, 143–7.

16. Chang, H. Y., Manning, E. G. and Metze, G. A. (1970) *Fault Diagnosis of Digital Systems*, pp. 55–63, Wiley Interscience, New York.

17. Chang, H. Y., Manning, E. G. and Metze, G. A. (1970) *Fault Diagnosis of Digital Systems*, Chapter 4, Wiley Interscience, New York.

18. Chang, H. Y., Manning, E. G. and Metze, G. A. (1970) *Fault Diagnosis of Digital Systems*, pp. 63–5, Wiley Interscience, New York.

19. Breuer, M. A. (ed.) (1972) *Design Automation of Digital Systems: Vol. 1 – Theory and Techniques*, pp. 383–8, Prentice-Hall International.

20. Putzolu, G. R. and Roth, J. P. (1971) A heuristic algorithm for the testing of asynchronous circuits. *IEEE Trans. Computers*, **C-20**(6), 639–47.

21. Bouricius, W. G., Hsieh, E. P., Putzolu, G. R. *et al.* (1971) Algorithms for the detection of faults in logic circuits. *IEEE Trans. Computers*, **C-20**(11), 1258–64.

22. Vaughn, G. D. (1976) CDALGO – A test generation program. *13th Design Automation Conference Proceedings*, June, 186–93.

23. Hsieh, E. P., Putzolu, G. R. and Tan, C. J. (1981) A test pattern generation system for sequential logic circuits. *Digest of Papers, 11th International Symposium on Fault Tolerant Computing*, 230–2.

24. Snethen, T. J. (1977) Simulator oriented fault test generator. *14th Design Automation Conference Proceedings*, June, 88–93.

25. Abramovici, M. and Menon, P. R. (1983) A machine for design verification problems. *Digest of Papers, International Conference Computer Aided Design*, 27–9.

26. Kramer, G. A. (1983) Employing massive parallelism in digital ATPG algorithms. *Digest of Papers, 1983 International Test Conference*, October, 108–14.

27. Breuer, M. A. and Friedman, A. D. (1980) Functional level primitives in test generation. *IEEE Trans. Computers*, **C-29**(3), 223–35.

28. Levendel, Y. H. and Menon, P. R. (1981) Test generation algorithms for non-procedural computer hardware description languages. *Digest of Papers, 11th International Symposium on Fault Tolerant Computers*, June, 200–205.

29. Levendel, Y. H. and Menon, P. R. (1972) Test generation algorithms for computer hardware description languages. *IEEE Trans. Computers*, **C-31**(7), 577–88.

30. Su, S. Y. H. and Hsieh, Y.-I. (1981) Testing functional faults in digital systems described by register transfer languages. *Digest of Papers, 1981 International Test Conference*, October, 447–57.

31. Lin, T. and Su, S. Y. H. (1985) The S-algorithm: A promising solution for systematic functional test generation. *IEEE Trans. Computer Aided Design*, **CAD-4**(3), 250–63.

32. Kawai, M., Shibano, M., Funatsu, S. *et al.* (1983) A high level test pattern generation algorithm. *Digest of Papers, 1983 International Test Conference*, October, 346–52.

33. Johnson, W. A. (1979) Behavioural level test development. *16th Design Automation Conference Proceedings*, June, 171–9.

34. Jain, S. K. and Agrawal, V. D. (1985) Modelling and test generation algorithms for MOS circuits. *IEEE Trans. Computers*, **C-34**(5), 426–33.

35. Reddy, M. K., Reddy, S. M. and Agrawal, P. (1985) Transistor level test generation for MOS circuits. *22nd Design Automation Conference Proceedings*, June, 825–8.

36. Chen, H. H., Mathews, R. G. and Newkirk, J. A. (1984) Test generation for MOS circuits. *Digest of Papers, 1984 International Test Conference*, October, 70–79.

37. Elziq, Y. M. (1981) Automatic test generation for stuck-open faults in CMOS VLSI. *18th Design Automation Conference Proceedings*, June, 347–54.

6
DESIGN-FOR-TESTABILITY TECHNIQUES

6.1 INTRODUCTION

Design for Testability (DFT) may be described as a means of easing the generation and application of test vectors to detect faults, in integrated circuits or systems, resulting from

1. the manufacturing process
2. 'wearout' due to normal device operation.

Design-for-testability techniques [1] range from a simple set of guidelines to a formal set of design rules.

The incorporation of DFT techniques into IC designs has become necessary to keep testing costs within acceptable limits, as device complexity increases through the advancements made in semiconductor fabrication techniques. Surveys [2]–[4] have shown that the increase in testing costs can be attributed to several factors:

(a) Increase in complexity

The amount of effort required for test pattern generation and fault simulation, on average increases as the square of the circuit complexity; that is, if advances in device fabrication techniques permit the basic line width to be halved, test generation times would increase by a factor of sixteen. Furthermore as circuit complexity increases, the sequential depth of the circuit increases, further increasing test generation costs. However, test generation and fault simulation costs in terms of time, are reduced to some extent by the continual increases in the performance of computer systems.

(b) Reduced accessibility

In the past when systems comprised small-scale and medium-scale integrated circuits, nodes internal to the circuit were more accessible; furthermore, testing could be carried out on individual modules. However, with the increase in the scale of integration access to the internal nodes has become greatly reduced. It is estimated that each time the scale of integration is increased the pin-to-gate ratio is reduced by a factor of ten. The essence of testing a circuit is the ability to control and observe signal values in a circuit, consequently if this ability is reduced the degree of difficulty in testing a circuit increases.

(c) Tester costs and performance

The performance of the devices within a VLSI circuit exceed those used in the test equipment; hence testing cannot be performed at full operational speed. When testing is performed at slower clock speeds certain faults, for example 'delay' faults within the circuit, can go undetected. Furthermore, with the increase in circuit complexity the volume of test data generated, i.e. input test patterns and output responses, also increases, requiring the test equipment to have the ability to store a vast amount of data and to be able to process it rapidly. These performance and storage requirements necessitate use of very expensive testers.

Consequently the objectives of DFT techniques are directed at reducing testing costs in general by one or more of the following:

1. Making the internal circuit nodes more accessible, that is improving the controllability and observability of signal values inside the circuit.
2. Transforming a sequential circuit, for the purpose of testing, into a combinational circuit by introducing some mechanism whereby the state of all internal storage elements can be monitored. The major problem in testing sequential circuits is trying to establish what 'state' a sequential machine is in at any given time; in the past much research effort was directed at the problem of generating 'homing sequences' whose objective was to drive the circuit into some known state independent of its present state. Consequently if the 'states' of a sequential machine can be controlled or observed, then the problems of testing a sequential circuit are reduced to that of testing a combinational circuit. DFT techniques which permit the states of a sequential circuit to be controlled and observed are said to enhance the 'predictability' of the circuit.
3. Making the circuit self-testing.
4. Reducing the amount of test data which is needed to test the circuit.

Although design techniques which improve the testability of a circuit are highly desirable, ultimately they incur some penalty [5] with respect to the effect they have upon the designer and the performance of the circuit.

If the particular design style, adopted to reduce testing cost, is to be accepted by the design community it must be easy to apply and must not be too constraining as to inhibit the ingenuity of the designer; there must also be adequate software support for a particular design technique, that is programs which will check a design for compliance to the design rules imposed by the design style. Furthermore, since most design-for-testability techniques involve either additional hardware and/or routing, the physical size of the circuit increases and subsequently reduces the yield; the additional hardware also introduces extra signal delays into the circuit which affect the performance of the circuit; the test hardware, itself, must be capable of being tested. There is also a requirement for extra input/output pins, increasing the cost of packaging. In view of these penalties it must be demonstrated that a particular design style produces a marked reduction in test generation costs with respect to CPU time and personal effort. The final choice, however, of whether to adopt DFT techniques or not is usually based on trade offs to establish the optimum amount of circuit testing required for a given circuit without endangering the reliability of the component or reducing profit margins to an unacceptable level. Manufacturers of high-volume components, in general, choose not to use DFT techniques, since although the initial cost of test pattern generation is high and test program development is long, the subsequent increase in area and reduction in yield through the use of design for testability is less acceptable; in some instances, however, to improve testability, whilst maintaining circuit yield constant, a slight reduction in circuit functionality is made. Conversely, the manufacturers of low-volume circuits in general incorporate DFT techniques, since design costs and time are considered more important than circuit size. However, regardless of whether low or high volume components are to be designed the manufacturer is obliged to produce reliable components, because at present the IC manufacturing industry is a buyers' market and customer satisfaction with respect to reliability is paramount. The cost to IC manufacturers for selling inadequately tested circuits is not only the loss of a customer who has had to expend time and money to diagnose faults in his systems which should have been detected by the IC manufacturer at chip level, but also the legal costs of liabilities resulting from system failure as a result of inadequately tested circuits. The legal liabilities resulting from defective circuits has, in some instances, become a barrier to manufacturers' describing any testability features incorporated in a circuit in the customer specification sheets, since these features could not be described adequately so as to avoid legal liabilities in the event of a fault in the circuit causing a system failure.

Designing for testability implies some modification to the circuit to ease the generation and application of test vectors to a circuit. The techniques to enhance testability have been categorized into three main groups:

1. *ad hoc* or retrofit methods
2. structured approaches
3. built-in self-test.

6.2 *AD HOC* OR RETROFIT METHODS [6]

These techniques employ additional hardware to improve the controllability/ observability of signal nodes within a circuit or to partition a circuit into simpler subfunctions. *Ad hoc* methods evolved through the necessity to solve a testing problem for a particular circuit rather than trying to solve the problem of testing complex circuits in general by using, for example, some formal design method. Consequently these methods require a degree of skill from the designer in knowing, for example, where test points can be placed to best advantage, although testability analysis tools are now available to identify areas of circuitry which, unless modified in some way, will have a low fault coverage. Several *ad hoc* methods for improving circuit testability include:

(a) Test point insertion

Test points are routed into the circuit to make certain internal nodes more accessible in order to observe or control a signal value on the node. If a signal is only to be observed the test point can be connected directly to the signal line, although the gate output drive capability may have to be increased to cater for the additional capacitance from the test point. However, if it is intended to use the test point to control a signal value on a line, it is necessary to insert a tri-state driver before the test-point connection on the signal line; the tri-state driver must be in the 'high impedance' state when the test point is used to control the signal value.

(b) Pin-amplification (funnelling)

In general, the number of pins available on a package for test purposes is minimal. The number of test pins, however, can be increased if some of the normal input/output pins are multiplexed to perform the additional function of acting as test inputs and outputs. However, the delays introduced by the multiplexers/demultiplexers will degrade the normal circuit performance.

(c) Blocking or degating logic

In this technique additional gates are incorporated into a design to inhibit data flow along certain paths in a circuit, for example, asynchronous feedback loops, thus partitioning the circuit into smaller modules for the purposes of testing. Blocking gates are simply two-input gates; one input is the normal data signal whilst the other is the blocking signal which can be controlled from a test input. During normal circuit operation the control signal is held at a non-dominant logic value for the type of blocking gate, that is, a logic 1 for an AND or NAND or a logic 0 for an OR or NOR gate.

(d) Control and observation switching

In this technique signal lines whose logic values are readily controlled and observed are identified in a circuit (using a testability analysis program) and are used in conjunction with a multiplexer/demultiplexer to improve access to difficult test nodes in close proximity to these lines. Test mode control signals to the multiplexers/demultiplexers determine whether the easy-to-control/ observe signal line transmits normal or test data.

(e) Test state registers

Test state registers are serial-in–parallel-out shift registers which are used to increase the number of test control signals which can be applied to a circuit at any given time. The input to the test state register is usually demultiplexed with the normal data inputs to the circuit. When the circuit is to be tested, test mode control inputs direct signals from the demultiplexer to the test state register; thereafter the test mode control signals are switched to normal mode so that the outputs of the demultiplexer, other than those connected to the test state register, propagate normal functional input signals or test patterns to the circuit.

The *ad hoc* methods for improving testability of a circuit have the advantage of not imposing severe constraints upon the designer but have the disadvantage that the techniques have no software support since they cannot be readily automated, although some testability analysis programs do explicitly suggest that test points should be added to a section of circuitry to improve its testability.

6.3 STRUCTURED APPROACHES

In contrast to the *ad hoc* methods, the structured approaches to design for testability are more formal and are incorporated into a design from the outset rather than introduced as an afterthought as with the *ad hoc* methods.

The objective of developing the structured approaches was to facilitate the testing of complex sequential circuits. The major difficulty in testing sequential circuits is determining the internal state of the circuit, since the output response of a sequential circuit depends not only upon the applied inputs but also upon the internal state of the circuit. Consequently structural DFT techniques were directed at improving the controllability and observability of the internal states of a sequential circuit. Thus, in having control of the internal states of a circuit the problems of testing a sequential circuit reduces to that of testing a combinational circuit.

Over the past few years several structured approaches have evolved, namely, scan path [7]–[8], level-sensitive scan design [9], scan/set [10] and

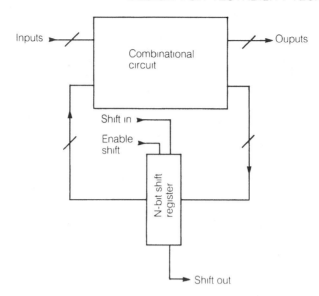

Fig. 6.1 Scan path principle: a classical sequential circuit using a shift register for storage, which enables the internal state of the circuit to be controlled or observed.

random access scan [11]; these techniques will now be discussed in more depth.

6.3.1 Scan path [7],[8]

The main objective of the scan path technique is to reconfigure a sequential circuit, for the purposes of testing, into a combinational circuit. This transformation is achieved by ensuring that all internal memory elements can be connected together, as shown in Fig. 6.1, to form a long serial shift register called a *scan path*. By means of this scan path the internal states of the circuit can be controlled and observed. Thus testing a sequential circuit containing a scan path is no more difficult than testing an equivalent combinational circuit, since the incorporation of the scan path essentially reduces the sequential depth of the original circuit to zero, by enabling any internal state of the circuit to be shifted (scanned) in or out.

When a circuit is designed with a scan path, it has two modes of operation, a normal mode and a test mode which configures the storage elements in the scan path giving access to the internal states. In order to control the internal state of the circuit it is necessary to set the storage elements to any desired value, which implies that some means of selecting the source of data to the flip-flops must exist, i.e. normal circuit input or test data input. The selection of the input source may be achieved by using a multiplexer on the data input or by using a two-port flip-flop which has two data inputs and two clocks; the input used by the flip-flop depends upon which clock is activated. As an example of a scan path configuration consider the Stanford scan path [7] shown in Fig. 6.2, which was the first published description of a scan path circuit. In this configuration the source of data input to the storage elements, which are D-type flip-flops, is selected using a multiplexer which is integrated into the flip-flop; this removes the circuit delay which would be incurred by simply adding a multiplexer directly to the input of the flip-flop. The

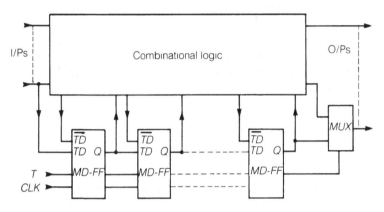

Fig. 6.2 Stanford scan path [7].

integrated multiplexer and D-type flip-flop is called an MD-flip-flop (multiplexed D-type). The method of testing a circuit with this scan path configuration is as follows.

When the test mode signal $T = 1$, the flip-flops accept data from the multiplexed input TD:

1. Set $T = 1$, verify the operation of the scan-path using shift and flush tests;
2. Shift/scan a test pattern into the flip-flops (controlling the internal state);
3. Apply a test pattern to the primary inputs of the circuit;
4. Set $T = 0$, allow the circuit to settle and monitor the primary output values;
5. Activate the circuit clock for one cycle (captures the next state of the circuit in the memory elements);
6. Set $T = 1$, scan out the contents of the registers, simultaneously scanning in the next pattern.

The storage elements can either be D, J–K, or R–S types of flip-flops; however, simple latches cannot be used in the scan path.

In addition to the scan path giving access to the internal states of a circuit, the scan path may be considered as a means of partitioning a circuit into less

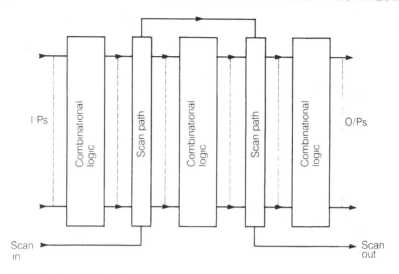

Fig. 6.3 Circuit partitioned by scan paths.

complex submodules, as shown in Fig. 6.3. Thus for the purposes of testing, the circuit is considered to comprise two complexes, namely blocks of combinational logic and a long shift register or scan path. The efficiency of the test pattern generation for the combinational blocks is greatly improved, since the combinational depth of the submodules is reduced, by the scan path, which may be used to apply test patterns and subsequently capture their responses. Consequently, before the scan path is used to apply test patterns and capture the response of the circuit, the scan path itself must be tested. Testing of the scan path is carried out using a set of *flush* and *shift* tests. The flush test comprises a string of all ones i.e. 111 . . . 11 or zeros, i.e. 000 . . . 0; the shift test comprises some pattern of ones and zeros, for example 00110, which exercises each register through all possible combinations of initial and next states.

6.3.2 Level Sensitive Scan Design (LSSD) [9]

LSSD incorporates two design concepts, namely level sensitivity and scan path.

The concept of level sensitive design requires that the operation of a circuit is independent of the dynamic characteristics of the logic elements, that is rise and fall times and propagation delays. Furthermore in a level sensitive design the next state of a circuit is independent of the order in which inputs change when a state change involves several input signals. This circuit characteristic implicitly places a constraint on what signal changes can occur in the circuit;

these constraints, which will be described later, are usually applied to the clocking signals.

The major element in a level sensitive design is the polarity hold shift register latch (SRL), shown in Fig. 6.4(a) which is used to implement all storage elements in the circuit. The SRL is similar to a master–slave flip-flop and is driven by two non-overlapping clocks which can be controlled readily from the primary inputs to the circuit. In test mode the SRLs are reconfigured to form a long shift register, permitting the internal states of the circuit to be controlled or observed as in the simple scan path technique.

The SRL is shown symbolically in Fig. 6.4(b); input D is the normal data input to the SRL, clocks CK_1 and CK_2 control the normal operation of the

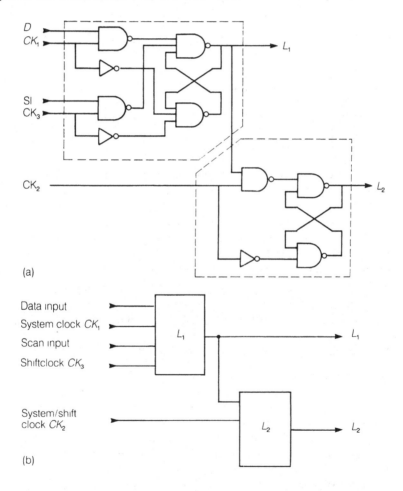

Fig. 6.4 (a) Polarity hold latch; (b) symbolic representation of a polarity hold latch.

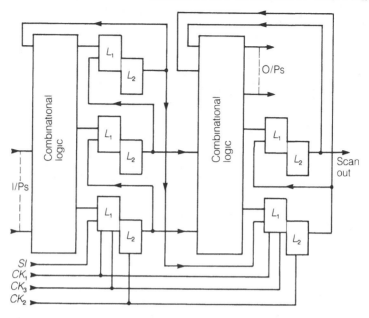

Fig. 6.5 Basic LSSD configuration.

element; input *SI* is the scan path input to the element; clocks CK_3 and CK_2 control the movement of scan path data through the SRL. The output of the SRL, in both normal and scan path modes of operation, is usually taken from L_2. The mode of operation depends upon which clocks are activated.

The basic LSSD configuration is shown in Fig. 6.5, where it is seen that for the purposes of testing the circuit is implicitly partitioned into subcircuits, comprising entirely combinational logic functions, by blocks of shift register latches which will be instrumental in applying test vectors to the subcircuits and subsequently capturing their responses. Thus, as outlined below, the first step in the procedure to test a circuit designed with LSSD is to verify that the SRLs are functioning correctly.

The procedure to test an LSSD circuit is as follows:

1. Verify the operation of the SRLs using a shift/flush test;
2. Preload the SRLs with a test pattern by applying the test pattern, serially, to the scan-in port, *SI*, and shifting it along the scan path by alternately pulsing clocks CK_3 and CK_2;
3. Apply a test vector to the primary inputs; after the circuit has settled apply a single pulse to the system clock CK_1. The response of the combinational subcircuits is now captured in the L_1 latches of the SRLs;
4. Pulse the system clock CK_2 to duplicate the contents of the L_1 latches into the L_2 latches;

5. Scan out the contents of the L_2 latches by systematically pulsing clocks CK_3 and CK_2. Whilst the contents of the L_2 latches are being scanned out a new test pattern can be scanned in.

The main advantages of the LSSD design technique are that it removes the need to perform a detailed timing analysis on the circuit since it is level sensitive; automatic test pattern generation is simplified since tests need only be generated for a combinational circuit, the scan path permitting internal states of the circuit to be controlled and observed; finally, as LSSD imposes a regimen on the designer the design can be checked for compliance to design rules.

The disadvantages of this technique are that the designer is constrained to implement the system as a synchronous sequential circuit; test application times are increased since the input and output data must be scanned serially and the system switched between normal and test modes. Although test pattern generation is simplified, test program development is complicated by the need to intersperse the test data with the appropriate clock pulses to switch from normal mode to test mode and to shift data; finally there is an increase in circuit area, nominally between 4% and 20%, and an increase in the number of input/output pins required for the scan-in/-out ports and clocks.

In the circuit configuration shown in Fig. 6.5, it is seen that the subcircuit outputs are taken from the L_2 latches; since there are two latches in the datapath this configuration is referred to as a *double latch* design. This configuration guarantees raceless circuit operation due to the master–slave action of the L_1 and L_2 latches; however, it does require two clock pulses in order to transfer data from one subcircuit to the next during normal circuit operation. Several modifications have been made to the SRL configurations to make better use of the test hardware in normal operation and also to improve circuit performance.

One variation on the basic LSSD configuration is called the *single latch* design, where the normal system outputs are taken from the L_1 latches, the L_2 latches only being used in the scan path. However, to ensure raceless operation of the circuit it is necessary to partition the circuit as shown in Fig. 6.6 into disjoint combinational blocks and impose the constraint that the output of an L_1 latch clocked by CK_1 can be applied only to a combinational block whose outputs terminate on L_1 latches activated by a second system clock, CK_4, and vice versa. Clocks CK_2 and CK_3 are still used to control the transfer of data through the scan path in test mode. The single latch design thus has the advantage of being faster than the double latch operation since only a single pulse of a system clock is required before the latch outputs can be used by the system.

The overhead of the redundant L_2 latch in the single latch configuration, as far as normal circuit operation is concerned, can be removed by modifying

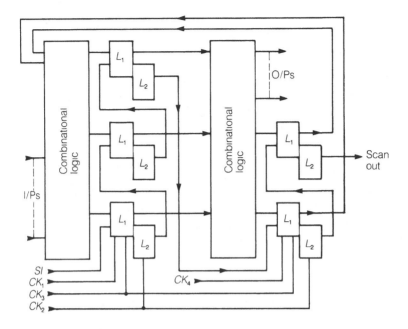

Fig. 6.6 Single latch configuration.

[12] the design of the L_2 latch so that it contains a system data input terminal and associated clock, as shown in Fig. 6.7. In this way both the L_1 and modified L_2 latches can be used in the single latch configuration. In this instance the circuit is partitioned such that the output of an L_1 latch, clocked by CK_1, can only be applied to a partition whose outputs terminate on the modified L_2 latches, clocked by CK_4, and vice versa. In test mode the L_1 and modified L_2 latches are configured in the normal way to form a scan path which is controlled by clocks CK_2 and CK_3.

In some instances it is necessary to interface LSSD and non-LSSD circuits. However, to prevent the non-LSSD structures affecting the LSSD circuits during testing a stable shift register latch (SSRL) [13] has been developed as shown in Fig. 6.8. The SSRL comprises a basic SRL and an additional latch L_3 activated by a clock P; the output of the L_3 latch is used as an input to the non-LSSD circuitry. When P is inactive the LSSD and non-LSSD sections are isolated, thus enabling the LSSD circuitry to be tested in the normal way. When clock P is activated the contents of the L_2 latch are transferred to the L_3 latch and hence to the non-LSSD circuitry; thus the scan path in the LSSD circuitry may be used to apply test patterns to the non-LSSD circuitry. Other applications of SSRLs are in 'on-line dynamic scan' for diagnosing faults in systems and 'on-line fault detection' in memories.

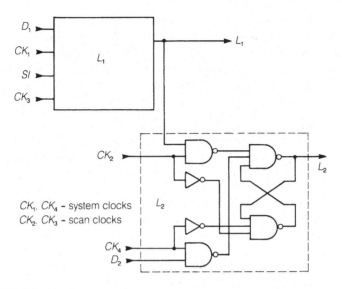

Fig. 6.7 Modified latch.

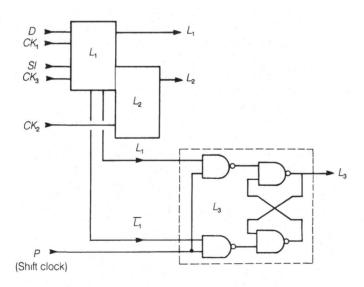

Fig. 6.8 Stable shift register latch (SSRL).

LSSD Design Rules [9]

The LSSD technique imposes certain rules on a design to ensure that the operation of the circuit is independent of the dynamic characteristics of the circuit and also that a scan path can be configured from the storage elements in the circuit. A basic set of design rules is outlined below; the first four rules ensure that the circuit is level sensitive and the last two ensure that a scan path can be configured:

1. All storage elements must be SRLs;
2. Each SRL must be controlled by two non-overlapping clocks such that
   (a) The output latch of one SRL may only feed the data input latch of another SRL provided both latches are not activated by the same clock;
   (b) A gated clock $C(G)$ may be derived from another clock C and the output of an SRL. The gated clock $C(G)$ may be used to activate the data input latch of another SRL, provided that the SRL output used to derive $C(G)$ is not controlled by clock C or any of its derivatives. This ensures that the data and clock inputs to a given latch are not controlled by the same clock source;
3. It must be possible to identify a set of SRL clocks which are directly controllable, i.e.
   (a) All clock inputs can be held inactive independently;
   (b) Any clock input can be activated whilst the others are inactive;
4. Clock inputs cannot feed data inputs to SRLs, either directly or indirectly through combinational circuitry;
5. All SRLs must be permanently connected into one or more shift registers, each of which has a scan-in and a scan-out port and shift clocks which are directly accessible from the primary inputs/outputs of the circuit;
6. There must exist a circuit configuration state, called the *scan state*, which is directly controllable from the inputs, such that
   (a) All SRLs are connected and available as a scan path(s);
   (b) All SRL clocks can be held inactive, except the shift clocks;
   (c) Any shift clock can be held inactive independently.

The above rules apply whether 'double' or 'single' latch LSSD configurations are implemented.

Design for testability rules checker [14]

If a design for testability scheme imposes rules or constraints upon a designer, any circuits implemented using the scheme must be checked, subsequently, for design rule violations; the checking procedure should be carried out automatically. One technique which has evolved for checking a circuit for design rule violations is based upon a DFT calculus which defines the types of signals and rules governing the transfer of signals through nodes in a circuit. The design rules which are checked by the DFT analyser are categorized into *basic design rules* and *test access rules*.

The basic design rules ensure, for example, that:

1. Latches can only have inputs from other latches if the latches are clocked from independent sources which are non-overlapping;
2. Data and clock inputs to a latch do not depend upon the same primary clock;
3. Asynchronous loops can exist only inside a block, unless disjoint clocked latches appear in the feedback path.

The test access rules ensure, for example, that:

1. Test input/output nodes are connected either directly to a test input/ output terminal or to a legal scan path;
2. Test point nodes can only belong to one test access path;
3. During testing all internal storage elements are controlled directly from the primary inputs to the circuit.

The advantage of having an automatic DFT rules checker is that having been constrained to obey certain rules, the designer is not burdened further by the onerous task of manually checking a design for rule violations.

The process of checking for design rule violations in a given DFT implementation requires that, first, a signal path related to a given design rule is traced out; second, the path is checked for violations. Path tracing is performed by a mechanism of signal set transfers through a node and the rule is checked by examining the contents of the signal set at the input and output of a node. To speed up the process of checking for design rule violations the logic function performed by a gate/module is usually ignored and only nodes which are inputs and outputs of storage elements or test points are examined.

The input to a DFT rules checker comprises a structural description of the circuit in terms of gates, storage elements and function blocks; the function blocks may or may not have been previously analysed. A declaration is required of the functions of the clocks, primary inputs/outputs and test data inputs/outputs. All global feedback loops are assumed to be open and a list of the resulting pseudo inputs/outputs defined.

The output from the DFT rules checker comprises, for example,

1. a list of violations and their locations
2. a DFT description set, which contains all the necessary information for path tracing and DFT rule checking, should the section of circuitry analysed be regarded as a functional block at a higher level of abstraction
3. a list of all test access paths and the clock and select signals necessary for their control
4. a list of all internal nodes which can be controlled or observed.

Implementation costs

The costs incurred by implementing designs with structured DFT techniques, in general, are:

1. The storage elements are more complex, in order to implement the scan path; hence require extra area;
2. Additional input/output pins are required for shift clocks, and scan-in/-out pins. The number of additional pins may be reduced if the test mode pins are multiplexed with the normal inputs/outputs of the circuit.
3. There is an increase in circuit area due to the additional logic and interconnect;
4. Normal circuit performance may be reduced due to additional gate delays resulting from the presence of the test mode logic;
5. Testing time is prolonged due to the process of shifting test data in and out of the scan path and switching between test and normal modes of operation. The testing time may be reduced if multiple scan paths are used. However, the overall testing time of the circuit will be much less than that required to test an unstructured sequential circuit performing the same function.

However, test generation costs are greatly reduced through the use of scan techniques. The area overheads may also be reduced if the test hardware can be used in the normal operation of the circuit. Furthermore with structured design techniques such as LSSD, since fault simulation and simulation for design verification are applied in general only to the combinational circuitry whose operation is independent of circuit delays, then compiled code types of simulators may be used which are much faster than, for example, a general purpose table driven simulator.

6.3.3 Scan/set logic [10]

In the scan/set technique shown in Fig. 6.9, the storage elements within the circuit are not used to implement a scan path; instead a separate register is added to the circuit whose sole function is to scan test data in and out of the circuit. Since all internal storage devices can neither be controlled nor observed the scan/set technique does not reduce the problem of testing sequential circuits to that of testing a combinational circuit. The decision on which circuit nodes should be connected to the scan/set register can be based upon the results derived from a testability analysis program; the nodes to be observed need not be limited to the outputs of storage elements. The main advantage of the scan/set technique is that test data can be scanned out of the system without interrupting its normal operation.

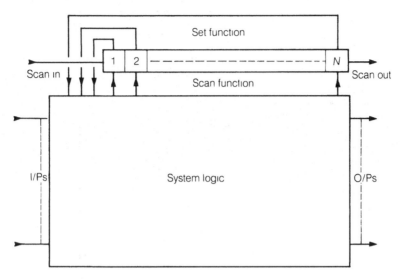

Fig. 6.9 Scan/set configuration.

6.3.4 Random access scan [11]

In this technique, as shown in Fig. 6.10, controllability and observability can be exercised over all buried latches in the circuit; consequently test pattern generation is reduced to that of generating tests for a combinational circuit.

In random access scan, the storage elements are not configured into a shift register but form a two-dimensional array with an associated addressing scheme which permits each element to be addressed and subsequently controlled or observed individually. The major disadvantage of this technique is the high overheads in terms of additional input/output pins required to implement the scheme.

6.4 BUILT-IN SELF-TEST TECHNIQUES [3],[15]–[17]

The motivation to develop Built-in Self-Test (BIST) techniques for VLSI circuits arose from foreseeable limitations in the current test strategies, even those employing structured DFT techniques, when applied to VLSI circuits. First, current test methods require the use of deterministic test patterns; the cost of test pattern generation for combinational circuits increases as the square of the number gates. Second, the volume of test data which must be processed during the testing phase is also increasing with circuit complexity. Finally, many VLSI circuits are used in high-speed applications, consequently there is

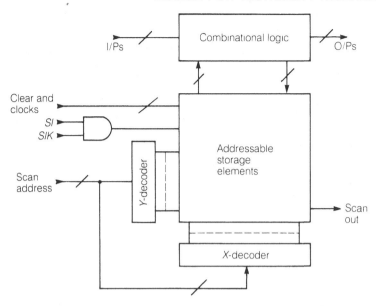

Fig. 6.10 Random access scan configuration.

a fundamental requirement to test these circuits at speed; unfortunately this requirement cannot be satisfied unless very expensive testers are employed. Although structured DFT techniques reduce test generation costs they do not reduce the volume of test data or permit the circuit to be tested at speed and have the disadvantage of prolonging test application times.

Consequently the objectives of BIST techniques are

1. to reduce test generation costs
2. to reduce the volume of test data
3. to produce an alternative to the costly automatic test equipment which will enable circuits to be tested at speed.

In BIST techniques these objectives are achieved by integrating the functionality of an automatic test system into a chip. Consequently a BIST technique must have the capability, not only of generating sequences of test vectors on chip, but also to compare efficiently the response of the circuit under test and indicate if it is faulty or fault-free. Functional self-test has been used with microprocessor systems for some time to test memory, instruction sets etc., the test program residing in an ROM and run off-line. Functional testing, however, has a low fault coverage; if this is to be improved through the use of deterministic tests the physical size of the stimulus and response ROMs limits the use of self-test to small circuits.

Several BIST techniques have been developed which not only enable large numbers of test vectors, which have a good fault coverage, to be generated efficiently on chip but also readily determine whether or not the circuit has responded correctly to the set of test vectors. Although BIST techniques have many advantages they also have several limitations, for example:

1. Not all circuits can be tested effectively using BIST techniques, although design techniques are available to make circuits more susceptible to BIST-generated test patterns;
2. The number of test patterns, generated using BIST, required to give a specified fault cover is usually greater than the number of deterministic test patterns required to give the same fault coverage. However, the cost of generating the BIST patterns is minimal and these patterns can be applied to the circuit at its normal operating speed (1 million patterns in one second with a 1 MHz clock);
3. It is difficult to estimate the fault coverage obtained from a BIST-generated test sequence; conversely, it is difficult to determine the test sequence length necessary to give a required fault coverage. Several techniques, however, have been developed which permit these estimations to be made without the use of extensive fault simulation.

Now that the advantages and disadvantages of the BIST technique have been discussed, the basic components required to implement the technique, in general, will now be described.

6.4.1 Stimulus generators for BIST systems [18]–[21]

The stimulus generators used in BIST systems can be classified as either static or dynamic. In the static system test patterns used to stimulate the circuit are stored in an on-chip ROM. These test patterns may have been generated by an automatic test pattern generation program or they may be a set of functional test patterns. The use of functional test patterns stored in ROMs has been widely used in BIST systems for microprocessors [3], where either a 'bootstrap' or 'architectural-checkout' approach is adopted. The major disadvantage, however, of the ROM-based system is that if the test patterns are generated automatically, large ROMs will be required to store the patterns in complex circuits. Alternatively, if functional test patterns are used test development times can be excessive – for example, 10 man months in order to obtain a 99% fault coverage – furthermore, the ROM-based approach negates one of the proposed advantages of BIST systems, namely the reduction of test pattern generation costs and the need to store large volumes of test data.

BIST systems which use the dynamic stimulus generation technique overcome the above problems by incorporating hardware on-chip to generate Pseudo Random Binary Sequences (PRBS) which are used to stimulate the

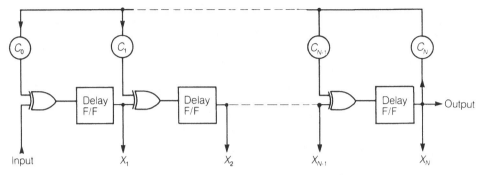

Fig. 6.11 Feedback shift register configuration to implement polynomial division [19].

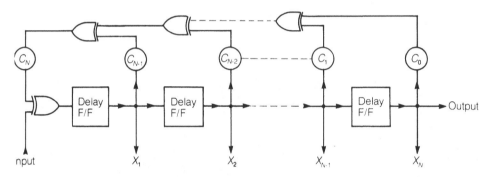

Fig. 6.12 An alternative feedback shift register configuration [19].

circuit; the PRBS generators may either be a separate entity on chip or it may be reconfigured from some of the system hardware. The PRBS generator is realized using a Linear Feedback Shift Register (LFSR) which is in essence a finite state machine comprising storage elements, modulo-two adders (EXOR gates) and binary constant multipliers which apply feedback to/from the individual stages. Two configurations of LFSR are shown in Figs 6.11 and 6.12, although the configuration in Fig. 6.12 is preferred since this minimizes the delay through the forward datapath in register. The transitions from state to state within the finite state machine are realized through the process of division by a polynomial called the 'characteristic' polynomial, which is defined by the coefficients of the constant multipliers; these can be assigned the values of 1 or 0 depending whether or not a connection between the feedback path and a given stage in the register exists. It should be noted that the constants of the polynomial are reversed in the configurations shown in Figs 6.11 and 6.12. This is a direct result of the position of the EXOR gates in the feedback loops; although both configurations will perform polynomial division correctly, the contents of the register in Fig. 6.11 will contain the remainder at each stage of the division process, whereas the register in Fig. 6.12 will not. If the LFSRs, regardless of the configuration, are initialized to a non-zero value and the serial input is held at a constant value a PRB sequence will be generated at the outputs $X_1, \ldots, X_n$. The length of the sequence produced by the generator will have the maximum value $2^n - 1$ (where n is the number of stages in register) provided the characteristic polynomial is primitive and irreducible; a table of irreducible polynomials can be found in Peterson [22]. The output sequence from the PRBS generator can be altered by changing the 'seed' or starting value of the register; this is used to advantage in some BIST systems which use a ROM, to store say 8 or 16 'seed' values, in conjunction with an LFSR to extend the number of patterns produced by the generator. It should be noted the LFSR must never be set to an all-zero 'seed' value since the register in this condition would take on a 'deadlock' state from which it can never recover. The normal configuration for a PRBS generator is shown in Fig. 6.13 where the feedback connection is returned directly to the first stage; this is equivalent to holding the input to the circuits shown in Figs 6.11 and 6.12 at a logic 0.

The polynomial used in the PRBS generator can readily be derived from the feedback tap positions in the following way. The movement of data through each stage of the shift register is controlled by the transfer function D which is defined [18] such that $X(t) = DX(t-1)$, i.e. D represents a delay of one time unit and successive multiplications by D represents delays of one, two, three etc. time units. Thus with reference to Fig. 6.13, the input to the PRBS can be written as $D^4X(t) + DX(t) + X(t)$ where $X(t)$ is the input data to the register, in this case a constant logic 0 (see Fig. 6.12). The feedback expression can be rewritten simply as $X^4 + X + 1$, where the multiplier coefficients $C_0 = 1$, $C_1 = 0$, $C_2 = 0$, $C_3 = 1$. Inserting these coefficients in the polynomial expression for the FSR shown in Fig. 6.11 yields the characteristic polynomial in the format given in the tables of irreducible polynomials, in this case $X^4 + X^3 + 1$. Conversely, given the irreducible polynomial the tap-off for PRBS generator configured as in Figs 6.12 and 6.13 can be derived by reversing the above procedure.

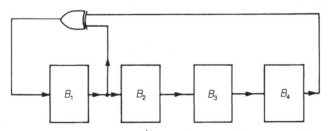

Fig. 6.13 PRBS Generator.

6.4.2 Non-linear Feedback Shift Registers (non-LFSR) [23]

When a designer uses a PRBS generator to produce test vectors for a circuit, he has no control over the order in which the vectors are applied; in the past this was of little consequence. However, in certain situations the order in which test vectors are applied is important: for example in the detection of stuck-open faults [24] in CMOS circuits or the required fault coverage may not be obtainable from the PRB sequences. A structured test set could be

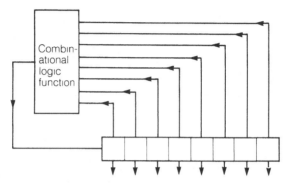

Fig. 6.14 Non-linear feedback shift register.

stored in an ROM; this however is uneconomical for complex circuits. Alternative, non-linear feedback registers can be used to generate ordered sets of test vectors without incurring the high area penalty resulting from the use of ROMs. The major difference between the LFSR and the non-LFSR is that the feedback lines are not connected through EXOR gates but through some combinational logic function realized using AND, NAND, OR and NOR gates as shown in Fig. 6.14.

In order to design a non-LFSR pattern generator it is necessary to order the test sequences so that a given test vector may be derived from k-shifts of a previous test vector; in some situations this is not possible and it is necessary to provide 'link' vectors in order to obtain the desired sequence. Once the test vectors have been ordered the feedback function for the generator must then be derived as shown below [23].

Consider that the test sequence to be generated is:

$$
\begin{array}{cccc}
1 & 0 & 0 & 0 \\
0 & 1 & 0 & 0 \\
0 & 0 & 1 & 0 \\
1 & 0 & 0 & 1 \\
& \vdots & & \\
\end{array}
$$

etc.

It is seen that each subsequent vector comprises $(n - 1)$ bits of the previous vector, shifted to the right; the first bit of each vector is generated by the feedback function; that is the first bit of each subsequent vector is some function of the last state of the register. In this instance the feedback function $F(X)$, where X represents the outputs, $x_1, x_2, x_3, \ldots, x_n$, from each register stage, is given by,

$$F(1\ 0\ 0\ 0) = 0; \qquad F(0\ 1\ 0\ 0) = 0;$$
$$F(0\ 0\ 1\ 0) = 1; \qquad F(1\ 0\ 0\ 1) = \text{etc.}$$

Subsequently, using some minimization technique a minimum complexity solution for the feedback function can be calculated.

When ordering the test vectors care must be taken to ensure the uniqueness of the next state function, consequently the ordered set of vectors must be scanned to locate repeated occurrences of test vectors which may have more than one different successor vector. If this situation arises procedures [23] must be invoked to remedy the situation; this involves extending each test vector with the first bit in the 'successor' vector. If the situation is still not resolved the process is repeated on the extended test vector list until a unique set of vectors is obtained. Each iteration of this procedure increases the size of the shift register by one bit. During testing, however, only the last n bits of the state vectors are applied to the circuit, that is the bits defined in the original ordered list of test vectors.

6.4.3 Data compression techniques for BIST systems [18], [19]

An important requirement of any BIST system is the ability to determine if the circuit is faulty or fault-free. The 'known' fault-free response to the input test vectors can be stored in an ROM; however, the physical size of the 'response' ROM would prohibit the use of this technique to small circuits. In order to overcome this problem data compression techniques have evolved which reduce the amount of output data which must be checked to determine if a given circuit is fault-free.

Early data compression schemes employed the technique of *transition counting* [25], [26] in which the number of transitions from a one to zero and vice versa occurring at each primary output in a circuit was counted and compared with the transition count derived from a simulated model of the circuit or a circuit which was known to be fault-free; if the counts differed the circuit under test was deemed to be faulty. It is possible, however, that when using this technique the faulty and fault-free circuits can produce the same transition count, as shown in Fig. 6.15, where an input to gate G_3 is considered to be stuck at 1. This situation is called *fault masking* and is a direct result of the compression process. If data compression techniques are to be acceptable

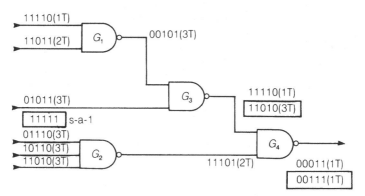

Fig. 6.15 Transition counting with fault masking [25].

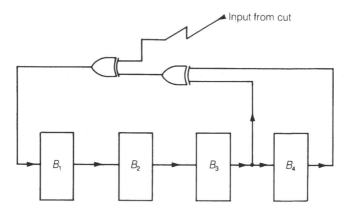

Fig. 6.16 Signature analysis register.

the probability of fault masking must be small. The data compression scheme currently used in BIST systems is called *signature analysis* [18] and was developed by Hewlett Packard in the late 1970s. The technique of signature analysis uses an LFSR similar to those used to generate pseudo-random binary sequences for test purposes. However, as shown in Fig. 6.16, instead of holding the serial input at a fixed logic value it is connected to some input source. In this instance the contents of the LFSR are no longer defined exclusively by the feedback taps, but are modified in a way determined by the signal coming from the other source. This modified bit pattern stored in the LFSR at any instant in time is called the 'signature' of the input source. In fact the signature is the last contents of the LFSR resulting from the division of the input waveform by the characteristic polynomial of the LFSR. Error detection relies on the fact that since the waveforms from the faulty and fault-free circuits differ the contents of the LFSR after division also differ. The error-detection capability of this technique is almost independent of the length of the input sequence and is mainly dependent upon the length of the register. It can be shown that if the length of the test sequence is L and the length of the register is r, the error coverage of this technique is given by

$$1 - \frac{2^{L-r} - 1}{2^L - 1} \, .$$

However, if the sequence length is long the error coverage can be approximated to $1 - 2^{-r}$, hence if the LFSR is 16 bits long the error coverage is 99.998%.

In practice, when using this technique of data compression, the signature register is initialized to some known pattern, the serial input is then connected to a given output in a fault-free circuit and the system is clocked for a preset number of cycles, whereupon the signature characteristic of the waveform observed at that output in the circuit will be stored in the signature register. The signature register is then reset to its original value and the procedure repeated with a circuit under test; if a fault exists in the circuit under test the signature generated will differ from that of the fault-free circuit.

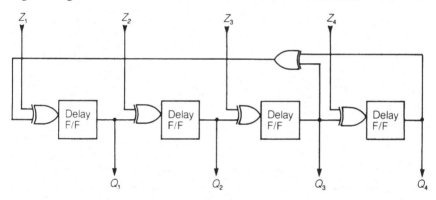

Fig. 6.17 Multiple input signature register.

Present-day integrated circuits have many outputs, consequently it would be impractical either to have a separate signature register attached to each output or to time multiplex one register between the outputs. To overcome this problem the Multiple Input Signature Register (MISR) [19] has been developed, and is shown in a simplified form in Fig. 6.17. If there are m inputs to the register, the input sequence, L, will be m bits wide and the error coverage, in this instance, will be,

$$1 - \frac{2^{mL-r} - 1}{2^{mL} - 1} \, .$$

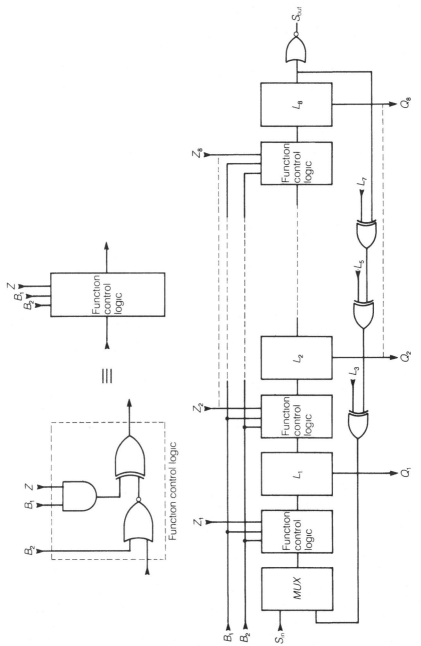

Fig. 6.18 BILBO multimode shift register.

If the sequence length is long the error coverage reduces to that of the serial input signature register, i.e. $1 - 2^{-r}$.

6.4.4 Built-in Logic Block Observer (BILBO) [27]

BILBO is a built-in test generation scheme which uses signature analysis in conjunction with a scan path. The major component in this technique is a multimode shift register, shown in Fig. 6.18, called a BILBO. The functions of the BILBO are controlled by two mode control signals B_1 and B_2. In practice the latches in the BILBO are the normal system latches in the circuit as shown in Fig. 6.19, where the Z-inputs are the outputs of the preceding combinational block and the Q-outputs are the inputs to the succeeding combinational block. The BILBO, in essence, partitions the circuit into simpler subfunctions for the purpose of testing, in a similar way to the SRL blocks used in LSSD.

The signal lines B_1 and B_2 control the function of the BILBO in the following way:

1. $B_1 = B_2 = 0$: the BILBO is configured as a long shift register forming a scan path;
2. $B_1 = B_2 = 1$: the BILBO functions as a set of system latches in which the Z-inputs are made available to the Q-outputs for normal circuit operation;

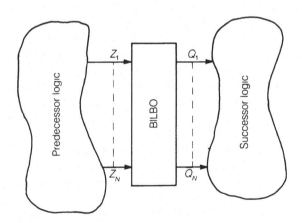

Fig. 6.19 BILBO in circuit.

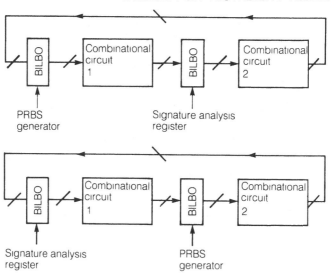

Fig. 6.20 BILBOs in a built-in test configuration.

3. $B_1 = 1$, $B_2 = 0$: the BILBO is configured into an LFSR with multiple inputs. If the inputs are held at a fixed value the LFSR operates as a PRBS generator; alternatively it operates as a signature analysis register.
4. $B_1 = 0$, $B_2 = 1$: the BILBO register is reset.

The technique of testing a circuit using BILBOs is shown in Fig. 6.20. A BILBO register is used as a PRBS generator to stimulate the combinational block under test; a second BILBO is then used as a signature analysis register to compress the test data response of the block under test, which after N cycles will contain the signature peculiar to that block (either faulty or fault-free). The BILBO configured as a signature register is now reconfigured into a scan path and the contents of the register are subsequently clocked out for analysis. The roles performed by the BILBOs are now reversed so that the next section of the circuit may be tested.

The advantages of this technique are that test pattern generation costs are eliminated and the volume of test data generated by, say, 10 000 test vectors has been reduced to two signatures of, for example, 16 bits each. The circuit can also be tested at operational speeds. The major problem, however, associated with test systems which employ signature analysis is determining the fault-free signature. This can be generated by simulation, although this approach is costly for complex circuits; alternatively the fault-free signature can be obtained from 'gold units'.

The BILBO technique is sometimes described as a 'simplex' method [19] of built-in self-test, since the BILBO can function both as a PRBS generator and

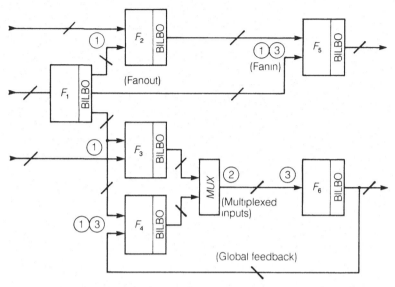

Fig. 6.21 An example of some problematic circuit configurations for the simplex method of built-in self-test [19]; (1) multiple sources may not be effective; (2) two signatures must be stored; (3) complex test scheduling required.

signature register. Simplex methods can be applied very efficiently to circuits which have a 'pipeline' architecture but their application can be problematic in circuits which have arbitrary interconnections between modules, as shown in Fig. 6.21 where, for example, all the outputs of one block may not be connected to the same successor block but split between several blocks, the inputs to a block may come from several sources, or the inputs to a block may be multiplexed from several sources. This situation is further aggravated in circuits which have reconvergent structures at functional level and also global feedback paths.

When a circuit has arbitrary interconnections between blocks, the above problems can be overcome, at the expense of chip area, by using the 'duplex' system [19] in which each functional block, essentially has its own test generator and signature register as shown in the simplified form in Fig. 6.22; for comparison the simplified form of the 'simplex' method is shown in Fig. 6.23, where the signature register and generator are essentially shared between consecutive blocks.

6.4.5 Problematic issues in the use of BIST systems

Although the use of BIST offers many advantages in reducing testing costs, several problems can arise in its use:

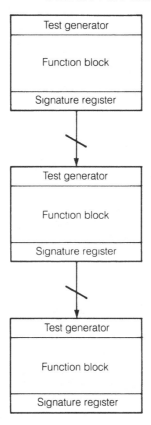

Fig. 6.22 Duplex method.

(a) Susceptibility of circuit under test to random patterns [28]

Not all circuits are susceptible to testing using random patterns. In particular circuits which have a high fanin: for example, to test for the output of a five-input NAND gate stuck-at-1 would require all inputs to be set to a logic 1, and the probability of this happening using random patterns is quite small, hence the probability of detecting this fault is low. For a similar reason PLAs are not susceptible to random pattern testing. It is possible, however, to modify the circuit, without altering the circuit function, to make it random pattern testable.

(b) Fault masking [29], [30]

In any data compression technique, as demonstrated in the technique of transition counting, the possibility of fault masking or 'aliasing' exists. In a

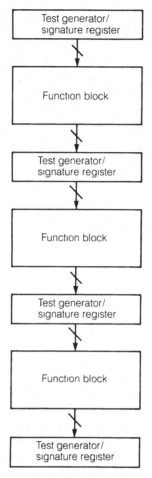

Fig. 6.23 Simplex method: N.B. The function performed by the test generator/ signature register module depends upon which block is to be tested.

BIST system the possibility of fault masking can be reduced to some extent by using the techniques of *multiple signature* or *split sequence* testing. In multiple signature testing the original test PRBS sequence, T, is divided into a number of subsequences and the signature derived from each subsequence is captured. In the split sequence technique the original test sequence, T, is again subdivided into several subsequences and these in turn are applied to the circuit and the resulting signatures captured. In this instance, however, the contents of the signature register are reset after the application of each subsequence. The error escape probability is identical for each technique and is given by 2^{-2r}, where r is the length of the signature register; this is an improvement on the error escape probability of the standard signature analysis technique, which is given by 2^{-r}. No additional hardware is required to implement either the multiple signature or split sequence technique although the test controller is more complex in that it has to capture several signatures. Multiple test sets may also be applied to the circuit under test; this

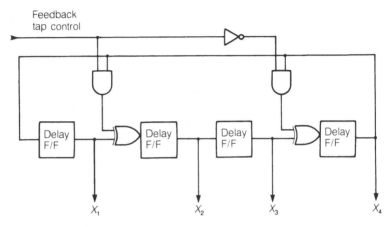

Fig. 6.24 Two-sequence PRBs generator.

technique is a more general form of split sequence testing in which each test set has a different fault coverage. In the multiple test technique each test set has the same fault coverage; in general two test sets are used, which may be generated as shown in Fig. 6.24 by adding some control logic to the PRBS generator so that it can realize a different characteristic polynomial; alternatively the original test set is simply run in reverse. If more test sets are used different patterns can be generated by loading a different 'seed' value from an ROM into the PRBS generator. In this case the error escape probability is, again, 2^{-2r}. Multiple data compression may also be used, when the outputs of the circuit under test are applied, simultaneously, to two signature registers having different characteristic polynomials; in this instance the error escape probability is again 2^{-2r}.

(c) Fault coverage/test length [31]–[36]

The most important characteristic of any test system is its fault coverage. Determination of fault coverage is usually achieved through the use of fault simulators implemented as either computer programs or hardware accelerators. However, in BIST systems which use PRBS generators and data compression, the problem of determining the fault coverage is quite difficult.

First, the number of test vectors used in a BIST system is usually much greater than a test system which uses deterministic test patterns. Second, the compression of test data into a signature also creates problems in fault simulation. For example, if the fault coverage is determined by the use of deductive fault simulation, which generates a list of all faults detected by a given input sequence at the outputs of a circuit, an enormous fault list can be produced (which may exceed the storage requirements of the computer system if unchecked) for each output since the signature for the data compressor is only available at the end of the test sequence, which may be several million test vectors long. The use of intermediate signatures can alleviate this problem to some extent by removing those faults detected by a given signature; although the fault coverage derived in this way is inaccurate, the reduction in CPU time obtained in deriving the fault coverage from intermediate signatures is quite considerable and in most cases the magnitude of the inaccuracy is negligibly small provided the error escape probability of the BIST system is small.

Another approach which has been used to improve the efficiency of determining fault coverage through simulation is the technique of *statistical fault simulation*, where instead of considering all faults in the circuit, a random selection of say 500 faults is selected; if the given test set detects an acceptable number of the faults, the test set is deemed to provide a sufficient fault coverage for the circuit.

The above approach to determining the fault coverage may be described as *a priori*, since the length of the test set is known.

The alternative *a posteriori* approach, however, is more useful to designers, since it essentially asks the question [34] 'What test length is required to achieve a given fault coverage?' This approach has attracted a vast amount of interest in view of the large amount of CPU time required to determine the fault coverage through simulation and the need to determine the fault coverage of a test set. One approach to determining the test length required to obtain a given fault coverage uses a fault detection probability profile. In generating this profile a 'cutting' algorithm [33] is used which enables hard-to-test faults to be identified. Circuit modification may be implemented to remove these hard-to-test faults, since the test length increases logarithmically with their number. From this profile, given an affordable test length it is possible to determine if a given fault coverage can be achieved; the detectability profile may be altered by raising or lowering the decision threshold on whether a particular fault is hard to test.

This approach, however, tends to suggest larger than required test lengths to achieve a given fault coverage since the analysis is based on true random pattern inputs and assumes that any pattern can and will be repeated; however, when the analysis is carried out considering the non-repeatability of the test patterns as occurs in BIST systems using PRBS generators a more accurate estimation of the required test length is obtained [35], [36].

6.4.6 Syndrome testing [37]

Although BIST systems which employ PRBS generators and signature analysis techniques offer many advantages in reducing the costs of testing circuits, the CPU time required to generate the fault-free signature can be excessive, particularly when MISRs are used in the signature register. Special-purpose hardware simulators, for example, the Yorktown Simulation Engine, capable of two billion gate evaluations per second, could be used to generate the fault-free signatures. Alternatively, special-purpose high-level simulators could be used for signature generation; for example, the SIGLYSER (SIGnature anaLYSER) [38], which has been developed to analyse the efficiency of MISRs in detecting faults. These techniques, however, are not widely available; consequently, as an alternative to long gate level simulation runs to develop fault-free signatures a very efficient although slightly suspect technique has evolved which uses 'gold units'. The essence of this technique is that a number of 'potentially' fault-free circuits, i.e. 'gold units', are identified, and these are clocked at, say, 1 MHz for one million or ten million cycles, and the output signatures from these devices are noted; the signature which is generated most often is considered to be the fault-free signature for the circuit against which all other fabricated circuits are tested.

In order to overcome this problem of fault-free signature generation the technique of *syndrome testing* was developed. The *syndrome* of a circuit is defined as $S = M/2^N$, where M is the number of minterms in the Boolean expression for the circuit and N is the number of inputs. For example, the syndrome of a three-input AND gate, which has only one minterm, is 0.125 $(1/8)$. Consider the function

$$W\bar{X}Y + \bar{W}X\bar{Y} + W\bar{X}Z + \bar{W}X\bar{Z} + WYZ$$

this function has six minterms and four inputs, so $S = 6/2^4 = 3/8$. The syndrome of the above function was generated solely from the Boolean expression of the function and is hence a functional characteristic of a circuit independent of its realization.

In general the structural description of a circuit is more readily available than its Boolean expression and the relationships outlined below permit the syndrome of a circuit described in terms of interconnected gates to be determined.

1. If the inputs to the connecting gates are disjoint and these inputs have syndromes S_1 and S_2, then the syndrome at the output of the connecting gate is given [37] by

Connecting gate	Overall syndrome
AND	$S_1.S_2$
NAND	$1 - S_1.S_2$
OR	$S_1 + S_2 - S_1.S_2$
NOR	$1 - (S_1 + S_2 - S_1.S_2)$
EXOR	$S_1 + S_2 - 2(S_1.S_2)$

2. The inputs to the connecting gates may, however, emanate from logic blocks, with known syndromes, which have shared inputs. In this instance, if the syndromes from the logic blocks are $S(A)$ and $S(B)$ respectively, then the overall syndrome at the output of the connecting gate is given by

Connecting gate	Overall syndrome
$AND - S(AB)$	$S(A) + S(B) + S(\bar{A}\bar{B}) - 1$
$OR - S(A + B)$	$S(A) + S(B) - S(AB)$
$EXOR - S(A \oplus B)$	$S(A\bar{B}) + S(\bar{A}B)$

A simple example illustrating the use of these relationships to calculate the output syndrome of a circuit is shown in Fig. 6.25.

The concept underlying syndrome testing is to exhaustively apply all input combinations to a circuit and to count the number of 1s at the output; the input test vector generator for syndrome testing simply comprises an N-bit counter eliminating the need to use involved test pattern generation algorithms to produce the input patterns. Furthermore the output data is compressed into a single word, that is the number of 1s appearing at a given output, which is stored in the output counter or syndrome register; the only difference between the number of ones counted and the syndrome is the

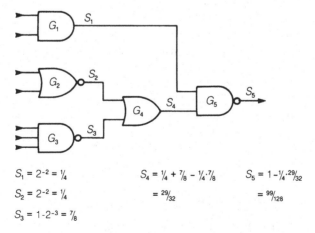

$S_1 = 2^{-2} = \frac{1}{4}$

$S_2 = 2^{-2} = \frac{1}{4}$

$S_3 = 1 - 2^{-3} = \frac{7}{8}$

$S_4 = \frac{1}{4} + \frac{7}{8} - \frac{1}{4}.\frac{7}{8}$
$= \frac{29}{32}$

$S_5 = 1 - \frac{1}{4}.\frac{29}{32}$
$= \frac{99}{128}$

Fig. 6.25 An example of calculating the output pattern of a circuit.

position of the binary point in the register. The basic configuration used to test a circuit using syndromes is shown in Fig. 6.26.

Not all circuits, however, are syndrome testable, for example, circuits containing EXOR gates or reconvergent fanout paths with unequal numbers of signal inversions. In these situations the circuit has to be modified to make it syndrome testable. As an example of a syndrome untestable circuit, consider the circuit which realizes the function, $F = xz + y\bar{z}$. The fault-free syndrome of this function is 1/2; however, if input z is stuck at 0, the syndrome of the faulty circuit is also 1/2. To make this type of circuit syndrome testable it is necessary to increase the size of the 'prime implicants' by introducing a control line w such that, in this instance $F = wxz + y\bar{z}$. In normal operation

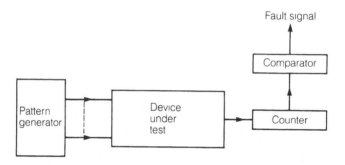

Fig. 6.26 Basic pattern test system.

the input w is held at a logic 1 and this has no effect on the logic function, but for the purposes of testing w is considered an input to the circuit. Algorithms [39] have been developed to determine the minimal number of control lines to be inserted into a circuit to make it syndrome testable.

The overheads, in terms of the number of additional inputs and the amount of control logic required to make a circuit syndrome testable, may be unacceptable. In such circumstances an alternative approach called *constrained syndrome testing* [40] may be used. In this technique parts of the circuit are desensitized during testing by holding particular inputs to the circuit at some logic value, in this way particular signal lines which make the complete circuit syndrome untestable are de-activated, the residual circuit is subsequently syndrome tested using all or some of the remaining inputs. Hence by performing multiple constrained syndrome tests on partitions in the circuit the complete circuit can be syndrome tested.

6.4.7 Syndrome testing applied to VLSI circuits [41]

The main disadvantage of applying syndrome testing to VLSI circuits is the necessity to exhaustively test the circuit, which results in long test times for circuits having more than twenty inputs. Consequently methods have had to be devised to partition the circuit into disjoint blocks which have a small number of inputs and which can be tested in parallel.

When partitioning the circuit into disjoint blocks a decision must be made whether to have many small blocks which can be tested quickly or several large blocks which require a longer testing time; the decision is usually based on the amount of additional hardware and input/output pins required to partition the system into disjoint blocks which can be tested in parallel. Furthermore the partitioned blocks, regardless of their size, will have multiple outputs. If each output is tested separately this requires many repetitions of the test inputs; however, it can be shown that outputs can be tested in parallel, each output producing the correct syndrome.

An important aspect of syndrome testing, when applied to VLSI circuits, is the size of the counter used to generate the test patterns. Intuitively it has an upper limit equal to the overall number of inputs to the circuit, and a lower limit equal to the maximum number of inputs to any subfunction being tested. It is essential to determine the minimal size of counter since this will affect the testing time. In order to minimize the size of the counter, as many subfunction inputs as possible must be connected to the same bit in the counter; inputs can only be connected to the same counter bit if they are 'non-adjacent' (inputs are said to be 'adjacent' if at least one output function is dependent upon both inputs). To determine the minimal counter size a non-adjacency graph [41] is drawn indicating the inputs which can be connected to the same counter bit; an attempt is then made to identify the maximal classes of inputs which can be connected and thereafter to find the minimum number of groups to include all inputs.

Another facet of self-test which is important is that of reducing the amount of reference data which must be stored and subsequently compared to determine whether a circuit is fault-free. When testing multi-output functions, instead of storing a syndrome for each output, a reduction in the amount of stored data can be achieved by using *weighted syndrome sums* [42]; care, however, must be taken in choosing the weighting factors since faults which are syndrome testable can become weighted syndrome sum untestable, or the amount of storage required for the weighting factors may become comparable to that required to store the individual syndromes.

The basic configuration for BIST using syndrome testing is shown in Fig. 6.27; it is assumed that the circuit has been designed using LSSD, each partition having a set of shift register latches on its inputs and outputs. In test mode the input shift registers are reconfigured as a counter to generate the test patterns for each partition, the multiplexer is used to select the outputs to be monitored during a given part of the test procedure, each block being tested in sequence and the syndromes compared with the reference values stored in the read-only store.

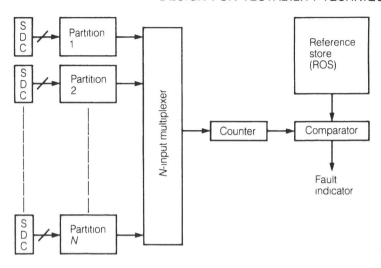

Fig. 6.27 Built-in pattern test architecture for VLSI [41].

6.4.8 Radamacher–Walsh coefficient testing method [43], [44]

The concept of using Radamacher–Walsh (RW) coefficients to test a circuit is also to create a signature for the circuit; if a fault is present this signature will be altered in at least one of the coefficients. The technique is similar to syndrome testing, which may be considered as an empirical verification of the zero-order RW coefficient, which is simply the number of ones appearing in the output function.

The RW spectral coefficients of a logic function define the function as uniquely as a truth table. If the number of inputs is n the number of coefficients is 2^n. In general the first $n + 1$ coefficients (primary coefficients) are sufficient to detect all distinguishable faults at the output of a circuit; however, recourse to the secondary coefficients is necessary in some situations when signal lines in the circuit are syndrome untestable. For any given function $F(X_1, X_2, X_3, \ldots, X_n)$ the spectral coefficients are obtained as follows

$$R = T_n F,$$

where R is the column matrix of spectral coefficients.

F is the column matrix of the output values of the circuit.

T_n is a $2^n \times 2^n$ transform matrix defined by

$$T_n = \begin{bmatrix} T_{n-1} & T_{n-1} \\ T_{n-1} & -T_{n-1} \end{bmatrix}; \qquad T_0 = 1.$$

Consider the function

$$F(X) = X_1\bar{X}_3 + \bar{X}_1\bar{X}_2 X_3:$$

$$
\begin{array}{c}
R \\
\begin{bmatrix}
r_0 \\
r_1 \\
r_2 \\
r_{12} \\
r_3 \\
r_{13} \\
r_{23} \\
r_{123}
\end{bmatrix}
\end{array}
=
\begin{array}{c}
T_n \\
\begin{bmatrix}
1 & 1 & 1 & 1 & 1 & 1 & 1 & 1 \\
1 & -1 & 1 & -1 & 1 & -1 & 1 & -1 \\
1 & 1 & -1 & -1 & 1 & 1 & -1 & -1 \\
1 & -1 & -1 & 1 & 1 & -1 & -1 & 1 \\
1 & 1 & 1 & 1 & -1 & -1 & -1 & -1 \\
1 & -1 & 1 & -1 & -1 & 1 & -1 & 1 \\
1 & 1 & -1 & -1 & -1 & -1 & 1 & 1 \\
1 & -1 & -1 & 1 & -1 & 1 & 1 & -1
\end{bmatrix}
\end{array}
\begin{bmatrix}
0 \\
1 \\
0 \\
1 \\
1 \\
0 \\
0 \\
0
\end{bmatrix}
=
\begin{array}{c}
F \\
\begin{bmatrix}
3 \\
-1 \\
1 \\
1 \\
1 \\
-3 \\
-1 \\
-1
\end{bmatrix}
\end{array}
$$

From the values of R the RW spectral response can be plotted.

To illustrate the effect of a fault on the RW coefficients, consider the function

$$F(X) = X_1(X_2 + X_3) + \bar{X}_1 X_4:$$

$$
\begin{bmatrix}
r_0 \\
r_1 \\
r_2 \\
r_{12} \\
r_3 \\
r_{13} \\
r_{23} \\
r_{123} \\
r_4 \\
r_{14} \\
r_{24} \\
r_{124} \\
r_{34} \\
r_{134} \\
r_{124} \\
r_{1234}
\end{bmatrix}
=
\begin{bmatrix} T_n \end{bmatrix} \cdot
\begin{bmatrix}
0 \\
0 \\
0 \\
1 \\
0 \\
1 \\
0 \\
1 \\
1 \\
0 \\
1 \\
1 \\
1 \\
1 \\
1 \\
1
\end{bmatrix}
=
\begin{bmatrix}
10 \\
-2 \\
-2 \\
1 \\
-2 \\
2 \\
-2 \\
2 \\
-4 \\
-4 \\
0 \\
0 \\
0 \\
0 \\
0 \\
0
\end{bmatrix}.
$$

If X is stuck at 1, so that $F'(X) = X_1 + \bar{X}_1 X_4$, the spectral coefficients become

$$
\begin{bmatrix}
r_0 \\
r_1 \\
r_2 \\
r_{12} \\
r_3 \\
r_{13} \\
r_{23} \\
r_{123} \\
r_4 \\
r_{14} \\
r_{24} \\
r_{124} \\
r_{34} \\
r_{134} \\
r_{234} \\
r_{1234}
\end{bmatrix}
=
\begin{bmatrix} \\ \\ \\ \\ \\ T_n \\ \\ \\ \\ \\ \end{bmatrix}
\begin{bmatrix}
0 \\
1 \\
0 \\
1 \\
0 \\
1 \\
0 \\
1 \\
1 \\
1 \\
1 \\
1 \\
1 \\
1 \\
1 \\
1
\end{bmatrix}
=
\begin{bmatrix}
12 \\
-4 \\
0 \\
0 \\
0 \\
0 \\
0 \\
0 \\
-4 \\
-4 \\
0 \\
0 \\
0 \\
0 \\
0 \\
0
\end{bmatrix}
$$

The coefficients comprising the RW spectra have the following meaning:

r_0 number of ones in the output function.

$r_1, r_2, \ldots, r_n$ a measurement of the correlation of the function with the input variables $X_1, X_2, \ldots, X_n$.

The other coefficients measure the correlation with particular EXOR functions, i.e. r_{12} compares the function $X_1 \oplus X_2$ and r_{123} compares $X_1 \oplus X_2 \oplus X_3$, the subscript indicating which variables are to be considered.

In the spectral response, a large positive value, except in r_0, indicates that the output is strongly dependent upon the positive value of that input, a large negative value indicates that the inverse of the input variable has a strong influence on the output. Consequently, when a fault exists, by observing which values are affected, it is possible to ascertain the location of a fault, since it will be on a node or a path in the circuit which contributes to that part of the spectral response. It should be noted that the alterations to the spectra brought about by some fault condition are unlikely to be arithmetic except in very restricted cases where the input variables are disjoint.

In general the primary coefficients are sufficient to detect all singly occurring distinguishable faults on the output. In some circuits, however, it may be found that certain primary coefficients are zero, implying that these lines are syndrome untestable; in this instance recourse must be made to the secondary coefficients. For example in a given circuit the primary coefficients r_2 and r_4 may be zero, implying that inputs X_2 and X_4 are not syndrome

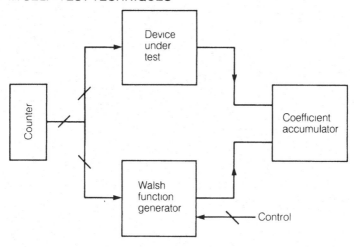

Fig. 6.28 Built-in Radamacher–Walsh coefficient test architecture [43].

testable; however, if r_{24} is not zero, the RW signature for the circuit would be (r_0, r_{24}). Since r_{24} does not equal zero it implies that X_2 is r_4-testable and X_4 is r_2-testable, that is a fault on X_2 or X_4 would be indicated by a change in the r_4 and r_2 coefficients respectively. In this instance the alternative spectral signature (r_0, r_2, r_4) could be used; from a hardware implementation aspect this latter signature would be preferred. It should be noted that not all RW coefficients are used in the spectral signature, a minimal subset is chosen so that all faults are detectable.

The hardware required to test a circuit by means of RW coefficients is shown in Fig. 6.28; it comprises a high-speed counter to generate all possible input combinations; a Walsh coefficient generator which is a simple multiplexer, if testing can be done using only the primary coefficients, i.e. $r_0, r_1, \ldots, r_n$, or if secondary coefficients are required an EXOR tree; and a coefficient accumulator. A separate test is run for each coefficient in the signature, which is carried out at full operational speed.

6.4.9 Autonomous testing [45]

The aim of autonomous testing is to partition a circuit into modules which are small enough to be exhaustively tested economically.

In autonomous testing all input sequences must be applied to the circuit; these input sequences are generated by a linear feedback shift register which has been modified so that it can generate an all-zero state without creating a 'deadlock' situation. The response of the circuit is checked by a multi-input signature analysis register.

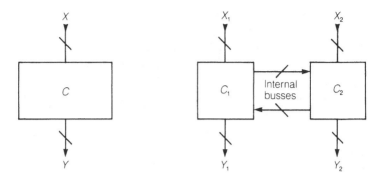

Fig. 6.29 Partitioning for autonomous testing.

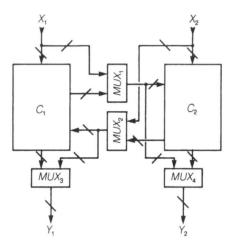

Fig. 6.30 Partitioned circuit with multiplexers added to control/observe internal busses [45].

The partitioning of the system into testable blocks can be accomplished either by using additional hardware or by 'sensitized' partitioning.

The simplest way to partition a system using hardware is to incorporate multiplexers into the system so that embedded inputs can be readily controlled and observed. As an example of the implementation of this technique consider the function block, C, in Fig. 6.29, with inputs X and output Y; this block is subsequently partitioned into two subfunctions C_1 and C_2 having inputs X_1, and X_2 and outputs Y_1 and Y_2 respectively. Multiplexers M_1–M_4 are then inserted, as shown in Fig. 6.30, to make the internal lines controllable and observable. In normal mode, multiplexers M_1 and M_2 permit the internal connections between the subfunctions to be maintained and M_3 and M_4

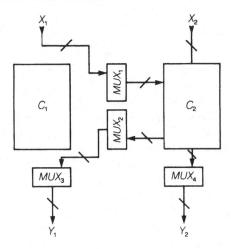

Fig. 6.31 Circuit configured to test subfunction C_2.

simply connect the blocks outputs to the external pins on the package. Figure 6.31 illustrates the configuration adopted in order to test subfunction C_2. The multiplexer M_1 permits the internal signals to C_2 to be controlled, whilst M_2 and M_4 permit the internal signals and main block outputs to be observed. This technique has been applied to a four-bit ALU having 14 inputs, which would require $2^{14} = 16K$ patterns if tested exhaustively; when the circuit was partitioned into two submodules having 10 and 5 inputs respectively only 1056 patterns were required. The introduction of the additional hardware, however, affected the circuit performance and increased the circuit area by 30%, this was due mainly to the additional interconnect.

The alternative to hardware partitioning is 'sensitized' partitioning, whereby certain patterns are applied to specific inputs, creating sensitized paths between primary inputs and embedded module inputs, and between the module outputs and the primary outputs. When this technique was applied to the four-bit ALU it was found that only 324 patterns were required to test the system.

6.4.10 Store-and-generate technique for built-in test [46]

Various schemes for built-in-test have been devised as possible solutions to the problem of testing VLSI circuits. However, the store-and-generate method treats other schemes as special cases and also permits a designer to tailor a technique to his own particular requirements, in that it offers flexibility in terms of test data generation, test data storage, output generation and storage.

The generalized model of a built-in test system comprises a test data complex, test results compressor, output data complex and a comparator together with the circuit under test. The way in which these functions are connected defines the method of self-test being used.

The function of the test data complex is to supply input stimuli to the device under test so that its response can be analysed. The test data may be stored in an ROM or it may be generated as required using counters or LFSRs for example. The output data complex produces reference values which are compared with the output response of the circuit under test. These responses may be held in a store or may be produced by a duplicate circuit. The test results compressor usually employs a technique similar to signature analysis in order to reduce the amount of test results which have to be analysed. The comparator simply compares the reference results with those responses obtained from the circuit under test.

Figure 6.32 illustrates the store-and-generate test scheme which uses both a test data store and a test data generator. The test data generator is a multi-input LFSR which has inputs coming from the test data store which is simply an ROM, permitting a larger number of test vectors to be produced than if either were used individually. The output data complex may simply be a store containing the fault-free circuit responses; however, for a multi-output circuit the amount of storage may be prohibitive. The alternative is to use a reduced version of the circuit under test as the output data generator. The duplicate circuit need only produce a response for the set of input test vectors; any other input patterns would produce a 'don't care' output pattern.

One important feature concerning built-in testing is that it must be self-checking. In the test data complex several faults can occur ranging from

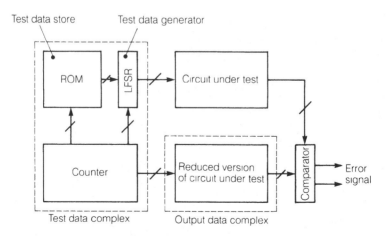

Fig. 6.32 Store-and-generate test scheme [46].

storage faults, address faults due to the address counter not being incremented properly or the LFSR not being initialized properly. The ROM could be replaced by a self-checking PLA, the counter could be periodically run through all its sequences, and the LFSR can be designed to be self-checking. In the output data complex, if the reference vectors are stored in an ROM then general coding theory techniques can be applied for error detection. If the reference data, however, is generated from a reduced version of the circuit under test, then a simple comparison between the actual circuit and the reduced version will check out this part of the system. The comparator can also be implemented as a self-checking function.

The store-and-generate approach to built-in testing is in fact a class of schemes whereby the designer can trade off additional hardware against off-line test pattern generation, by interchanging generator modules with storage modules; in this way he has sufficient flexibility to produce the built-in test system most suited to his requirements.

6.4.11 Examples of standard components with built-in test features [47], [48]

Until recently, design techniques to enhance testability could only be incorporated, readily, into full custom-designed circuits. However, several companies have now developed standard cell components and gate arrays which include features to enhance the testability of complete designs.

In response to the requirements of the system designer for a micro-controller which could be customized for a specific application, Intel produced a SuperCel Design System which comprises a microcontroller supercell whose function can be made application specific by modifying the peripheral logic, which includes standard cells (basic gates), high-level cells (registers, decoders, etc.) and high density array cells (ROM, RAMs, PLAs). In order to test the microcontroller core independently of the peripheral customizing logic some test hardware was incorporated into the micro-controller supercell in the form of an 'isolation' ring: during test mode the microcontroller can be disconnected from the peripheral logic and tested as a stand alone component; thereafter the microcontroller core is used to test the peripheral logic.

CDC have also included some built-in test features into their 6K gate array, based on the BILBO technique. The array contains three additional registers for testing purposes, that is an input register, an output register and a control register. The input and output registers are used as either a data source or sink for nodes between the primary inputs/outputs of the circuit and the internal gate array. The control register simply regulates the function of the input/output registers, which are essentially in parallel with the primary input/output lines to the logic array; consequently there is no performance penalty by incorporating the built-in test logic on the chip. The circuit can be configured into a self-test mode in which the primary inputs are disabled and the input register is converted to a pseudo-random number generator which supplies the test vectors for the logic array. The output register is configured into a check sum register and accepts outputs directly from the logic array or the output buffers. The additional test hardware on the chip can also be used to perform interconnect tests between chips on a printed circuit board, since patterns can be loaded into the output register of one chip and transmitted to the input registers of the other chips on the board to which it is connected, thus permitting open and shorted interconnections on the printed circuit board to be located.

It is generally accepted that some sort of DFT technique must be incorporated into a design if testing costs are to be reduced. However, with the large number of methods available, each with its own advantages and disadvantages, the designer now has the added complication of identifying the technique most suitable to a given design. In an attempt to alleviate this problem several expert systems have evolved to guide the designer's choice; the application of expert systems to this problem is discussed in Chapter 11.

6.5 REFERENCES

1. Williams, T. W. and Parker, K. P. (1983) Design for testability – a survey. *Proc. IEEE*, **71**, 95–122.
2. Eichelberger, E. B. and Lindbloom, E. (1983) Trends in VLSI testing. VLSI '83, pp. 339–48, Elsevier Science Publishers.
3. Maunder, C. (1985) Built in test – a review. *Electronics and Power*, March, 204–8.
4. Sharad, C. S. and Agrawal, V. D. (1985) Cutting chip testing costs. *IEEE Spectrum*, April, 38–45.
5. Segers, M. T. M. (1982) Impact of testing on VLSI design methods. *IEEE Journal of Solid State Devices*, **SC-17**(3), 481–6.
6. Grason, J. and Nagle, A. W. (1980) Digital test generation and design for testability. *Design Automation Conference Proceedings*, June, 175–89.
7. McCluskey, E. J. (1984) A survey of design for testability scan techniques, *VLSI Design*, **5**(12), 38–61.
8. McCluskey, E. J. (1985) Built-in self test structures. *IEEE Design and Test of Computers*, **2**(2), 29–36.
9. Eichelberger, E. B. and Williams, T. W. (1977) A logic design structure for LSI testability. *14th Design Automation Conference Proceedings*, June, 462–8.
10. Stewart, J. H. (1977) Future testing of large LSI circuit cards. *Digest of Papers, 1977 Semiconductor Test Symposium*, October, 6–15.
11. Ando, H. (1980) Testing VLSI with random access scan. *Digest of Papers, Compcon '80*, Spring, 50–2.
12. Bennetts, R. G. (1984) *Design of Testable Logic Circuits*, Chapter 3, Addison-Wesley.
13. Das Gupta, S., Walther, R. G., Williams, T. W. and Eichelberger, E. B. (1981) An enhancement to LSSD and some applications of LSSD in reliability, availability and serviceability. *Proc. 11th Fault Tolerant Computing Symposium*, 32–4.

14. Bhavsar, D. K. (1983) Design for test calculus – an algorithm for DFT rule checking. *20th Design Automation Conference Proceedings*, June, 300–307.
15. McCluskey, E. J. (1985) Built-in self test techniques. *IEEE Design and Test of Computers*, **2**(2), 21–8.
16. Williams, T. W., Walther, R. G., Bottoroff, P. S. and Das-Gupta, S. (1985) Experiment to investigate self testing techniques in VLSI. *IEE Proceedings*, **132**, Part G(3), 105–7.
17. Totton, K. A. E. (1985) Review of built-in test methodologies for gate arrays. *IEE Proceedings*, **132**, Parts E and I(2), 121–9.
18. Frohwerk, R. A. (1977) *Signature Analysis: A New Digital Field Service Method.* Hewlett Packard Application Notes, Number 222-2, 9–15.
19. Bhavsar, D. K. and Heckelman, R. W. (1981) Self testing by polynomial division. *Digest of Papers, 1981 International Test Conference*, October, 208–16.
20. Tang, D. T. and Chen, C. L. (1983) Logic test pattern generation using linear codes. *Digest of papers, 13th International Symposium on Fault Tolerant Computing*, June, 222–6.
21. David, R. (1980) Testing by feedback shift register. *IEEE Trans. Computers*, **C-29**(7), 668–73.
22. Peterson, W. W. and Weldon, E. J. (1972) *Error Correcting Codes.* The MIT Press, Cambridge, Massachusetts.
23. Mucha, J. and Daehn, W. (1981) Hardware test pattern generation for built-in testing. *Digest of Papers, 1981 International Test Conference*, October, 110–13.
24. Wadsack, S. L. (1978) Fault modelling and logic simulation of CMOS and MOS integrated circuits. *Bell System Technical Journal*, **75**, 1449–74.
25. Hayes, J. P. (1976) Transition count testing of combinational logic circuits. *IEEE Trans. Computers*, **C-25**(6), 613–20.
26. Breuer, M. A. and Friedman, A. D. (1977) *Diagnosis and Reliable Design of Digital Systems*, pp. 152–6, Pitman.
27. Koenemann, B., Mucha, J. and Zwiehoff, G. (1979) Built-in logic block observation techniques. *Digest of Papers, 1979 Test Conference*, October, 37–41.
28. Eichelberger, E. B. and Lindbloom, E. (1983) Random pattern coverage enhancement and diagnosis for LSSD logic self test. *IBM J. Research and Development*, **27**, 265–72.
29. Bhavsar, D. K. and Krishnamurthy, B. (1984) Can we eliminate fault escape in self testing by polynomial division (signature analysis). *Digest of Papers, 1984 International Test Conference*, October, 134–9.
30. Hassan, S. Z. and McCluskey, E. J. (1984) Increased fault coverage through multiple signatures. *Digest of Papers, 14th International Symposium on Fault Tolerant Computing*, June, 354–9.
31. Agrawal, V. D. (1978) When to use random testing. *IEEE Trans. Computers*, **C-27**(11), 154–5.
32. Williams, T. W. (1984) Sufficient testing in a self testing environment. *Digest of Papers, 1984 International Test Conference*, October, 167–72.
33. Savir, J., Ditlow, G. and Bardell, P. H. (1983) Random pattern testability. *Digest of Papers, 13th International Symposium on Fault Tolerant Computing*, June, 80–9.
34. Savir, J. and Bardell, P. H. (1983) On random pattern test length. *Digest of Papers, 1983 International Test Conference*, October, 95–106.
35. Chin, C. K. and McCluskey, E. J. (1987) Test length for pseudorandom testing. *IEEE Trans. Computers*, **C-36**(2), 252–6.
36. Wagner, K. D., Chin, C. K. and McCluskey, C. J. (1987) Pseudorandom testing. *IEEE Trans. Computers* **C-36**(3), 332–43.

37. Savir, J. (1980) Syndrome-testable design of combinational circuits. *IEEE Trans. Computers*, **C-29**(6), 442–51. (Corrections: **C-29**(11), 1012–13 (1980).)

38. Sridhar, T., Ho, D. S., Powell, T. J. and Thatte, S. M. (1982) Analysis and simulation of parallel signature analysers. *Digest of Papers, 1982 International Test Conference*, November, 656–61.

39. Markowsky, G. (1981) Syndrome testability can be achieved by circuit modification. *IEEE Trans. Computers*, **C-30**(8), 604–6.

40. Savir, J. (1981) Syndrome testing of 'syndrome-untestable' combinational circuits. *IEEE Trans. Computers*, **C-30**(8), 606–8.

41. Barzilai, Z., Savir, J., Markowsky, G. and Smith, M. G. (1981a) VLSI self testing based on syndrome techniques. *Digest of Papers, 1981 Test Conference*, October, 102–9.

42. Barzilai, Z., Savir, J., Markowsky, G. and Smith, M. G. (1981b) The weighted syndrome sum approach to VLSI testing. *IEEE Trans. Computers*, **C-30**(12), 996–1000.

43. Muzio, J. C. and Miller, D. M. (1982) Spectral techniques for fault detection. *Digest of Papers, 12th International Symposium on Fault Tolerant Computing*, June, 297–302.

44. Susskind, A. M. (1983) Testing by verifying Walsh coefficients. *IEEE Trans. Computers*, **C-32**(2), 198–201.

45. McCluskey, E. J. and Bozorgui-Nesbat, S. (1981) Design for autonomous test. *IEEE Trans. Computers*, **C-30**(11), 866–75.

46. Agarwal, V. K. and Cerney, E. (1981) Store and generate built-in-testing approach. *Digest of Papers, 11th International Symposium on Fault Tolerant Computing*, June, 35–9.

47. Koehler, R. (1983) Designing a microcontroller 'super cell' for testability. *VLSI Design*, October, 44–6.

48. Resnick, D. R. (1983) Testability and maintainability with a new 6K gate array. *VLSI Design*, March/April, 34–8.

7
HYBRID DESIGN-FOR-TESTABILITY TECHNIQUES

7.1 INTRODUCTION

When using structural design for testability techniques, such as scan path, several problems can arise:

1. In order to efficiently use the structural techniques it is necessary to generate the test patterns for the combinational blocks associated with each scan path;
2. The test equipment must be capable of storing long sequences of serial data for shifting into the registers. This usually means that two or three probes on a tester must be associated with a large amount of memory;
3. The test patterns cannot be applied to the device under test at full clock speed. This is caused by the stop/go form of testing used by the scan path; that is the system must be stopped while the test patterns are shifted into the scan paths, the circuit is then placed in a 'go' condition to apply the pattern.

Although the BILBO [1] technique can eliminate most of these difficulties it has a major problem of high area overhead because its registers are multifunctional. To overcome these problems a class of Built-In Self-Test (BIST) schemes has evolved which use both scan path and signature analysis in a hybrid form. These techniques have acquired the generic name of S^3 (Self-test using Scan path and Signature analysis) [2] techniques. The methods used can be divided into two categories – internal or external. As this discussion is concerned with built-in test methods only the internal structures will be considered. The internal category can be further subdivided into centralized, distributed and mixed mode.

In the centralized schemes buffer registers are added to the primary inputs and outputs. These registers are used to form a pseudo-random bit-sequence generator and a signature analysis register respectively. Figure 7.1 illustrates the basic configuration for this scheme. The technique has two advantages in

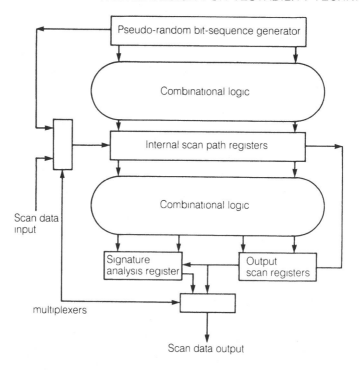

Fig. 7.1 Internal centralized S³ technique [2].

that the area overhead required to implement the technique is minimal and the test hardware is very easy to incorporate into existing circuits especially if registers are already present within the design. Additional multiplexers may be required on the inputs and outputs of the scan registers. A major disadvantage of this technique is that the testing time can be long [3] since each vector has to be shifted through the entire scan register.

The problem of long test times associated with the centralized scheme can be overcome by the use of a distributed scheme where each storage element is modified so that it can perform either the generator or the analysis function. The scheme is capable of testing a system at near operational speeds. However, the input and output pins are not tested, and the final performance of the design will be degraded by the extra test hardware. Another problem is that the extra area required by the modified storage latches can be quite high.

To overcome the problems of large area overhead and long testing times associated with the two previous techniques, the mixed-mode method has been proposed [4] (Fig. 7.2). In this scheme the pattern generator and signature analyser are again centralized and are formed from a group of registers at the beginning and end of the scan path chain, respectively. In this

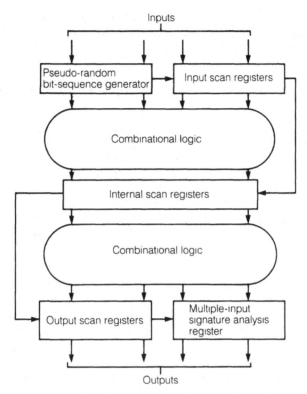

Fig. 7.2 Internal mixed-mode S^3 technique [2].

case, however, the registers are capable of accepting both parallel and serial test data. Thus during test operation the pattern generator not only applies patterns to the inputs of the combinational block to which it is attached but also serial data is fed into the start of the shift register chain, providing the necessary patterns for self-testing the remaining logic. Similarly the signature analysis register is capable of accepting parallel data from the combinational logic to which it is attached and also serial data can be input from the scan register; therefore no extra hardware is required to compact the fault information. The advantages of this technique are that the area overhead is minimal since the generation/analysis hardware is not distributed throughout the system, and testing times are short since there is no need to shift data through the complete scan path. As the testing environment and functional environment are very similar it is possible to detect some performance and delay problems. A basic discussion of the S^3 philosophy is given by Komonytsky [3].

The following sections of this chapter will look at three hybrid techniques in detail. These are HILDO [4], LOCST [5] and SASP [6]. In the case of the final method a detailed implementation study is undertaken comparing the BILBO, scan path and SASP techniques.

7.2 THE HIGHLY INTEGRATED LOGIC DEVICE OBSERVER (HILDO) [4]

The basic HILDO technique is illustrated in Fig. 7.3. As can be seen from the figure this technique does not use an independent pattern generator and signature analyser; instead these operations are combined into the HILDO register. During test operations the HILDO register is first initialized with a seed value. The pattern input to the circuit is then taken from the HILDO register via the multiplexer on the input. The output pattern is compressed in the HILDO to form the next input pattern. These operations will continue until either the pattern sequence begins to repeat itself or all the faults have been detected. Two methods can be used to determine the pattern length produced by the HILDO register. The first method involves determining mathematically the characteristic polynomial of the circuit under test. This method is however very difficult. The second method uses more traditional simulation methods. The circuit being tested effectively becomes part of the feedback logic of the signature analysis register; therefore by simulating the complete circuit it is possible to determine the length of the test sequence that will be produced. Once the patterns that the circuit will receive during the test operation have been calculated a fault simulation can be performed to determine the number of faults detected. Obviously the number of patterns produced from the HILDO register cannot be greater than the number

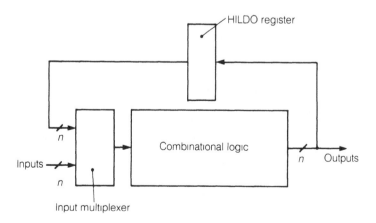

Fig. 7.3 HILDO technique [4].

generated by the maximal length polynomial used in a BILBO register of the same number of bits; indeed a much smaller set of patterns will probably be generated. If the number of test patterns generated is inadequate for the desired fault coverage then the number of patterns can be increased by altering the inputs to the analysis register. In addition to detecting faults in the combinational logic this technique can also detect stuck-at faults in the multiplexer and HILDO register. Since the signature analysis and pattern generation can be performed in real time the circuit can be tested at full speed; therefore it is possible to detect timing faults.

The HILDO register is constructed from a modified D-type flip-flop and has four operating modes [4]; these are:

(a) Normal mode

In this mode data from the circuit is passed through the flip-flop, some signal propagation delay will be incurred during normal operation.

(b) Test mode

In this mode the data present at the input of the flip-flop is Exclusive-ORed with the output data of the preceding flip-flop in the register. This is similar to the operation of a multiple input signature analysis register.

(c) Shift mode

In this mode the data in the register can be shifted out for comparison off chip. Alternatively data can be shifted in so that the seed value for any test sequence can be set.

(d) Initialization mode

In this mode the required seed value for the test can be set automatically in the registers by using the Preset and Clear inputs of the D-type flip-flop.

If, as is often the case, the number of inputs and outputs on a circuit are not the same then the HILDO register must be modified so that the correct number of bits can be fed back to the inputs. If the number of inputs to the circuit are less than the number of outputs, then the output bits must be compressed. The compression of the output bits can be achieved by Exclusive-ORing together bits that are not of the same type, that is bits that do not belong to the same group of signals (thus in an ALU the output bits should not be Exclusive-ORed together). If the number of inputs is greater than the number of outputs then the HILDO register must be extended by the required number of bits. These extra bits do not receive any output bits but simply act as a pattern generator for the inputs.

In trials performed by Beucler and Manner [4] on three logic circuits to compare the performance of the HILDO register and BILBO devices it was found that in general both registers had approximately the same fault detection capability. However the area overheads associated with the HILDO technique were almost half of those of the BILBO method.

7.3 LSSD ON-CHIP SELF-TEST (LOCST) [5]

The LOCST scheme was developed by IBM in order to reduce the complexity, in terms of input patterns required and output patterns stored, of their test environments. Currently four test areas are considered: manufacture test, card test, system level test and field/repair testing. However, the LOCST scheme is directed mainly at the last three test areas since more rigorous tests are required in manufacturing testing although LOCST can be used to provide rapid initial screening of the chips.

The basic block diagram of the LOCST method is shown in Fig. 7.4. The technique itself is based on the LSSD [7] structures proposed by IBM to perform normal scan testing. Also included in the scheme is the On-Chip Monitor (OCM) used to control the operation of the on-chip test hardware.

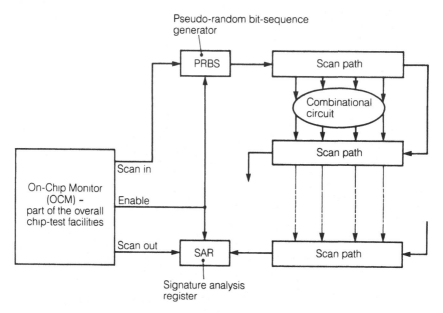

Fig. 7.4 The LOCST test scheme (adapted from [5]).

To complete the test scheme it is necessary for the latches to be able to access the input and output pins; this is sometimes known as 'boundary scan'. If this is implemented, then it is possible to control, for the purposes of testing, non-LSSD chips (these are devices which have been designed without incorporating any LSSD latches; consequently the LSSD devices must be capable of controlling the pins of the non-LSSD chips, this is usually achieved by the boundary latches on the LSSD chip) which are incorporated onto the same board as the LOCST devices – or the interconnections between LSSD devices can be tested. In self-test mode the first 20 latches of the scan path in the LOCST chip are reconfigured into a maximal length Linear Feedback Shift Register (LFSR) to produce pseudo-random test patterns; this is called the Pseudo-Random Bit-Sequence generator (PRBS) in Fig. 7.4. The feedback polynomial for the LFSR is chosen so that a maximal length output can be obtained while keeping the number of feedback points to a minimum so that the area overhead for the device can be reduced. The reconfiguration of the PRBS can be controlled by an enable signal from the OCM. If different random test patterns are required then the initial 'seed' value in the PRBS can be altered. To collect the responses from the scan path the last 16 latches are configured into a Signature Analysis Register (SAR), again under the control of the OCM.

In order to perform self-test using the LOCST scheme the OCM first enables the PRBS and SAR, the internal shift register latches are configured into a scan path and the seed values are scanned into the scan registers (PRBS and SAR). The scan clocks are then operated to fill the scan path with pseudo-random patterns. At the same time the previous data in the scan path is compressed in the SAR. Once the data has been set the system clocks are enabled to allow the test pattern stimuli to activate the combinational logic. This complete cycle is repeated until the required number of test patterns have been applied to the device. The signature in the SAR is then scanned out for comparison with the 'good' signature. The 'good' signature can be obtained by two routes. Either the device can be simulated to produce the signature or alternatively 'gold' units can be used to generate the signature. Currently the 'gold' unit approach is the most cost effective and has been used to generate the good signatures for other chips. During self-test mode any boundary latches must be inhibited or known input data must be applied to their inputs so that signature generation is not affected.

The overall fault coverage of this technique depends on the distribution of 'exterior' and 'interior' logic [5]. 'Interior' logic is defined as logic whose inputs and outputs are fed by scan latches. 'Exterior' logic is defined as logic which only has scan latches at its inputs or outputs but not both. LOCST is capable of testing 'interior' logic but not 'exterior' logic. Also LOCST cannot be used to fully test the OCM or any embedded RAMs in the circuit.

When this technique was applied to three signal processing chips it was found that the fault coverage for the interior logic was greater than about 97% with fewer than 5000 random patterns. When deterministic patterns were applied the fault coverage was greater than 99%.

A major disadvantage of this technique occurs when the scan rate is slow or the scan path is long, as the test time can be increased from a few milliseconds to several minutes. However, in order to overcome this problem the PRBS could be reconfigured to load several individual scan paths in parallel. The output responses can then be compacted in a multiple input signature register. This method of operation is used in the Signature Analysis and Scan Path (SASP) technique.

7.4 SIGNATURE ANALYSIS AND SCAN PATH (SASP) [6]

A block diagram of the hardware configuration for the SASP technique is shown in Fig. 7.5. The design consists of three major test blocks:

(a) Pseudo-Random Bit-Sequence (PRBS) generator

This device is simply a linear feedback shift register which is capable of producing pseudo-random output sequences.

(b) Signature Analysis REGister (SAREG)

This is an LFSR designed to receive multiple inputs and generate a signature. Both the PRBS and SAREG are 16 bits in length.

(c) Scan paths

These are simply pseudo-dynamic latches that can be reconfigured into a shift register to form the familiar scan paths.

In addition to these major blocks certain ancillary circuitry is also required, such as the multiplexers (SEL_1, SEL_2, SEL_{out}) which direct data flow during the test operations.

To apply the technique the circuit is first partitioned into blocks by the scan path registers. The scan outputs of the scan path registers are then connected to the multiplexer, SEL_1, which is used to select one of the scan path register outputs in order to send it to an external pin via SEL_{out}. The output of the scan path also connects to one of the inputs of the SAREG; the output of SAREG can also be directed to an external pin by the multiplexer SEL_{out}. Consequently either the SAREG or a scan path output can only be directed to an external pin at any one time, which reduces the number of extra pins required for test purposes. The inputs to the scan paths come from the multiplexer

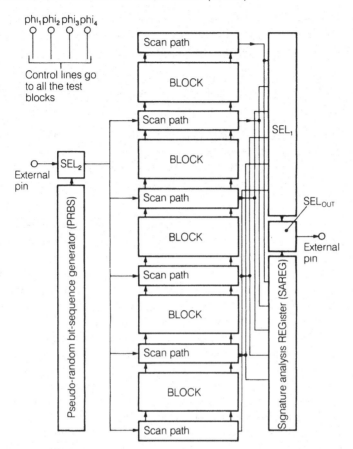

Fig. 7.5 The SASP schema [6].

SEL$_2$. The input to SEL$_2$ can come from either an external test pin or the output of the PRBS.

The procedure for self-testing a circuit using the SASP technique is now as follows:

1. The inputs to all the scan paths are connected via SEL$_2$ to the output of the PRBS generator. The scan paths are also initialized to a starting value. A pseudo-random pattern is then clocked into each of these scan paths in parallel;

2. The test pattern is then applied to the combinational logic blocks and the output response is collected in a scan path register, the original test pattern being overwritten;

3. The outputs from the scan paths are then clocked simultaneously into the multiple-input signature analysis register. At the same time another random pattern is shifted into the scan registers.

This procedure is repeated for say N test patterns, at the end of this period the signature is shifted out of SAREG and compared to a 'good' signature.

The advantages of this technique are:

1. The combinational logic blocks are tested in parallel. Long shifts register sequences are therefore avoided;
2. The advantages of BILBOs on-chip test-pattern generation have been achieved without the use of the multifunction register required by the BILBO technique.

The major disadvantage, however, of all the techniques which require a signature for test comparison is that the generation of the fault-free signatures is very difficult [8], particularly for multiple-input signature analysis registers, as used in this technique. The SASP technique attempts to solve this difficult problem by allowing the designer access to the short internal scan paths and therefore the ability to test each block individually, as may occur in the prototype stage of a design. Once each block has been verified as operating correctly, the system can then be switched to internal self-test and the signature of the system generated using the actual hardware as a 'gold' unit. The signature produced can then be used as the comparison signature in the production testing of the devices. This method of signature generation is certainly possible with the BILBO technique, however it is likely that more than one signature will be required and also the size of the BILBO registers will be shown to be extremely large in comparison to SASP in Section 7.4.1.

In order to improve the area overhead requirements of SASP several modifications can be made to the original SASP concept. The first is to remove the PRBS on-chip generator and place it in the tester that is controlling the test; this would reduce the test area overhead if required, but it does mean that the test equipment is more complicated. This is undesirable if a high-speed technology is being tested, on-chip test would then be preferable. Alternatively the signature analyser could be used to generate the patterns as demonstrated by the HILDO technique (Section 7.2). This also has disadvantages since the test coverage prediction may be difficult.

One useful application of the SASP technique is in the testing of PLAs. The testing of PLAs using a modified BILBO to generate pseudo-random patterns has been discussed by Daehn and Mucha [9]. However, PLA structures are usually very tightly packed, consequently the pitch between the bit lines is very narrow. This makes it very difficult to build the relatively large BILBO structure onto these bit lines for testing without extensive modifications to the PLA. However, a shift register can easily be inserted onto these lines because of its smaller size. Therefore the SASP technique could be used to break up the PLA structure with scan paths and still allow on-chip pattern generation and data compression. Obviously several PLAs could be linked to one SAREG and PRBS.

7.4.1 Comparative study

In order to decide which test method is the most economical to implement in a system it is necessary to have some criteria on which to base the decision. This section will compare three different structural test techniques based on the following test criteria:

(a) The total area occupied by the test structure

It will be assumed that no hardware is available from the circuit to be tested, this will give a worst case value of area overhead, since in reality it may be possible to use blocks from the circuit to be tested.

(b) Time taken to apply the test

A nominal value for the number of test patterns required to test a device to a given fault coverage will be chosen. In order to give a realistic time for the test a clock speed of 1 MHz will be assumed.

The three structural methods that will be compared are the scan path, BILBO and SASP techniques. In order to estimate the area of the test structure it is necessary to actually lay out all the required circuits in the desired technology. The necessary pieces of the test structure were designed using the PLAP [10],[11] suite of programs in 6 μm 5 V NMOS technology.

Two main blocks of logic are required to implement all of the test methods; these are a latch and an Exclusive-OR gate (for the feedback paths in the LFSR). The two blocks designed to carry out these tasks are shown in Fig. 7.6 and Fig. 7.7. As can be seen the latch is of the pseudo-dynamic type. Each block was constructed so that it is possible to use abutment [12] in the layout of the final designs. Figure 7.8 illustrates this layout technique for an 8-bit scan path block; the figure also shows two other blocks that are necessary to implement the shift register – a control block and a buffer. The control block is used to implement the clocking scheme, shown in Fig. 7.9, for the scan paths and the LFSR blocks. Four control lines are required by the structure: phi$_1$ and phi$_2$ act as clock signals and phi$_3$ and phi$_4$ are used to control the holding or latching of data respectively. The transparent mode allows the latches to pass data straight through without storing it, which is useful if the scan path is no longer required to separate the blocks in normal system operation; however,

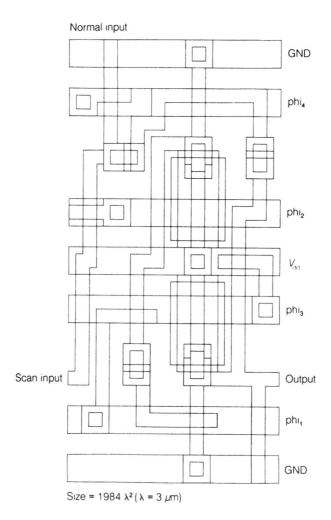

Normal input

GND

phi$_4$

phi$_2$

V_{dd}

phi$_3$

Scan input   Output

phi$_1$

GND

Size = 1984 λ^2 (λ = 3 μm)

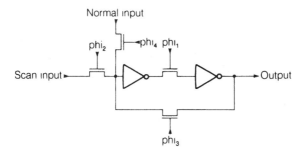

Normal input

phi$_2$ phi$_4$ phi$_1$

Scan input   Output

phi$_3$

Fig. 7.6 Dynamic shift register layout (DSR).

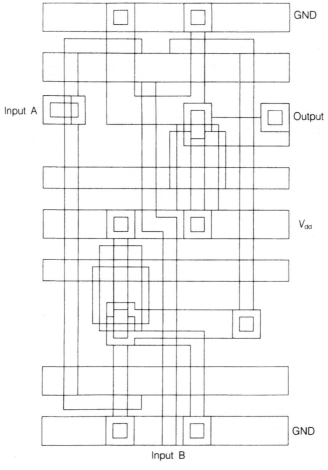

GND

Input A

Output

Vdd

GND

Input B

Size = 2170λ^2 (λ = 3 μm)

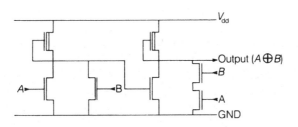

Fig. 7.7 Exclusive-OR gate layout (EORG).

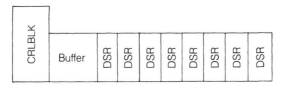

Fig. 7.8 Eight-bit scan path block.

Register type	Operation	Clock		control	
		phi$_1$	phi$_2$	phi$_3$	phi$_4$
Scan path	Shift	Clock		L	L
	Hold	H	L	H	L
	Copy	H	L	L	H
	Transparent	H	L	L	H
LFSR	Shift	Clock		L	L
	Hold	H	L	H	L
	Reset	H	L	L	H
	Transparent	H	L	L	H

Fig. 7.9 Clocking schemes for LFSR and scan path.

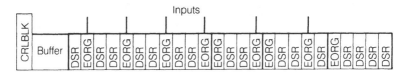

Fig. 7.10 SAREG.

there will be an extra inverter pair delay in any signal path. The buffer is used so that long scan paths do not degrade the control signals.

Figure 7.8 illustrates the method of constructing a scan path from the blocks discussed. Figure 7.10 shows the layout of a 16-bit 6-input multiple-input signature analysis register; the Exclusive-OR gates are used to produce the feedback paths and as the inputs for the signature-forming bits. A pseudo-random bit-sequence generator can also be formed in this way; however, the input Exclusive-OR gates are not required. The feedback points for the PRBS and SAREG were obtained from the maximal length polynomials discussed by Peterson and Weldon [13]. A BILBO register can be formed using the signature analysis type of structure, except that two extra control bits will be required to control the functionality of the compound registers.

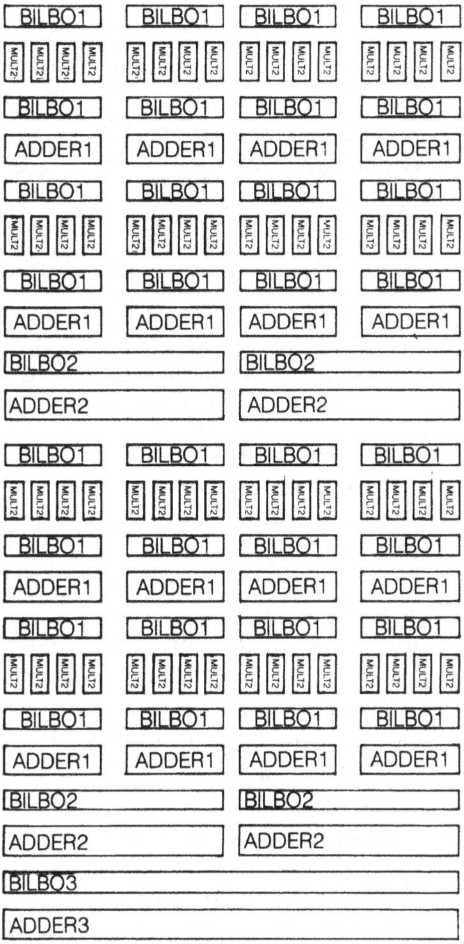

Fig. 7.11 Breakdown of BILBO sequences.

To obtain a comparison between these techniques a typical design was chosen, in this case an $N \times N$-bit parallel multiplier structure [14]. The multiplier is constructed from two basic building blocks, a 2×2-bit multiplier and a full adder. This simple arrangement allows the extension of the multiplier from its minimum 4×4-bit configuration to a 32×32-bit configuration. As the multiplier is built up recursively, 'natural' breaks occur in its structure and allow for easy partitioning of the design. Figure 7.11 schematically shows how this was achieved for the BILBO technique using the 32×32-bit multiplier. The block labelled MULT2 is the 2×2-bit multiplier and the block labelled ADDER ($n = 1, 2, 3$) is the other basic

Table 7.1 Sizes of the various test structures

Multiplier	Size (λ^2)	Scan path (%)	BILBO (%)	SASP (%)
4 × 4	420 000	12	39	51.5
8 × 8	1 963 300	15	38.9	24.7
16 × 16	8 433 500	16	38.5	24
32 × 32	34 911 100	17	38.2	23.9
Extra pins required		6	6	11

4 pins are required by the control lines
2 pins are required for input/output of data
5 pins are required by the SASP technique to control the multiplexers etc.

building block of the multiplier. The scan paths were placed at the same points as the BILBOs, for both the SASP and the normal scan path technique.

Four stages in the hierarchical development of the multiplier were chosen as test vehicles; these were the 4 × 4-, 8 × 8-, 16 × 16- and the 32 × 32-bit designs. The area of the multiplier circuit at each of these points was then calculated, and Table 7.1 gives a breakdown of these figures. The sizes are given in lambdas as used by Mead and Conway [12] in this case 1 λ = 3 μm. The area required by each of the test strategies was then calculated. This was done by totalling the number of latches required in each case and then for the BILBO technique the number of Exclusive-OR gates required was also calculated. For the SASP technique the maximum number of registers which are allowed on any one SAREG is 16, therefore every 16 scan path blocks will require a new SAREG block, in a worst-case scenario. Using this information and the sizes of the latches, exclusive-OR gates etc. already designed, an estimate of the area for the three techniques was produced for each of the four multiplier configurations. The area of the test structures was then calculated as a percentage increase above the total area of the multiplier. The results in Table 7.1 are shown graphically in Fig. 7.12. It can be seen that the scan path shows the least area requirement, as might be expected. The SASP technique improves its performance as the size of the devices on which it is implemented increases; this is to be expected since the design is virtually a scan path implementation, with extra area for the PRBS and SAREG. The BILBO shows the worst area usage probably because BILBO registers are inherently large, as they can perform scan path and LFSR functions. The selection circuits for SASP were not considered, because of their small size.

The next parameter of interest is the length of time required to perform the actual tests. For the scan path a set of deterministic patterns was produced which would give approximately 100% fault coverage. The BILBO and SASP techniques both use random patterns to test the device. As the multiplier was

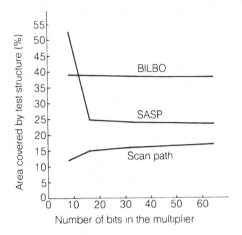

Fig. 7.12 Size of multiplier test structures.

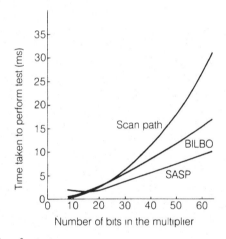

Fig. 7.13 Time taken for tests.

made from non-concurrent PLAs, which are difficult to test using random patterns, 255 random patterns were used in the BILBO and SASP techniques to test the devices, the BILBO and scan paths for SASP were built up using 8-bit registers since this is the number of input bits required by the minimum multiplier configuration. The 255 pattern test represents an exhaustive test for the lower 4×4-bit multiplier blocks which make up the majority of the design; therefore the fault coverage should be comparable with that of the deterministic scan path patterns. The graph in Fig. 7.13 shows the times taken to perform the tests in milliseconds for the various test methods. The scan path method shows an exponentially increasing test time; this is due to the

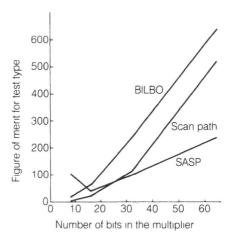

Fig. 7.14 Figure of merit for tests.

length of the scan path increasing in a similar fashion as the number of bits in the multiplier design increases. The BILBO technique shows an almost linear time increase since the patterns are generated internally without any need to scan in/out patterns, except for the signatures at the end of the test. The SASP technique shows quite a large test time initially; however, the ability to scan in/out in parallel greatly reduces the test time as the designs become larger.

To compare the three designs a third value, the *figure of merit*, was calculated from the product of the testing time and the percentage area increase. The figures of merit for BILBO, scan path and SASP are shown plotted in Fig. 7.14 (the smaller the figure of merit the better the test performance). It is seen that the SASP technique exhibits an improved performance characteristic in terms of the area-testing time trade-off over scan path or BILBO techniques, particularly as the circuit complexity increases.

7.5 CONCLUSIONS

This chapter has presented several hybrid designs for testability techniques and shown that they can be viable and worthwhile alternatives to the pure scan path and BILBO design philosophies.

7.6 REFERENCES

1. Koenemann, B., Mucha, J. and Zwiehoff, G. (1979) Built-in logic block observer. *Digest of Papers, 1979 Test Conference, IEEE*, October, 37–41.

2. El-Ziq, Y. M. (1983) S³: VLSI self testing using signature analysis and scan path techniques. *ICCAD '83*, September, 73–6.
3. Komonytsky, D. (1982) LSI self test using level sensitive scan design and signature analysis. *Digest of Papers, 1982 Test Conference, IEEE*, 414–24.
4. Beucler, F. P. and Manner, M. J. (1984) HILDO: The Highly Integrated Logic Device Observer. *VLSI Design*, June, 88–96.
5. Le Blanc, J. J. (1984) LOCST: A built-in self test technique. *IEEE Design and Test*, November, 45–52.
6. Sayers, I. L., Kinniment, D. J. and Russell, G. (1985) New directions in the design for testability of VLSI Circuits. *Proc. ISCAS 1985*, 1547–50.
7. Eichelberger, E. B. and Williams, T. W. (1977) A logic design structure for LSI testability. *Proc. 14th Design Automation Conference*, 462–8.
8. Resnick, D. R. (1983) Testability and maintainability with a new 6K gate array. *VLSI Design*, March/April, 34–8.
9. Daehn, W. and Mucha, J. (1981) A hardware approach to self testing of large programmable logic arrays. *IEEE Trans. Computers*, **C-30**, 829–33.
10. Russell, G., Kinniment, D. J., Chester, E. G. and McLauchlan, M. R. (1985) *CAD for VLSI*, Van Nostrand Reinhold.
11. VLSI design tools (1983) Dept. Electrical and Electronic Eng., Newcastle University.
12. Mead, C. and Conway, L. (1980) *Introduction to VLSI Systems*, Addison-Wesley.
13. Peterson, W. W. and Weldon, E. J. (1972) *Error Correcting Codes*, 2nd edn, MIT Press.
14. Yung, H. C. and Allen, C. R. (1984) Part I – VLSI implementation of a hierarchical multiplier. *Proc. IEE*, Pt G., **131**(4), 56–60.

8
HARDWARE FAULT TOLERANCE USING REDUNDANCY

8.1 INTRODUCTION

The reliability of a computing system has always been an important consideration from the earliest days of electronic calculations. Initially this concern arose due to the high failure rates of the components used in the first computers, which prevented even the simplest calculations from being completed without interruption. However, today, reliability is an important issue since so many critical applications involve the use of computers or digital circuits. These applications can range from flight electronics on passenger airliners to non-maintainable control computers in unmanned space satellites. Consequently the pursuit of more reliable digital systems will consume a large proportion of chip and computer design effort in the future.

To increase the reliability of a computer system there are two entirely different approaches [1]: fault intolerance (fault avoidance) or fault tolerance. Fault intolerance methods attempt to achieve high reliability by the use of careful design rules and the use of already highly reliable components. However, even the use of such techniques will not stop faults occurring; hence the system will eventually fail, although the rate of failure may be acceptably low. Fault tolerance, on the other hand, expects faults to occur within the logic. To counteract these faults and thus maintain a viable system redundancy is employed, that is additional facilities are included in the system which allow the system to continue to produce correct results even in the presence of faults. These additional facilities include extra information bits with the data, extra hardware or extra time to perform the operation more than once. Although a fault-tolerant system may cost more, in terms of hardware requirements etc., than a non-fault-tolerant system this difference in cost is diminishing, consequently the use of fault tolerance in even the simplest computer system could become the norm rather than the exception. The first part of this chapter will concentrate on the redundancy techniques that have been used to build fault-tolerant circuits.

One possible method of introducing fault tolerance is to use information redundancy, that is to add extra bits to the data used in the system so that the integrity of the system can be checked. This is achieved through the use of coding techniques in which the data is encoded into a codeword in such a way that any of the errors expected in the system will cause a codeword to become a non-codeword, therefore the error can be detected. However, the usual method of detecting an error is to use a checking circuit to continuously check for valid codewords. This checker circuit is not immune from faults; consequently, the question arises of 'who checks the checker?' To overcome this problem an alternative approach using Totally Self-Checking (TSC) circuits has been developed for certain types of codes. The TSC checkers are capable of not only detecting errors in the code bits applied to their inputs but also faults that may occur internally; therefore faults in the whole system can be detected. In Section 8.3 these circuits will be discussed along with the design of TSC checkers for many of the codes that can be used in fault-tolerant designs.

8.2 REDUNDANCY TECHNIQUES

In order to detect or mask faults in a system during operation it is necessary to use some form of on-line error detection technique. Faults within a system usually manifest themselves as errors in the outputs of the system. Several techniques are available to perform the on-line detection of errors. These include hardware redundancy, such as Triple Modular Redundancy (TMR) [2], information redundancy, such as Single Error Correcting/Double Error Detecting (SEC/DED) Hamming codes [3] and time redundancy where an operation may be repeated several times on the same piece of hardware to detect faults. All three techniques can be used to mask errors within the system by producing correct outputs even if a fault exists internally in the circuit. The three methods will be discussed in detail, with extensions to the basic techniques also presented.

8.2.1 Hardware redundant techniques

The simplest, conceptually, form of hardware redundancy, Triple Modular Redundancy (TMR) which was originally proposed by von Neumann [4] in 1956, is illustrated in Fig. 8.1. The logic circuits labelled L are triplicated and fed into a majority voting element, M. The logic circuits can range in complexity from a simple gate to a microprocessor. The voting element generates the majority function $M_v = XY + XZ + YZ$. Therefore if one circuit is faulty the other two will override the error and mask the fault, but a second circuit failing in a similar way could produce an incorrect result. The reliability of the

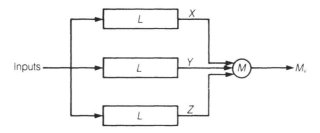

Fig. 8.1 A simple triple modular redundant system.

TMR circuit in comparison to the single circuit is very easy to calculate. If the reliability of a single circuit is given by R_L, i.e. R_L is the probability that the circuit will still be operating after a time t, then the overall reliability of the TMR circuit is given by the following equation:

$$R_{TMR} = R_L^3 + 3R_L^2 (1 - R_L)$$
$$= \text{probability of three systems surviving until time } t$$
$$+ \text{probability of two systems surviving until time } t$$
$$= 3R_L^2 - 2R_L^3$$

Obviously this does not include the reliability of the voting circuit. However, this value would only be a constant multiplier and since the voting circuit is usually less complex than the triplicated circuits its reliability will be much greater. The value of R_L is an exponential function with time, i.e. $e^{-\lambda t}$, where λ is the failure rate of the device. If the values of R_{TMR} and R_L are plotted then the graph shown in Fig. 8.2 is produced. As can be seen from this graph the reliability of the TMR system is greater than that of the simplex system up to a time T, at which point the simplex system becomes more reliable. However, for times less than T the TMR system is more likely to

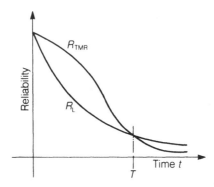

Fig. 8.2 Graph of the functions R_L and R_{TMR}

survive, therefore in situations where a system must be highly reliable for short periods of time, e.g. flight computers, then the TMR system is more suitable. The time T is normally selected so that it is very much greater than the actual mission time. One interesting point to note is that if a simplex system is inherently unreliable then placing it in a TMR configuration will not increase its reliability; for example, if $R_L = 0.5$ then $R_{TMR} = 0.5$.

The reliability equation derived for the TMR System is very pessimistic since compensating failures may occur. That is, two circuits could fail but they may not produce faulty outputs for the same input patterns. Thus any input pattern may only activate one faulty circuit. Alternatively the two circuits may have failed in such a way that the outputs produce inverse values; thus the correctly operating device will have the casting vote, so the correct values will always be generated. To achieve a better prediction for the reliability of a TMR System, York et al. [5] have introduced extra terms into the reliability equation which include the probability that there will be compensating failures. York et al. also provide an algorithm to calculate these probabilities for combinational circuits containing only basic logic gates. The algorithm assumes that the circuit in question will only have single stuck-at faults and does not contain any reconvergent fanouts. If multiple faults and re-convergent fanouts are present then the algorithm must be modified. The technique used simply calculates the probabilities that any branch in a circuit will compensate another fault in the circuit. The calculations are based on fault dominance and equivalencing in the basic gate functions. For every circuit these calculations will produce different results. However, it is possible to calculate an upper and lower bound on the probability of a compensating failure. Therefore extra terms must be added to the reliability equation which include the probability of compensating failures for two and three modules failing.

In a TMR system if one system should fail then the overall reliability of the system will be less than that of the single system. Therefore schemes have been proposed where the TMR system is operated in simplex mode once a single failure has occurred [6].

The reliability of the voter circuit was ignored in calculating the reliability of the whole system, since it was assumed to have negligible effect on the overall reliability, nevertheless if the voter circuit does fail then the whole system will fail. To overcome this problem the Triplicated TMR [2] scheme, illustrated in Fig. 8.3, was introduced. If the reliability of the voters is less than 1 then this scheme is better; however, the initial TMR is preferable if the reliability of the voter is very high since fewer pieces of hardware are necessary: this could ultimately have an effect on the yield of a device in a VLSI environment.

One further problem associated with TMR schemes is to synchronize the three circuits so that they produce the results of their computations at the same time for comparison. Although this can be easily achieved by using a single clock applied to all three systems it can cause problems if the clock fails since the complete system will then fail. To overcome this difficulty the clocking circuits must also be made fault tolerant. Lewis [7] has achieved this by using standby sparing. In this scheme two clock circuits are run together; the 'primary oscillator' and the 'secondary oscillator'. If a fault is detected in the primary oscillator by two of the driven circuits then a switchover to the secondary oscillator takes place. However, because this switchover will take a finite amount of time some of the faulty clock signals could reach the circuits with disastrous results. Therefore the primary clock is fed through a delay line before reaching the circuits. Thus on switchover there is sufficient time for the

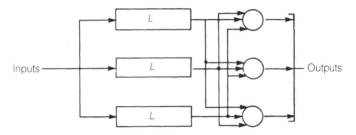

Fig. 8.3 A triplicated TMR system.

new clock signal to take over. The clock signal itself is monitored for the normal stuck-at condition, also the clock is checked for out-of-tolerance variations in the period, duty cycle and amplitude. Any variations in these factors may cause internal timing problems to occur in the clocked circuits, so measurement of these parameters is quite important. Smith [8] attempts to keep these values in tolerance by using a slightly different approach. In this scheme independent oscillators each send their signal to a voting element where the majority clock signal is generated. The clock signal produced is then compared with the input signals and the phase error is sent back to the oscillator to control its frequency. Quad redundancy is used in this scheme – that is, the clock system is replicated four times – therefore single failures can be tolerated.

The TMR concept is the most basic form of fault masking; however, it is possible to extend the scheme to N-Modular Redundancy (NMR) where N identical circuits are used to generate the output from the circuit. Again if the voting elements were perfect then this scheme could tolerate $n = (N + 1)/2$ individual circuit failures before it failed completely. A scheme where the modules never change as they fail is called static redundancy; as can be noted from Fig. 8.2, as the device ages then the reliability will decrease. In the dynamic redundancy scheme the failed modules are replaced by new modules as they fail. Before switching out a suspect module a retry of the last operation may be performed to determine if the fault was transient or permanent. This could save switching out a module that was not really faulty. In some systems the spare modules may be left unpowered possibly because their reliability is higher in that condition or alternatively it may be a method of saving power in a satellite for instance. In some systems the spare units may perform exactly the same functions as the operational units so that they can be switched in quickly; this is especially true for computer systems. Some reliable computer designs run in duplex mode. In this mode one computer is used to perform the tasks while the standby spare performs exactly the same operations and also checks the outputs of the working computer. If an error is detected and it is determined to be the working computer that is faulty then the spare system can be switched in to take over the computations. Several systems have been designed to operate in this mode, for example the Bell Electronic Switching System (ESS) [9].

A scheme which attempts to combine the attributes of both static and dynamic redundancy is hybrid redundancy. A hybrid redundant system uses a normal NMR or TMR system 'with spare units' at the heart of the design. However in this design if an error occurs in one of the modules then the outputs are generated correctly, a signal is also generated that points to the circuit which produced the faulty output. This signal is then used to reconfigure the system by switching in a circuit which has not as yet been connected to the voting circuits. As the spare units are used up the system will eventually resort to a TMR configuration in an NMR system. A hybrid redundancy scheme based on this system of switching in spares has been proposed by Siewiorek and McCluskey [10]. This scheme uses TMR to form the core of the system with two modules as spares to be switched in when one of the modules in the core fails. An iterative array is used to select the next fault-free module when a module fails. A problem could arise if two modules become faulty in the TMR core, which would indicate that the good module was faulty and cause it to be switched out. Therefore, providing the spares were fault free they would also be exhausted as none of their outputs would produce a valid comparison. The probability of two systems failing in the same way is, fortunately, very low. However, if the switching circuit fails then it may be impossible to switch in a good module, so the number of possible spares is reduced. Alternatively, it may not be possible to switch out a faulty circuit, so the system will operate in duplex mode and is really only capable of detecting errors.

8.2.2 Information redundancy

Error detecting/correcting codes are another means of improving the reliability of a system. Coding techniques however increase the information needed to represent data; consequently they are referred to as information redundant techniques. There are many different types of code available; however, they all essentially perform the same basic operation of mapping the normal output vector space of a device onto an extended code space, thereby allowing errors in the data to be detected and corrected. Probably the best known codes are the Hamming-type codes [3], therefore the use of these codes in the case of binary numbers will be used to illustrate information redundant techniques.

First a simple parity code will be discussed. For this code one extra bit, known as the check bit, is added to the data to be checked. For the odd (even) parity code the check bit is defined so that the total number of 1 bits occurring in the data is odd (even). For example if $I = 0110$ the information bits, then in this case for even parity the check bit $P_e = 0$ and for odd parity the check bit $P_o = 1$. Any single bit error is easily detected since the parity of the data will change from odd (even) to even (odd). This ability to detect errors can be quantified by the Hamming distance, which is the minimum number of bits in which any two codewords differ. For the simple parity code the Hamming distance $d = 2$. Therefore it is capable of detecting single errors but not correcting these errors. If two errors occurred this scheme would not be able to detect the fault. In order to be able to mask faults, the requirement of any fault tolerant system, it is necessary to have a Hamming distance of at least 3. This Hamming distance can be achieved by adding further check bits to form a single error-detecting code. In general any code of distance $(d + 1)$ is capable of detecting d errors. If a code has a distance $(2c + 1)$ then it is capable of correcting c or less errors; hence if the distance is 3 then c equals 1. In order to determine the number of extra checkbits required the Hamming relationship can be used; this is given by

$$2^k \geq m + k + 1.$$

In this case $m =$ number of information bits and $k =$ the number of extra check bits required. Therefore if we wanted to produce a single error-correcting code which had 8 bits, then $k = 4$ ($m = 8$). There are two possible methods of representing this code: either each code word can be individually enumerated in a table or alternatively a parity check matrix can be formed. This matrix can be used to generate the code and gives a very compact definition of the code bits. The matrix is made up of r rows, called the rank of the matrix, and $(m + k)$ columns. For example the parity check matrix for the single bit even parity code ($r = 1$) would be a vector of all 1s, i.e. $H = [1 \ 1 \ 1 \ \ldots \ 1]$, containing $(m + 1)$ bits ($k = 1$ in this case). Further material on the parity check codes and a wide range of other codes can be obtained from Peterson and Weldon [11]. A code has a minimum distance of three if no two columns of the parity check matrix are linearly dependent; in the binary case this means that no column should be zero and no two columns should be equal to each other. The construction of a Hamming code for $m = 8$ will now be examined.

If $m = 8$ then $k = 4$, and there will be 12 bits in all. The parity check matrix shown in general form is then as follows:

$$H = \begin{array}{cccccccccccc} 1 & 2 & 3 & 4 & 5 & 6 & 7 & 8 & 9 & 10 & 11 & 12 \\ C_1 & C_2 & I_1 & C_3 & I_2 & I_3 & I_4 & C_4 & I_5 & I_6 & I_7 & I_8 \\ \begin{bmatrix} 1 & 0 & 1 & 0 & 1 & 0 & 1 & 0 & 1 & 0 & 1 & 0 \\ 0 & 1 & 1 & 0 & 0 & 1 & 1 & 0 & 0 & 1 & 1 & 0 \\ 0 & 0 & 0 & 1 & 1 & 1 & 1 & 0 & 0 & 0 & 0 & 1 \\ 0 & 0 & 0 & 0 & 0 & 0 & 0 & 1 & 1 & 1 & 1 & 1 \end{bmatrix} \end{array}$$

The check bits (C_1-C_4) occur at the bit positions which are a power of 2; the information bits (I_1-I_8) are simply the other bits. Each check bit is then related to other information bits by the following equations (where $\oplus$ is the Exclusive-OR operation).

$$C_1 = I_1 \oplus I_2 \oplus I_4 \oplus I_5 \oplus I_7$$
$$C_2 = I_1 \oplus I_3 \oplus I_4 \oplus I_6 \oplus I_7$$
$$C_3 = I_2 \oplus I_3 \oplus I_4 \oplus I_8$$
$$C_4 = I_5 \oplus I_6 \oplus I_7 \oplus I_8$$

i.e. the information bits chosen to form the check bits are simply those that have a 1 in the same row of the check matrix as the check bit. By using the scheme it is easy to generate a syndrome which indicates where the faulty bit is in the data. This is best illustrated by the following example case:

Let $I_1-I_8 = 0\ 1\ 1\ 0\ 1\ 0\ 1\ 0$; then $C_1 = 1, C_2 = 0, C_3 = 0, C_4 = 0$.

Therefore the complete word with parity is

$$V = 1\ 0\ 0\ 0\ 1\ 1\ 0\ 0\ 1\ 0\ 1\ 0.$$

Now if this data were generated by some circuit in a fault-tolerant system and information bit 6 was stuck at one, then the faulty code word would be

$$V' = 1\ 0\ 0\ 0\ 1\ 1\ 0\ 0\ 1\ \overset{*}{1}\ 1\ 0.$$

To check this received code word the error syndrome is generated this simply involves an Exclusive-OR operation between the received check bits and the regenerated check bits from the faulty data which in this case gives $(C_1^R - C_4^R$

are the check bits as received for checking):

$$E_1 = 0 = C_1 \oplus C_1^R$$
$$E_2 = 1 = C_2 \oplus C_2^R$$
$$E_3 = 0 = C_3 \oplus C_3^R$$
$$E_4 = 1 = C_4 \oplus C_4^R$$

If the received data was correct then the error syndrome E_1–E_4 would be zero, but in this faulty case it represents the bit that is in error which in the example case is bit 10 (1010) or information bit 6, so this bit simply needs to be inverted in order to be corrected. The generation of a non-zero syndrome could indicate that either a single or double (possibly multiple) error had occurred. As maximum likelihood decoding is used by the error-detecting code (that is the most likely information vector is chosen when an error occurs) then it is impossible to differentiate, using this scheme, between single and double errors. Therefore a double error would go undetected and the codeword would be wrongly corrected. In order to make a code that is capable of detecting such errors it is necessary to form the above code into a distance 4 code. This can be easily achieved by adding a row of all ones and an extra column to the parity check matrix, which is effectively the parity check over the whole data word:

$$C_0 = C_1 \oplus C_2 \oplus C_3 \oplus C_4 \oplus I_1 \oplus I_2 \oplus I_3 \oplus I_4 \oplus I_5 \oplus I_6 \oplus I_7 \oplus I_8;$$

therefore there will now be 5 check bits. Thus if a double error had occurred C_0 would equal zero but the error syndrome generated would not be equal to zero, hence a double error could be detected. In the case of a single error if it occurs within the normal data (i.e. not C_0) then C_0 will indicate an error and the error syndrome will give the position of the incorrect bit. If, however, C_0 is incorrect, then an error syndrome of zero will refer to the overall parity check bit C_0. If C_0 is correct and the error syndrome is zero then 'no error' has occurred. The use of a Hamming code in the design of a self-checking data path circuit is discussed in Chapter 9. The value of the syndrome can be represented as $V.H^T$, where H^T is the transpose of the parity check matrix. The general form of H as a systematic code is $H = [P_{m \times k}^T : I_k$ where I_k represents the identity matrix and $P_{m \times k}^T$ represents the transpose of the parity matrix (m and k are as previously defined).

As with the hardware redundancy method the whole operation of this scheme depends on the integrity of the checkers and syndrome generators associated with the code. If these are generating incorrect data, then the whole scheme will fail to be fault tolerant. The next section will discuss the production of check circuits that are designed to overcome this problem. As well as correcting errors when data is in storage or after transmission it is also possible to detect errors when data undergoes a transformation of some type [12]–[19]. The basic requirement behind all these techniques is illustrated in Fig. 8.4. In this diagram there is a module performing the function such that $M*N = R$. There is also a second independent block operating on the check bits in order to produce C_R, the check bits for the result R. The value of R can then be compared with C_R; if the comparison indicates that no errors have occurred then the computation is assumed to have been completed correctly; otherwise the computation has failed and the output must be corrected using

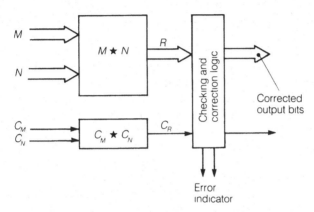

Fig. 8.4 Fault tolerance using a code scheme.

the coding scheme. However, there are two problems associated with this view of the computation:

1. Building the code checkers so that they themselves can be checked for correct operation. This style of design is usually called totally self-checking;
2. Finding a code that is useful for all types of operations, both arithmetic and logical. As early as 1959 Peterson and Rabin [14] pointed out that codes for checking non-trivial logical operations could not be used without complete duplication. However, in 1972 Monteiro and Rao [15] demonstrated the use of a residue code for checking both logical and arithmetic operations. A wider discussion of this problem can be found in Chapter 9. Also in 1972 Pradhan and Reddy [20] proposed the design of fault-tolerant logical and arithmetic processors using Reed–Muller codes. Since Reed–Muller codes can be applied to memory error detection as well, this gives a single method that is applicable to the whole range of computational elements.

8.2.3 Time redundancy

The previous redundancy schemes have relied upon extra hardware or space in which to perform their operations. That is, to implement these schemes extra circuits are required to perform the majority voting function or check and correct the codewords of the chosen code. This extra hardware causes an increase in the complexity of the device and possibly a decrease in the overall yield of the chips. To overcome this problem time redundancy has been proposed. This technique relies on the ability of the circuit to recompute the values on different parts of the hardware so that they may be compared and corrected if necessary. Laha and Patel [21] have used this scheme to perform error correction in an adder circuit. In this scheme the adder circuit is divided into three k-bit slices. Each bit slice then receives the operands split into k-bit sections. The addition operation is performed three times each time the operands are rotated by k bits, so that each adder will eventually add all three sections of the operand. Using this scheme it is possible to detect one faulty adder and perform a correction operation. The algorithm for detecting a faulty adder segment proceeds as follows. Initially the operands are added unshifted and the result stored, including the carry bits. The values to be added are then rotated k bits and the addition operation repeated. A comparison is made between the stored values and the recomputed values, and any differences are noted and stored. The operands are again rotated by k bits and the comparison operation repeated, noting any disagreeing bits. If no disagreements occur then the results produced in the first step can be used as the output of the adder. However, if disagreements occur then it is possible to determine which bit slice is faulty. With this information a series of re-computation steps can be performed so that the faulty element is eliminated and the result is generated using those elements of the adder which have been deemed to be functional. This technique can also be applied to multipliers that use partial product formation followed by the addition of these values. Faults other than the stuck-at faults can be considered by this method.

The adder example is quite a simple illustration of the time redundancy method. However, Choi and Malek [22] have used this technique to implement a fault-tolerant Fast Fourier Transform (FFT) processor. In this scheme the fault detection is achieved by using 'recomputation by an alternate path'. In order to achieve fault tolerance it is necessary to alter the structure of the FFT processor so that it can be made symmetrical. This involves the addition of switches to the computation elements (butterflies) so that input and output mappings can be altered. Also comparators are added to the outputs to compare the output results. To detect an error a duplicate computation is performed, delayed by one cycle and through an alternative path. From the information obtained at the output is is possible to determine which butterfly element is faulty and reorganize the computation accordingly. This scheme is capable of detecting both transient and permanent faults.

8.3 TOTALLY SELF-CHECKING CIRCUITS

The previous section has concentrated on designs which are said to be 'self-checking', that is the circuits have the ability to detect and mask errors during the course of their normal operations. An interesting subclass of these circuits is called 'Totally Self-Checking' (TSC) circuits. When this type of circuit is used to produce a code checker it has the capability of not only detecting errors in the code values but also of easily detecting errors that occur internally, thereby solving the problem of 'who checks the checker' and eliminating the 'hard core' fault problem. The TSC circuits have been widely covered in the literature [23]–[32]. The features which distinguish TSC circuits were originally defined by Anderson [33] in 1971 and later consolidated by Anderson and Metze [23] in 1973. The two main requisites of a TSC circuit are that it must be self-testing and fault secure. The following definitions are used to distinguish TSC circuits (extracted from [33]):

Definition 1

A circuit is self-testing for a set of faults F, if for every fault in F the circuit produces at least one non-code output for at least one code input.

Definition 2

A circuit is fault secure for a set of faults F, if for every fault in F the circuit never produces an incorrect code output from a code input.

Definition 3

A circuit is totally self-checking if it is both self-testing and fault secure.

The typical structure of a totally self-checking design is shown in Fig. 8.5. In any TSC design it is assumed that the input vector space to the circuit is actually a subset of all the possible inputs and similarly the output vector space is actually a subset of all possible outputs. This restriction usually means that the inputs and outputs are coded in some way. Hence if a non-code output is generated then it will be detected by the code checker and an error reported. In order to illustrate these points consider a circuit that has a code input space S, that is, any input $x \in S$. Also consider the output space to consist of two sets of outputs: Y, the valid codewords and Y' the invalid codewords ($Y \cup Y' =$ the complete output space). Now consider a set of faults F which can occur in the self-checking circuit. Let input $x \in S$ generate the code output $y \in Y$ in the absence of any faults in the self-checking circuit. If the input x now generates the output $y' \in Y$ in the presence of fault $f \in F$, then $y' = y$, this is the fault secure property. This means that if a codeword is produced it must be the

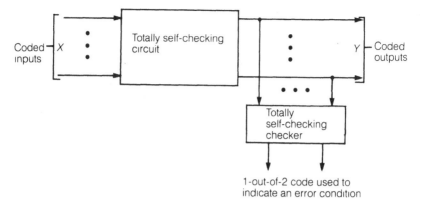

Fig. 8.5 Typical totally self-checking circuit.

correct codeword for input x. The self-testing property guarantees that for any fault $f \in F$ there is at least one input $x \in S$ which generates a non-codeword output $y' \in Y'$. Therefore this property effectively guarantees that the fault must be detected. Although this class of circuits has some very interesting properties it may be possible for one particular fault to be only detected by one input pattern. Consequently if this fault occurs, and another fault occurs before it is detected by the input pattern, the possibility exists that any tests for these faults may become invalid: this is the so-called *error latency problem*. This highlights another area of difficulty with TSC circuits in that it must be possible to apply all the fault-detecting patterns to a circuit to make sure that a fault does not go undetected when it first occurs.

The design of TSC circuits is not a very straightforward task. However, to assist in the design of TSC systems Smith and Lam [25] have proposed a method of verifying that a system is indeed TSC when it is built up from TSC blocks. The method proposed involves representing the circuit as an interconnection graph. Each arc in the graph is then used to represent the properties of the block of TSC logic through which it passes. Three values are represented on the arc: the time required to produce a result; whether or not the block will transmit non-code inputs; and the length of time required to transmit the non-code input. From this information it is possible to determine if a circuit will be TSC, therefore it is also possible to alter the system so that it has the required properties. Since the graph may become quite difficult to deal with for large systems it is possible to verify that the system is TSC hierarchically. This technique can also be used to assist in the optimal placement of checkers within the system. This can be very useful for making a circuit achieve the TSC goal. It can also assist in reducing the checker overhead by only adding a minimum number of checkers. It is also possible to use this system to determine where contaminated data may have reached when an error is indicated. This feature is useful if the situation must be recovered before computation can be continued. Another useful attribute of this system of representing a TSC is its ability to assist in the isolation of faulty circuits within the complete system. A systematic approach to the design of TSC circuits has also been suggested by Smith and Metze [31].

In general, however, because of the difficulty of building TSC systems a large majority of the design effort directed towards TSC circuits has been employed in the production of TSC checkers for various codes. The checker is designed to map non-codeword inputs into non-codeword outputs and code-word inputs onto codeword outputs; this type of circuit is said to be *code disjoint*. In order to represent the error information the checker usually generates a two-rail code at its outputs. The values of 01 and 10 are used to represent a valid output (therefore a valid input) and 11 and 00 represent an invalid output (therefore an invalid input). A TSC checker for two-rail codes is shown in Fig. 9.4. To obtain an arbitrarily sized checker for two-rail codes is simply a matter of cascading the checkers in a tree structure as shown in Fig. 8.10, the mod b adders being replaced by the two-rail encoder circuit. So far TSC checkers have been designed for m-out-of-n codes, 1-out-of-n codes, Berger codes, parity codes and arithmetic codes. Since the TSC checkers form an important class of circuits the next few sections will take a close look at their design.

In order to perform comparisons between self-checking circuits of the same type it is necessary to have some form of quantitative measure of their self-checking ability; Lu and McCluskey [34] have proposed just such a technique. The method used is based on measuring the self-testing and fault secure properties of a circuit. This is achieved by defining two quantitative measures, the Testing Input Fraction (TIF) set and the Secure Input Fraction (SIF) set. For each fault in the circuit the TIF value is the fraction of the number of code inputs that produce non-code outputs, i.e. detect the fault, and the SIF value is the fraction of code inputs that do not produce non-code outputs in the presence of that fault. Currently only stuck-at fault conditions are considered in order to restrict the fault set to manageable proportions. Once each value of TIF and SIF have been evaluated for every fault in the fault set a weighted geometric mean is produced of the TIF and SIF values to give an overall indication of circuit properties, the weight being determined by the probability of the fault. Geometric means are used instead of arith-metic means so that zero values of TIF are taken into account. In fact the geometric means are subdivided into three categories, depending on whether TIF > 0 and SIF > 0 or TIF = 0 and SIF > 0 or TIF = 0 and SIF = 0. From these measurements it is possible to evaluate the performance of a self-checking circuit. The values of SIF and TIF can be calculated either from a simulator or by analytical means, depending on the complexity of the circuit. To demonstrate the technique Lu and McCluskey used different types of self-checking Linear Feedback Shift Registers (LFSR). From the values of SIF and TIF calculated in each of the three groups for the three different LFSRs it is possible to determine which register would have the optimum performance in terms of circuit complexity and self-checking properties. Obviously this is extremely useful in evaluating new self-checking logic circuits. In the case of a totally self-checking circuit TIF > 0 and SIF = 1 for the case of a stuck-at fault set.

So far the description of totally self-checking circuits has concentrated on purely combinational designs; however, these techniques are equally applicable to sequential circuits [35]. The sequential circuit is self-testing if for every fault from an assumed fault set a non-code-space output is produced for an input received during the normal operation of the device. Similarly the fault secure property of a sequential machine is defined so that the circuit never produces an incorrect code space output for any normal input; again this is over the prescribed fault set.

Many different techniques have been proposed for the design of self-checking sequential machines. Diaz *et al.* [36] presented a scheme for designing Moore-type sequential machines. In this case an m-out-of-n code is used to represent the state information and an m-out-of-$2m$ code is used to represent the inputs and outputs since they are assumed to be associated with two wires. The problem is to make the circuit self-testing for faults that occur within the structure. Ozguner [37] has proposed a similar technique for Mealy-type asynchronous and synchronous sequential machines, using 1-out-of-n or $(m-1)$-out-of-m codes to represent state information and inputs. Viaud and David [38] suggested the design of Sequentially Self-Checking circuits (SSC). These circuits in fact form a superset of the totally self-checking machines and the strongly fault secure machines.

8.3.1 Totally self-checking checker design

Although there has not been a great deal of work carried out on the design of general TSC combinational circuits the same is not true for checker circuit design. The checker circuits are generally used to detect errors in coded data values; consequently it is imperative that they are designed so that they are tested by the normal code inputs and can detect errors in their own operations. These properties are all covered by the TSC circuits. Therefore this section will look at some of the techniques that are available to design TSC checkers for the most popular codes.

(a) M-*out-of*-N *codes*

The m-out-of-n codes contain n information bits of which m must be ones and the rest are zeros. Therefore a checker for this code must be able to detect m

ones. The m-out-of-n codes are useful because they can be used to detect unidirectional errors, that is faults within a circuit can cause ones to be changed into zeros or vice versa, but not both at the same time. For a fixed value of n an m-out-of-n code has the maximum number of codewords when $m = \lceil n/2 \rceil$. One of the first attempts to explore the design of TSC m-out-of-n checkers was by Anderson and Metze [23] in 1973.

The scheme proposed by Anderson and Metze concentrated on the design of checkers for k-out-of-$2k$ codes using AND–OR logic. The use of this logic is preferred since it makes possible the detection of unidirectional errors within the checker itself, whereas with NAND and NOR logic it is possible to detect only single faults because of the inversion produced by these gates. The design procedure uses the majority function $T(k \geq i)$ which has the value 1 when the number of ones, k, in the codeword is greater than or equal to the value of i. The checker circuit is constructed from a realization of the majority function generating a two-rail code at the output. The method of constructing an m-out-of-n code checker is very straightforward. However, it is only for the k-out-of-$2k$ case that a totally self-checking checker can be easily designed, so it is necessary to produce this checker before that for the m-out-of-n code. The design of such a checker proceeds as follows. The k-out-of-$2k$ code is split into two groups, $A(X_1, \ldots, X_k)$ and $B(X_{k+1}, \ldots, X_{2k})$, both containing k inputs; let k_a and k_b be the number of ones that occur in A and B respectively. If the checker has two outputs E_1 and E_2, then the sum-of-products form of these outputs can be specified using majority functions as follows:

$$E_1 = \sum_{i=0}^{k} T(k_a \geq i) . T(k_b \geq k - i) \qquad (i \text{ odd})$$

$$E_2 = \sum_{i=0}^{k} T(k_a \geq i) . T(k_b \geq k - i) \qquad (i \text{ even})$$

This design method produces a circuit that is totally self-checking. To illustrate the method, a 2-out-of-4 code checker will be designed. In this case $k = 2$, so there are four inputs $\{ X_1, X_2, X_3, X_4 \}$ and the two outputs E_1 and E_2 are defined as follows:

$$E_1 = T(K_a \geq 1) . T(K_b \geq 1).$$

The majority function $T(k_a \geq 1)$ is true for $(X_1 + X_2)$. Similarly $T(k_b \geq 1)$ is true for $(X_3 + X_4)$; therefore

$$E_1 = (X_1 + X_2) . (X_3 + X_4).$$

In the case of E_2

$$E_2 = T(K_a \geq 0).T(K_b \geq 2) + T(K_a \geq 2).T(K_b \geq 0);$$

using the same principle as before

$$E_2 = 1.(X_3.X_4) + (X_1.X_2).1$$
$$= (X_3.X_4) + (X_1.X_2)$$

The circuit diagram for this function is shown in Fig. 8.6. The test set for the k-out-of-$2k$ checker consists of only 2^k codewords. Therefore for the 2-out-of-4 checker only 4 patterns are required to test it completely.

To build a checker for m-out-of-n codes it is necessary to use a code disjoint translator. The translator takes the m-out-of-n code as input and produces a k-out-of-$2k$ code as output which can then be checked by a circuit produced from the above design method. One of the most frequently used m-out-of-n codes is when $m = 1$, i.e. 1-out-of-n codes. To check this code it must first be converted into a k-out-of-$2k$ code by a single level of OR gates. Of course the translator must itself be TSC and code disjoint. To produce the translator to perform the check on an m-out-of-n code it is necessary to convert it into a 1-out-of-C_m^n code first. The method used for checking the 1-out-of-n code can be used to convert the code into a k-out-of-$2k$ code. The only codes which cannot be checked using this method are the case when $n = 3$ and $n = 7$, with $m = 1$.

The design of checkers in the above case is limited to k-out-of-$2k$ codes with m-out-of-n codes simply being an extension of the method. However Marouf and Friedman [24] present four procedures for the following types of codes:

(i) $(2m + 2) \leq n \leq 4m$
(ii) $n = 2m + 1$
(iii) $n > 4m$
(iv) $n < 2m$.

The method used to generate a $(2m + 2) \leq n \leq 4m$ code will be described to illustrate the approach taken by Marouf and Friedman. The first step is to partition the inputs into two sets A and B so that the number of bits in A is $N_a = \lfloor n/2 \rfloor$ and B is $N_b = \lceil n/2 \rceil$. Two functions E_1 and E_2 are then defined as used by Anderson and Metze:

$$E_1 = \sum_{i=1}^{m-1} T(k_a \geq i).T(k_b \geq m - i) \qquad (i \text{ odd})$$

$$E_2 = \sum_{i=1}^{m-1} T(k_a \geq i).T(k_b \geq m - i) \qquad (i \text{ even}).$$

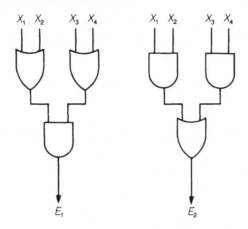

Fig. 8.6 A 2-out-of-4 checker circuit.

The two sets A and B are then each subdivided into further subsets A_1, A_2 and B_1, B_2 so that $N_{x_1} = \lceil N_x/2 \rceil$ and $N_{x_2} = N_x - N_{x_1} = \lfloor N_x/2 \rfloor$ (where $x = a$ or b). This then gives four further functions

A-case:

$$E_3 = \sum_{i=m-n_{a_2}}^{n_{a_1}} T(k_{a_1} \geq i).\,T(k_{a_2} \geq m-i) \qquad (i \text{ odd})$$

$$E_4 = \sum_{i=m-n_{a_2}}^{n_{a_1}} T(k_{a_1} \geq i).\,T(k_{a_2} \geq m-i) \qquad (i \text{ even})$$

B-case:

$$E_5 = \sum_{i=m-n_{b_2}}^{n_{b_1}} T(k_{b_1} \geq i).\,T(k_{b_2} \geq m-i) \qquad (i \text{ odd})$$

$$E_6 = \sum_{i=m-n_{b_2}}^{n_{b_1}} T(k_{b_1} \geq i).\,T(k_{b_2} \geq m-i) \qquad (i \text{ even}).$$

These functions are all generated separately using majority detection circuits which consist of AND–OR gates. A translator is then used to take this 1-out-of-6 code and produce a 2-out-of-4 code which can be checked by an appropriate checker circuit (see Fig. 8.6). The circuit produced is a totally self-checking checker for the selected code. To illustrate this technique the design of a 2-out-of-8 ($m = 2$) code circuit will be demonstrated.

The set of inputs for the code is

$$\{X_1, X_2, X_3, X_4, X_5, X_6, X_7, X_8\}.$$

Therefore when partitioned this gives the two sets

$$A = \{X_1, X_2, X_3, X_4\}, \qquad B = \{X_5, X_6, X_7, X_8\}$$

$$n_a = n_b = 4.$$

The majority functions E_1 and E_2 are as follows:

$$E_1 = T(k_a \geq 1) . T(k_b \geq 1)$$
$$= (X_1 + X_2 + X_3 + X_4) . (X_5 + X_6 + X_7 + X_8)$$
$$E_2 = 0 \text{ since there are no even terms.}$$

The subdivision of A and B produces

$$A_1 = \{X_1, X_2\}, \quad A_2 = \{X_3, X_4\}, \quad B_1 = \{X_5, X_6\}, \quad B_2 = \{X_7, X_8\}$$
$$n_{a_1} = n_{a_2} = n_{b_1} = n_{b_2} = 2.$$

Therefore the majority functions for E_1, E_2, E_3 and E_4 are

$$E_3 = T(k_{a_1} \geq 1) . T(k_{a_2} \geq 1)$$
$$= (X_1 + X_2) . (X_3 + X_4)$$

$$E_4 = T(k_{a_1} \geq 0) . T(k_{a_2} \geq 2) + T(k_{a_1} \geq 2) . T(k_{a_2} \geq 0)$$
$$= (X_3 . X_4) + (X_1 . X_2)$$

$$E_5 = T(k_{b_1} \geq 1) . T(k_{b_2} \geq 1)$$
$$= (X_5 + X_6) . (X_7 + X_8)$$

$$E_6 = T(k_{b_1} \geq 0) . T(k_{b_2} \geq 2) + T(k_{b_1} \geq 2) . T(k_{b_2} \geq 0)$$
$$= (X_7 . X_8) + (X_5 . X_6)$$

As can be seen from this derivation when $m = 2$ this translator will produce $E_2 = 0$; therefore a 1-out-of-5 code is produced which must be translated into the 2-out-of-4 code. All single and unidirectional faults in the checker can be detected if this design method is employed. The faults can be detected with T codewords, where

$$
T = \sum_{i=1}^{m-1} \max[C_i^{na}, C_{m-i}^{nb}] + \sum_{i=m-n_{a_2}}^{n_{a_1}} \max[C_i^{na_1}, C_{m-i}^{na_2}]
$$

$$
+ \sum_{i=m-n_{b_2}}^{n_{b_1}} \max[C_i^{nb}, C_{m-i}^{nb_2}].
$$

Therefore for the case of the 2-out-of-8 code shown

$$
T = \sum_{1}^{1} \max[C_i^4, C_{m-i}^4] + 2 \sum_{0}^{2} \max[C_i^2, C_{m-i}^2]
$$

$$
= C_1^4 + 2(C_0^2 + C_1^2 + C_0^2)
$$

$$
= 4 + 2(1 + 2 + 1)
$$

$$
= 12 \text{ patterns.}
$$

Figure 8.7 shows a circuit diagram for this particular checker. In general the checkers produced for m-out-of-n codes by this scheme are more efficient than those generated by the previous method. The scheme for producing $n = 2m + 1$ checkers is very similar except that only the B partition is subdivided further. This will give only 4 majority functions. Thus a 1-out-of-4 code will be produced which must be converted into the 2-out-of-4 code for checking. The method used for $n > 4m$ involves subdividing the inputs initially as before. Then a recursive algorithm is used to produce the final result.

The cases for $n = 2m$ and $m = 1$ are not considered by these algorithms. This algorithm can produce designs which are 70–97% less complex than the method proposed by Anderson and Metze.

Reddy [29] has considered the design of checkers for k-out-of-$(2k + 1)$ and $(k + 1)$-out-of-$(2k + 1)$ codes. Also the design of multilevel k-out-of-$2k$ checkers $(k > 1)$, which only require $2k$ test patterns to verify the hardware is working correctly, is considered. With this technique the design of a 3-out-of-6 checker is produced which can be used to check a 1-out-of-7 code. Again the philosophy behind checking k-out-of-$(2k + 1)$ codes is to convert them to 1-out-of-C_k^{2k+1} codes and then into a 1-out-of-2 code for checking. Unidirectional and stuck-at faults can then be detected within the checker.

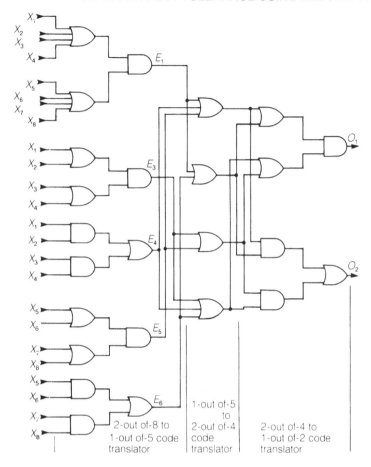

Fig. 8.7 A TSC checker for a 2-out-of-8 code.

All of the above checking schemes are unable to produce TSC checkers for the 1-out-of-3 code using AND or OR gates, i.e. a combinational circuit. However, David [39] suggests a method of producing such a checker for this code by using a sequential circuit which takes as its input the 1-out-of-3 code and produces a 1-out-of-4 code which can be checked by a TSC checker. The sequential circuit is only used on one input bit and generates two bits in the 1-out-of-4 code. This circuit is totally self-checking for all single stuck-at faults. Another scheme by Golan [40] converts the 1-out-of-3 code into an $(m + 1)$-out-of-$(n + 3)$ code by combining it with an m-out-of-n code. The new code generated is then translated into a 1-out-of-p code for which a checker can be designed. To date no TSC combinational checker circuit has been designed to check this code directly.

The 1-out-of-n code occurs quite frequently in digital circuits; for instance address decoders for non-concurrent Programmable Logic Arrays (PLAs) (see Chapter 10), consequently checkers for these types of codes are quite important. Khakbaz [30] has suggested a method of converting a 1-out-of-n code into a two-rail code for which a checker already exists. The checker is designed in two stages. First the 1-out-of-n code is converted to a two-rail code. In this scheme any type of logic could be used; however, the original technique was implemented on NMOS NOR gates. The translator takes n inputs and has $2x$ outputs, where $x = \lceil \log_2(n) \rceil$. Figure 8.8(a) shows a translator for $n = 4$. (The dots represent transistors connected across the lines to form NOR gates.) The outputs of the translator represent a two-rail code. The positions of the devices on the intersection of the input and J lines are determined by the binary value of that line. If the bit value for that line is zero then there is no device; a one indicates a device is present. The K lines on a particular input are simply the inverses of the J lines. Thus, for example, input three is '11' in binary, indicating that both the J lines attached to this input

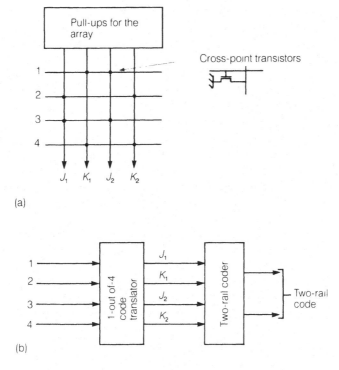

(a)

(b)

Fig. 8.8 Code translator for a 1-out-of-4 code [30]: (a) code translator; (b) complete checker circuit.

have devices. If the number of inputs is a power of two then the last line, number 4 in this case, is represented by the all-zero case. The two-rail encoder used to check the inputs of the translator is simply a PLA with $2x$ inputs, 2^x product terms and two outputs to act as the two-rail code to indicate whether or not there has been an error in the two-rail inputs to the PLA checker. The two-rail checker PLA is designed to be self-testing. The complete scheme is shown in Fig. 8.8(b). The design constitutes a TSC checker, since if the 1-out-of-n translator receives an all zero input, then all the two-rail code outputs will be one; similarly if more than one input is one then at least one of the outputs will become all zero (i.e. $J = 0$ and $K = 0$). Therefore both of these error conditions will be detected and signalled by the two-rail checking circuit. In order to test the circuit for single faults it is necessary to apply all possible 1-out-of-n code inputs. This will check the 1-out-of-n translator and also apply all possible two-rail code inputs to the checker circuit, which is therefore also tested. However, this is only true if n is a power of 2. If n is not a power of 2, then all possible input patterns to the two-rail checker from the 1-out-of-n code translator may not be applied. In order to overcome this problem and allow the same test patterns to be used it is necessary to rearrange the two-rail checker so that the line $J_1 K_1$ and $J_x K_x$ are fed to a two-input two-rail checker and all the other lines enter a separate checker. The outputs of these two checkers are then fed to another two-rail checker before generating the final output. Overall, by making use of this design technique, instead of the one suggested by Anderson and Metze [23], substantial savings on the area required for their implementation of 1-out-of-n checkers can be made so long as $n < 1025$. Again this system does not solve the problem of 1-out-of-3 code checking.

Crouzet and Landrault [28] have made a comparative study of three types of k-out-of-$2k$ checkers. The types chosen are the designs by Anderson and Metze [23], Reddy [29] and Smith [41]. The scheme proposed by Anderson and Metze is based directly on the majority function as discussed previously, whereas the schemes by Reddy and Smith use cellular realizations, which are easily testable, to form the majority function. It was found that as k increased the number of transistors required by the Anderson and Metze method increased exponentially, an undesirable feature where VLSI designs are concerned. The Smith and Reddy schemes show no such deterioration in size requirement. However, it was found that the Reddy technique uses less transistors and is therefore a better choice for integration, although the Smith method would be better for circuits implemented as single gates.

(b) Berger codes

The Berger codes [42] are separable codes unlike the m-out-of-n codes, that is, the information and check bits can be formed as two separate parts of the

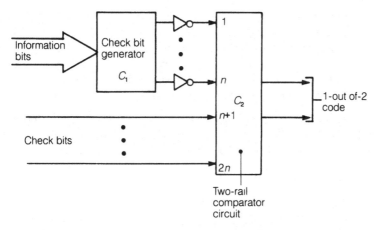

Fig. 8.9 TSC checker for the separable codes [44].

code word. The Berger codes can detect unidirectional errors like the m-out-of-n codes. The Berger codes require the fewest number of check bits amongst all the separable codes that detect unidirectional errors. However, they are more redundant than m-out-of-n codes. A Berger code of length n has I information bits and k check bits, where $k = \lceil \log_2(I + 1) \rceil$; the codeword therefore has $n = I + k$ digits. The k check bits are formed from a binary number which corresponds to the number of ones in the information bits being complemented bit by bit. For example consider the data 110, in this case $I = 3$ hence $k = \lceil \log_2(3 + 1) \rceil = 2$. The number of ones in the data is 2 ($= 10$ binary) therefore the check bits are 01. Consequently the complete codeword $n = 11001$. If $I = 2^k - 1$, then this is termed a Maximal Length Berger Code (MLBC) [43]. In the above example $I = 3$, therefore this is a maximal length Berger code; if $I = 4$ then the code is not MLBC since $k = 3$.

One of the first schemes to produce totally self-checking checkers for Berger codes was proposed by Ashjaee and Reddy [44] in 1976. Their scheme uses the design of a checker circuit shown in Fig. 8.9. This checker however is only suitable for maximal length Berger codes. Therefore since non-MLBC are more likely to occur a method is necessary to convert this code into a complete separable code which can be checked by the proposed checker.

A more formal method of generating MLBC checkers has been demonstrated by Marouf and Friedman [43]. The procedure developed essentially consisted of the C_1 checker design (shown in Fig. 8.9) using full adders with ripple carry between the adders. This generator produces the binary value of the number of ones which is fed into the two-rail checker circuit along with the check bits, which are the inverse of the generated bits; therefore there is no need for any inverters in this circuit, consequently the design can also detect unidirectional errors internally. The two-rail checker is very straightforward to design for any number of bits.

For a Berger code with $I = 3$ and $k = 2$ the check bit generator consists of a full adder with the sum and carry bits generating the two check bits. Therefore the adder circuits are used simply to generate the check bits. The procedure for generating MLBC checkers for $I > 3$ is as follows [43]:

(i) Let the set of information bits $I = \{X_1, X_2, \ldots, X_{2^k-1}\}$; also let $m = k$ and $J = 1$.

(ii) The set I is now partitioned into three sets A, B and E as follows:

$$A^J = \text{The leftmost } P \text{ bits } (P = (2^{m-1} - 1))$$

$$B^J = \text{The next } P \text{ bits}$$

$$E^J = \text{The right most bit.}$$

Let $a^J = (a^J_{m-1}, \ldots, a^J_1)$, $\mathbf{b}^J = (b^J_{m-1}, \ldots, b^J_1)$ and $\mathbf{e}^J$ be the binary representation of the number of ones in each of the partitioned sets A^J, B^J and E^J respectively.

(iii) The binary value for the number of ones in I can then be obtained from the following addition, using $(m - 1)$ full adders with ripple carry between them:

$$\mathbf{g}^J = \mathbf{a}^J + \mathbf{b}^J + \mathbf{e}^J.$$

This will produce the values $\mathbf{g}^J = (g^J_m, g^J_{m-1}, \ldots, g^J_1)$ where g^J_m is the carry generated by the addition operation.

(iv) If $m > 2$ let $m = m - 1$, $L = J$ and the set $I = A^J$, $J = J + 1$, then repeat steps (ii) and (iii) to generate $\mathbf{a}^L = \mathbf{g}^J$. Then $\mathbf{b}^L$ can be generated in the same way as for $\mathbf{a}^L$. Otherwise go to step (v).

(v) Finish.

This procedure for generating the circuit is obviously recursive, producing a binary tree structure where every node in the tree is a ripple carry adder. To test the circuit produced by this method only requires 8 patterns which are again generated by a recursive procedure. To illustrate the design of this type of checker the development of an MLBC checker for $I = 7$ and $k = 3$ will be demonstrated. Each of the stages presented above is illustrated below for this code.

(i) $I = \{X_1, X_2, X_3, \ldots, X_7\}$, $J = 1$, $m = 3$.

(ii) $P = 3$, therefore

$$A^1 = \{X_1, X_2, X_3\}$$

$$B^1 = \{X_4, X_5, X_6\}$$

$$E^1 = \{X_7\}.$$

Therefore

$$\mathbf{a}^1 = \{a_2^1, a_1^1\}, \qquad \mathbf{b}^1 = \{b_2^1, b_1^1\} \qquad \text{and} \qquad \mathbf{e}^1 = X_7.$$

(iii) Now to produce the sum of the number of one bits to give $\mathbf{g}^1$. Therefore

$$
\begin{array}{r}
a_2^1\, a_1^1 \\
+ b_2^1\, b_1^1 \\
+\quad X_7 \\
\hline
g_3^1\, g_2^1\, g_1^1 \\
\hline
\end{array}
$$

This can be easily generated by using a two-stage ripple carry adder.

(iv) Since the value of $m > 2$ it is necessary to repeat steps (ii) and (iii) to calculate A^2 and B^2, by setting $m = 2$ and $L = 1$, $J = 2$ and $I = \{X_1, X_2, X_3\}$, i.e. A^2.

(v) Therefore $(P = 1)$

$$A^2 = \{X_1\}, \qquad B^2 = \{X_2\} \qquad \text{and} \qquad E^2 = \{X_3\}.$$

This will give

$$\mathbf{a}^2 = a_1^2 = X_1, \qquad \mathbf{b}^2 = b_1^2 = X_2, \qquad \mathbf{e}^2 = X_3.$$

(vi) Now to produce the sum of the number of one bits to give $\mathbf{g}^2$ requires

$$
\begin{array}{r}
X_1 \\
+ X_2 \\
+ X_3 \\
\hline
g_2^2\, g_1^2 \\
\hline
\end{array}
$$

(vii) $m = 2$ for A^2; therefore we can now generate values for $I = \{X_4, X_5, X_6\}$, i.e. B^2, which in this case will be the same as A. Since $\mathbf{a}^1 = \mathbf{g}^2$, $g_1^2 = a_1^1$ and $g_2^2 = a_2^1$, these values can be generated by a full adder circuit.

The final circuit consists of two full adders generating the sum and carry bits for $\{X_1, X_2, X_3\}$ and $\{X_4, X_5, X_6\}$. The sum bits of these adders are then fed to a second level adder with X_7 as a carry input; this generates g_1^1. All three carry bits from the adders are then fed to the final adder to produce g_2^1 and g_3^1. This will produce the necessary check bits for the $I = 7$ and $k = 3$ code.

This generator and all the generators designed by this method can be tested by only eight patterns. These patterns will detect multiple faults in a single full adder module and any single faults. The eight patterns required for the above code are shown below. Four of the patterns are unique, the other four are simply their complements.

	1	2	3	4	5	6	7	8
X_1	1	1	1	1	0	0	0	0
X_2	1	1	0	0	0	0	1	1
X_3	1	0	1	0	0	1	0	1
X_4	1	1	0	0	0	0	1	1
X_5	1	1	1	1	0	0	0	0
X_6	1	0	0	1	0	1	1	0
X_7	1	1	1	1	0	0	0	0

A procedure for the generation of checkers for Berger codes where $I \neq 2^k - 1$ is also presented. This scheme uses both full adder and half adder modules. Consequently the test generation phase is slightly more complicated and may require more patterns. Another problem is caused by the fact that only a subset of the 2^k possible check bits is available to the two-rail comparator. If two-level realization of the comparator circuit is used, then the circuit will not be self-testing. To overcome the problem it is necessary to construct the comparator from a tree of two-input comparator circuits.

Piestrak [45] has proposed the conversion of Berger codes for which $I = 2^k - 2$ and $I = 2^k - 1$ into m-out-of-n codes to produce fast self-testing circuits. Again all unidirectional errors can be detected. Dong [46] has proposed the *modified Berger code*, which has two sets of check bits related to the information bits, instead of the usual single set of check bits. One set of check bits is an altered Berger code formed from the information bits while in the example the second set of check bits is the complement of the first set of check bits, i.e. the first and second set of check bits form a two-rail code (in fact any code could have been chosen for the second set of check bits). This will detect any unidirectional errors within the check bits; therefore a two-rail checker can be used to check this part of the codeword. The Berger code for the first check symbol is produced by taking the binary count of the number of zeros or ones in the word and then dividing it modulo $(m + 1)$ to give the check bits $(1 < m < I, I =$ number of information bits). This will give a code that can detect up to m unidirectional errors. Only $\lceil \log_2(m + 1) \rceil$ check bits are needed for this code. For example if $I = 8$ then normally $k = 4$; however, if m is chosen to be seven (i.e. seven unidirectional errors will be detected) then only

3 check bits are required. The checkers can be produced in a similar manner to those of Marouf and Friedman given earlier. However the checkers are generally less expensive in terms of hardware and are faster than for the normal Berger codes. Although this code has a lower error-detection ability than the normal Berger code it does have fewer check bits. The 'error-detection ability' will remain constant with changes in the number of information bits.

(c) Parity codes

The design of checkers for the parity codes is very straightforward [47], [35]. The checker for the parity code simply consists of a tree of Exclusive-OR gates. To form a self-checking design is simply a matter of splitting the tree into two parts so that it generates a two-rail code output; an inverter must be added to one of the outputs if an even parity code is used. However, if during normal operation any of the input lines or lines within the Exclusive-OR tree do not receive 0 and 1 transitions, then the self-testing properties can be lost. To overcome this problem Khakbaz [47] has suggested an algorithm for swapping the input lines of the tree between different subtrees until every line receives a 0 and 1 transition. This algorithm can also be applied to the design of checkers for Hamming SEC/DED codes as well as the design of two-rail checkers which are also formed from trees. A further discussion of parity checks and Hamming codes can be found in Chapter 9.

(d) Residue codes

The residue codes are arithmetic codes that are capable of detecting multiple unidirectional errors. They also have the ability to undergo arithmetic operations while maintaining the code values; therefore they have the capability of checking the arithmetic operations carried out inside a processor [19], [48], [49]. As well as arithmetic operations, these codes are capable of checking logical operations [15]. Since the properties of these codes are covered extensively in Chapter 9 only the outline of the checker design will be discussed in this section for completeness.

The check bits for the residue code can be formed by dividing the information bits by a base, b. The remainder from this division gives the check bits that will be used. For example, consider a number I, this is equal to $I = kb + r$, k is a constant and r is the remainder after division ($0 \leq r < b$), therefore $r = I$ modulo b. The number of check bits for a code is given by $\lceil \log_2(b) \rceil$. For example if $b = 7$ then the number of check bits required is 3. If $I = 011111$ then $r = 011$ (since this is equivalent to 31 modulo 7 = 3). The most important class of residue codes are the Low Cost Residue (LCR) codes. In this code the check base $b = 2^a - 1$ ($a \geq 2$). Therefore the above example

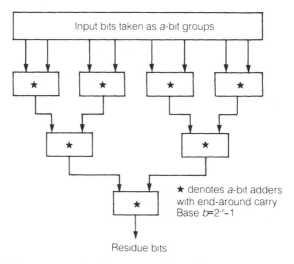

Fig. 8.10 Check bit generator for a low cost residue code.

of $b = 7$ is an LCR code. The main advantage of the LCR code is the ease with which the residue can be formed. Instead of performing a complicated division on the information bits it is simply necessary to produce the residue by a tree of modulo $(2^a - 1)$ adders, as shown in Fig. 8.10. The modulo adders can be formed from a-bit adders with end around carry. A totally self-checking checker for the LCR code can be formed as shown in Fig. 8.9 [44]. The check bit generator C_1 is produced from the circuit shown in Fig. 8.10. The inverted residue bits are compared with the check bits of the codeword using the two-rail comparator circuit as for the Berger code example described above.

8.3.2 Strongly fault secure circuits [27]

The totally self-checking circuits are capable of detecting faults by producing non-code outputs in the presence of the fault. However, certain restrictions are placed on the design: the faults must come from an assumed fault set; and the faults occur one at a time, the time interval between two faults being sufficiently long so that all the code input will have been applied to the circuit.

Using the same assumptions as for the TSC circuits it is possible to define a broader class of circuits that can still achieve the TSC objective. The self-testing property of these circuits is, however, essential. Consider the circuit that conforms to the fault secure property for a fault set F. If the circuit is not self-testing for the fault f_1, from F, then it will never produce a non-code output should this fault occur; therefore the fault will not be detected. If a second fault now occurs, f_2, then there will be two faults present in the circuit.

If the new fault condition $f_1 \cup f_2$ is not in F, then it may or may not be detected. Therefore a circuit should be fault secure for an initial subsequence of faults and self-testing for a combination of faults from this sequence of faults.

Definition (extracted from [27])

Given a fault sequence $\langle f_1, f_2, \ldots, f_m \rangle$, let k be the smallest integer for which the above sequence generates a non-code output when a code input is applied. If no such k exists, then set $k = m$. The circuit is said to be Strongly Fault Secure (SFS) if for the given fault sequence either a correct code output or a non-code output is generated when a code input is applied. The circuit is strongly fault secure with regard to the fault sequence if it is strongly fault secure with regard to all the member fault sequences of F. It is obvious from the above definition that if $k = 1$ then the circuit is TSC. Therefore the TSC circuits are a subset of the SFS circuits. It is possible to produce a TSC circuit from an SFS circuit by a process of 'reduction' [50].

Nicolaidis and Courtois [51] have produced a series of rules governing the layout of NMOS circuits so that they may satisfy the SFS fault hypothesis. These rules cover such things as the number of inversions that a signal can undergo in propagating from the primary inputs to the primary outputs. This allows faults to produce unidirectional errors at the outputs. Other rules govern the layout of power-supply lines so that any unidirectional errors created inside a circuit can produce unidirectional errors at the outputs of the circuit. Also the problems of shorts between lines is considered. This problem is solved by placing signal lines on layers in the layout in such a way that shorts cannot occur.

Three different designs are used to demonstrate the technique in operation. The SFS design technique is applied to Programmable Logic Arrays (PLA), Read Only Memories (ROM) and Arithmetic Logic Units (ALU). The PLA and ROM are augmented with codes in order to detect faults. The ALU is designed in a bit slice fashion using PLA structures. Each bit slice cell takes in a dual rail code and produces a dual rail code as output. A complete ALU is produced by cascading the bit slice cells together. In all of the designs the basic layout rules for producing an SFS circuit are followed and verified.

8.3.3 Partially self-checking circuits

The 'partially self-checking' circuits were defined by Wakerly [35], [52]. The partially self-checking circuits are self-testing for a set N of normal inputs. However, they are fault secure only for a subset I of the normal inputs N. Therefore when the inputs are from I the circuit is totally self-checking as defined before. This mode of operation is termed secure mode. If, however, the inputs are from a set I' ($I' = N - I$) of the inputs, then the circuit is not fault secure and this is termed insecure mode. In secure mode non-code values will not be transmitted and any fault from a chosen fault set is detected. This is the totally self-checking property of a circuit. However, in insecure mode it is possible that an error may be undetectable. The probability of an error being undetected depends on the frequency that the circuit operates in insecure mode. If this is low then the probability of an error being detected in secure mode will be relatively high. The possibility still exists, however, that a transient fault could occur in insecure mode and cause an error. Therefore partially self-checking circuits are really only useful in situations where the

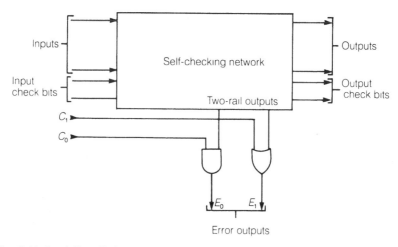

Fig. 8.11 Partially self-checking circuit [52].

reliability is not critical and the less expensive hardware of a partially self-checking design is more economical.

Wakerly defines three types of partially self-checking circuit. The basic outline of the type I and II circuits is shown in Fig. 8.11. The basic circuit consists of the normal self-checking hardware, which produces a two-rail code output to indicate an error. These error signals are fed into an AND and OR gate controlled by C_0 and C_1 respectively. Therefore if $C_0 C_1 = 10$, the error indicator will be enabled; this condition is associated with secure mode. However, if $C_0 C_1 = 01$ then the output is forced to 01 indicating no error, and this is associated with insecure mode. Obviously in insecure mode the output of the circuit may or may not be a code value. The type III networks are more complicated in that the check bits are regenerated by a check bit generator. In this circuit the control lines $C_0 C_1$ operate a multiplexer, which either directs the output of the functional circuit to the code outputs in secure mode or directs the outputs of the code bit generator to the outputs in insecure mode. Therefore in insecure mode the generated check bit outputs which are used in the comparison operation will be checked with themselves; consequently a good comparison will always be produced. One of the problems with this circuit is that in insecure mode a faulty output may be used before it is detected. Wakerly proposed these circuits, in general, to deal with the problem of checking logic operations. He also proposed the design of a partially self-checking ALU [52] which operated in secure mode during arithmetic operations and insecure mode during logical operations, thus avoiding the problem of producing a code to check logical operations.

8.4 VLSI PROCESSOR DESIGNS

In the past the cost of adding fault tolerance to a design would have meant increases in component count and possibly degradation in performance. However, with the advent of VLSI the opportunity now arises for the introduction of fault-tolerant techniques into the chip. This has brought a change in the philosophy of fault tolerance, as it is quite feasible to add or even increase fault tolerance on the chip without greatly affecting circuit performance, and it may even be possible to upgrade the performance of the device. This is a direct result of the cost of adding complexity to the chip not increasing rapidly with the number of discrete devices used. The advantages to be gained in using VLSI for increasing fault tolerance have resulted in a completely new spectrum of possible computer designs.

One of the first major studies of fault-tolerant practice in the design of computer systems was carried out by Avizienis *et al.* [53] with the design of the Self Testing And Repairing (STAR) computer. This study lasted for several years and performed most of the ground work for later designs. However the use of self-checking techniques was originally suggested in 1968 by Carter and Schneider [54].

In 1980 Sedmak and Liebergot [55] proposed the design of a VLSI processor using duplicate complementary logic to perform the checking operation internally to the chip. To detect errors on inner chip networks an error-correcting code or a simple parity check was suggested for most data lines. However, single or 'non-homogeneous' lines would use redundancy. Included in the overall specification an 'error-handling chip' was designed to co-ordinate the fault information available from the self-checking chips. This chip is capable of being used in a tree to produce compressed fault information for use in fault recovery procedures. Using this scheme it is possible to pinpoint the fault to the chip in question, and recovery from transient faults is possible as well as the detection of permanent faults.

The use of strongly fault secure circuits has been suggested to produce a self-checking MC68000 [56] (see also [51]). Along with the SFS design rules a set of layout rules for NMOS circuits is also proposed. The data path section of the self-checking 68000 processor was checked using a parity code, whereas the PLAs in the control section were checked using a Berger code since this code is capable of producing SFS PLAs. The overall increase in area required by this technique was 48%; however, the class of faults covered is comprehensive, including stuck-at faults, shorts and open lines. Nanya and Kawamura [57] have also suggested the use of strongly fault secure circuits in the design of self-checking processors. In the discussion they present the design of an SFS microprocessor with a similar instruction set to the Intel 8080 8-bit microprocessor.

Totally self-checking circuits have been proposed for the design of a self-checking MIL-STD-1750A microprocessor [58]. The arithmetic units are designed using totally self-checking PLAs which can be cascaded to produce any size of adder. The microprogram controller is designed to detect 100% of all single faults with only a 15–30% overhead in self-checking hardware, depending on the encoding scheme chosen for the microinstruction. To achieve this objective the ROM microprogram store uses an m-out-of-n code for its address logic. The selected address is recoded for checking with the input address selected to detect address selection logic faults within the ROM. The overall self-testing hardware required by this scheme is only 40–60% extra, depending upon whether a parity or two-rail code is used to check the registers in the data path.

The Intel 432 [59] is a processor which is capable of producing a fault tolerant system with varying levels of fault tolerance. The 432 has very comprehensive facilities for detecting both transient and permanent errors in the processor, busses and memories. The detection is performed totally within the VLSI circuit; therefore there is no need for extra self-checking logic or software to produce test diagnostics for the system. Simply by increasing the number of 432 processors it is possible to upgrade the bus bandwidth and processor performance. Along with this increase, the fault-tolerance capabilities of the system are also increased. Thus it is possible to select the tolerance level at which the processor will operate. This increase in performance, however, is transparent to the software running on the system. Therefore the system can be configured to support any level of fault-tolerant requirement.

A different approach to the design of a self-checking system has been taken by Crouzet and Landrault [60]. This scheme involves the design of a detection processor to produce self-checking systems from commercially available integrated circuits such as the M6800 series of devices. The circuit is designed to allow the implementation of duplicate systems, error-correcting coded memory techniques, duplicate peripheral circuits and rollback methods.

8.5 SUMMARY

This chapter has presented many schemes to implement fault-tolerant systems. These techniques range from extra hardware to extra information bits which can be incorporated into the design of a system. With the advent of VLSI circuits it is possible to incorporate many of these techniques at very little extra hardware cost. Therefore as VLSI matures into WSI circuits these techniques will prevail. Consequently continued research into fault-tolerant and fault masking methods will eventually provide solutions to the problems of transient faults which are becoming the major source of errors in VLSI circuits as densities increase.

8.6 REFERENCES

1. Siewiorek, D. P. (1984) Architecture of fault tolerant computers. *IEEE Computer*, August, 9–18.
2. Breuer, M. A. and Friedman, A. D. (1977) *Diagnosis and Reliable Design of Digital Systems*, Pitman.
3. Hamming, R. W. (1950) Error detecting and error correcting codes. *The Bell System Technical Journal*, **26**(2), 147–60.
4. Von Neumann, J. (1956) *Probabilistic Logics and Synthesis or Reliable Organisms from Unreliable Components*, Automata Studies, Annals of Mathematical Studies, No. 34, Princeton University Press, 43–98.
5. York, G., Siewiorek, D. P. and Zhu, Y. X. (1985) Compensating faults in Triple Modular redundancy, *Proc. 15th Fault Tolerant Computing Symposium*, 226–31.
6. Ball, M. and Hardie, H. (1969) Majority voter design considerations for TMR Computers. *Computer Design*, 100–4.
7. Lewis, D. W. (1979) A fault tolerant clock using Standby Sparing. *Proc. 9th Fault Tolerant Computing Symposium*, 33–9.
8. Smith, T. B. (1981) Fault tolerant clocking system. *Proc. 11th Fault Tolerant Computing Symposium*, 262–4.
9. Toy, W. (1978) Fault tolerant design of local ESS processors. *Proc. IEEE*, **66**(10), 1126–45.
10. Siewiorek, D. P. and McCluskey, E. J. (1973) An iterative cell switch design for hybrid redundancy, *IEEE Trans. Computers*, **C-22**, 290–7.
11. Peterson, W. W. and Weldon, E. J. (1972) *Error Correcting Codes*, 2nd edn, MIT Press.
12. Langdon, G. G. and Tang, C. K. (1970). Concurrent error detection for group look-ahead binary adders. *IBM J. Research and Development*, September, 563–73.
13. Khodadad-Mostashiry, B. (1979) Parity prediction in combinational circuits. *Proc. 9th Fault Tolerant Computing Symposium*, 185–8.
14. Peterson, W. W. and Rabin, M. O. (1959) On codes for checking logical operations. *IBM J. Research and Development*, **3**(2), 163–8.
15. Monteiro, P. and Rao, T. R. N. (1972) Residue checker for arithmetic and logical operations. *Proc. Fault Tolerant Computing Symposium*, 8–13.
16. Garcia, O. N. and Rao, T. R. N. (1968) On the methods of checking logical operations. *2nd Conference Information Sciences and Systems*, 89–95.

17. Garner, H. L. (1966) Error codes for arithmetic operations. *IEEE Trans. Electronic Computers*, **EC-15**(5), 763–70.
18. Ashjaee, M. J. and Reddy, S. M. (1976) On totally self-checking checkers for separable codes. *Proc. 6th Fault Tolerant Computing Symposium*, 151–6.
19. Avizienis, A. (1973). Arithmetic algorithms for error-coded operands. *IEEE Trans. Computers*, **C-22**(6), 567–72.
20. Pradhan, D. K. and Reddy, S. M. (1972) A design technique for synthesis of fault tolerant adders. *Proc. 2nd Fault Tolerant Computing Symposium*, 20–23.
21. Laha, S. and Patel, J. H. (1983) Error Correction in arithmetic operations using time redundancy, *Proc. 13th Fault Tolerant Computing Symposium*, 298–305.
22. Choi, Y. H. and Malek, M. (1985) A fault tolerant FFT Processor. *Proc. 15th Fault Tolerant Computing Symposium*, 266–71.
23. Anderson, D. A. and Metze, G. (1973) Design of totally self-checking check circuits for m-out-of-n codes. *IEEE Trans. Computers*, **C-22**(3), 263–8.
24. Marouf, M. A. and Friedman, A. D. (1978) Efficient design of self-checking checker for any m-out-of-n code. *IEEE Trans. Computers*, **C-27**(6), 482–90.
25. Smith, J. E. and Lam, P. (1983) A theory of totally self-checking system design. *IEEE Trans. Computers*, **C-32**(9), 831–43.
26. Rao, K. V. S. S. P. and Basu, D. (1983) Design of totally self-checking circuits with an unrestricted stuck at fault set using redundancy in the space and time domains. *IEEE Trans. Computers*, **C-32**(5), 464–74.
27. Smith, J. E. and Metze, G. (1978) Strongly fault secure logic network. *IEEE Trans. Computers*, **C-27**(6), 491–9.
28. Crouzet, Y. and Landrault, C. (1980). Design of self-checking MOS-LSI circuits: application to a four bit microprocessor. *IEEE Trans. Computers*, **C-29**(6), 532–7.
29. Reddy, S. M. (1974) A note on self-checking checkers. *IEEE Trans. Computers*, **C-23**(10), 1100–2.
30. Khakbaz, J. (1982). Totally self-checking checker for 1-out-of-n codes using two rail codes. *IEEE Trans. Computers*, **C-31**(7), 677–81.
31. Smith, J. E. and Metze, G. (1977) The design of totally self-checking combinational circuits. *Proc. 7th Fault tolerant computing symposium*, 130–4.
32. Clary, J. B. and Sacane, R. A. (1979) Self testing computers. *Computer*, October, 49–59.
33. Anderson, D. A. (1971). *Design of Self-checking Digital Networks Using Coding Techniques*. Co-ordinated Sci. Lab. Rep. R-527, Univ. Illinois, Urbana, PhD Thesis.
34. Lu, D. J. and McCluskey, E. J. (1984) Quantitative evaluation of self-checking circuits. *IEEE Trans. Computer Aided Design*, **CAD-3**(2), 150–5.
35. Wakerly, J. (1978) *Error Detecting Codes, Self-checking Circuits and Applications*, North-Holland.
36. Diaz, M., Azema, P. and Ayache, J. M. (1979) Unified design of self-checking and fail-safe combinational circuits and sequential machines. *IEEE Trans. Computers*, **C-28**(3), 276–81.
37. Ozguner, F. (1977). Design of totally self-checking asynchronous and synchronous sequential machines. *Proc. 7th Fault Tolerant Computing Symposium*, 124–9.
38. Viaud, J. and David, R. (1980) Sequentially Self-Checking circuits. *Proc. 10th Fault Tolerant Computing Symposium*, 261–8.
39. David, R. (1978). A totally self-checking 1-out-of-3 checker. *IEEE Trans. Computers*, **C-27**(6), 570–2.
40. Golan, B. (1981) A new totally self-checking checker for 1-out-of-3 code. *Proc. 4th International Conference on Fault Tolerant System Diagnostics*, 246–7.

41. Smith, J. E. (1977) The design of totally self-checking check circuits for a class of unordered codes. *Design Automation and Fault Tolerant Computing*, October, 321–43.

42. Berger, J. M. (1961) A note on error detection codes for asymmetric channels. *Information and Control*, **4**, 68–73.

43. Marouf, M. A. and Friedman, A. D. (1978) Design of self-checking checkers for Berger codes. *Proc. 8th Fault Tolerant Computing Symposium*, 179–84.

44. Ashjaee, M. J. and Reddy, S. M. (1976) On totally self-checking checkers for separable codes. *Proc. 6th Fault Tolerant Computing Symposium*, 151–6.

45. Piestrak, S. J. (1985) Design of fast self testing checkers for a class of Berger codes. *Proc. 15th Fault Tolerant Computing Symposium*, 418–23.

46. Dong, M. (1984) Modified Berger codes for detection of unidirectional errors. *IEEE Trans. Computers*, **C-33**(6), 572–5.

47. Khakbaz, J. (1982) Self testing embedded parity trees. *Proc. 12th Fault Tolerant Computing Symposium*, 109–16.

48. Avizienis, A. (1971) Arithmetic error codes: Cost effectiveness studies for application in digital system design. *IEEE Trans. Computers*, **C-20**(11), 1322–31.

49. Massey, J. L. (1964) Survey or residue coding for arithmetic errors. *ICC Bulletin*, **3**(4), 195–209.

50. Smith, J. E. (1976) *The Design of Totally Self-checking Combinational Circuit*. Co-ordinated Science Laboratory, Rep. R-737, University of Illinois, Urbana.

51. Nicolaidis, M. and Courtois, B. (1986) Design of NMOS strongly fault secure circuits using unidirectional error detecting codes. *Proc. 16th Fault Tolerant Computing Symposium*, 22–7.

52. Wakerly, J. F. (1973) Partially self-checking circuits and their use in performing logical operations. *Proc. 3rd Fault Tolerant Computing Symposium*, 65–70.

53. Avizienis, A., Gilley, G. C., Mathur, F. P. *et al.* (1971) The STAR (Self Testing And Repairing) Computer: An investigation of the theory and practice of fault tolerant computer design. *IEEE Trans. Computers*, **C-20**(11), 1312–21.

54. Carter, W. C. and Schneider, P. R. (1968) Design of dynamically checked computers. *IFIP Congress 68*, **2**, 878–83.

55. Sedmak, R. M. and Liebergot, M. L. (1980) Fault tolerance of a general purpose computer implemented by very large scale integration. *IEEE Trans. Computers*, **C-29**(6), 492–500.

56. Marchal, P., Nicolaidis, M. and Courtois, B. (1984) *Microarchitecture of the MC68000 and the Evaluation of a Self-checking Version*. NATO Advanced Study Institute on Microarchitectures of VLSI Computers.

57. Nanya, T. and Kawamura, T. (1985) Error secure/propagating concept and its application to the design of strongly fault secure processors. *Proc. 15th Fault Tolerant Computing Symposium*, 396–401.

58. Halbert, M. P. and Bose, S. M. (1984). Design approach for a VLSI self-checking MIL-STD-1750A microprocessor. *Proc. 14th Fault Tolerant Computing Symposium*, 254–9.

59. Johnson, D. (1984) The Intel 432: A VLSI architecture for fault tolerant computer systems. *IEEE Computer*, August, 40–8.

60. Crouzet, Y. and Landrault, C. (1980) Design specification of a self-checking detection processor. *Proc. 10th Fault Tolerant Computing Symposium*, 275–7.

9
THE DESIGN OF SELF-CHECKING CIRCUITS USING INFORMATION REDUNDANCY

9.1 INTRODUCTION

In an ideal world a VLSI system would always produce valid outputs. However in the real world design faults, fabrication defects and temporary errors (see Chapter 2) can cause a system to occasionally produce invalid outputs. Built-in test techniques such as scan path [1] and BILBO [2] can enhance the testability of a VLSI circuit by increasing the observability and controllability of various internal points. However, these test techniques are only capable of detecting design faults and fabrication defects they cannot detect temporary errors. It has been predicted in Chapter 2 that the temporary errors are likely to increase and become more dominant in integrated circuits as device geometries decrease in size. The inability of scan path and BILBO to detect temporary errors arises from the way these 'classical' fault-detection techniques are used in testing. Generally these test methods are used only once during initial testing or possibly to diagnose a system after it has failed. If the fault causing the error is permanent then it can be detected; however, if it is temporary then it may go undetected. Therefore other techniques must be considered which will allow a system to monitor its own operations. The self-checking methods, such as error codes and triple modular redundancy, are capable of detecting errors while the system is on-line. The main aim in applying self-checking techniques to VLSI and other hardware is to assure the user that a particular system has a very low probability of failing due to temporary faults while it is operating; it may also assist in increasing the Mean Time Between Failures (MTBF) and reducing the Mean Time To Repair (MTTR) the failed device, both of which are very important considerations in the design of any new piece of digital hardware.

The probability of two systems being faulty in a way which produces identically bad outputs is usually negligible. Therefore self-checking approaches such as triplication (TMR) and duplication, followed by majority vote taking and matching, have been extensively studied. However, such techniques demand a large increase in the amount of silicon area required for their implementation.

The use of an error-detecting/correcting code does, however, promise adequate fault detection without the large area overheads; this chapter will therefore explore the necessary requirements for implementing such a system. The separable codes (e.g. residue codes), which have the information and check symbols as two distinct parts of the codeword, are the most applicable to VLSI self-testing schemes, offering the following advantages:

(1) The checking hardware can be implemented almost independently of any logic which will manipulate the information bits;
(2) Unlike the non-separable codes (e.g. m-out-of-n codes) they do not require decoders. The decoders introduce extra delays, increase the module count in the device and also create the problem of checking the decoder outputs [3].

A possible code for use in the design of self-checking systems is the residue code. Residue codes are particularly suitable for error detection in digital systems as they can be applied easily to the checking of arithmetic [4], [5], logical [6] and other operations; they also require very few extra check bits. Error-detecting codes also have the ability to provide checking across chip boundaries. The Hamming code [7] may also be able to provide this self-checking capability, although at a much higher cost as will be shown later.

This chapter will concentrate on the application of residue codes to self-checking VLSI designs. Initially the properties of residue codes will be discussed. The necessary hardware to perform residue code checks will also be presented. The result of applying the residue code to various VLSI structures will then be discussed, and the problems that may be encountered when this code is applied to certain arithmetic operations will be outlined. A case study of the application of residue codes for concurrent error detection in a microprocessor data path will be used to demonstrate the technique.

9.2 GENERAL THEORY OF THE LOW COST RESIDUE CODES

The residue number system and its attributes are well documented [8] and will not be discussed in detail. However, before discussing some of the properties of residue codes applicable to self-checking circuits a few informal definitions are required:

Definition 1

N is a separable code if and only if the information and check digits can be obtained without any decoding of the codeword, e.g. N is of the form $N = \{X : X = IC\}$, where I = information digits and C = check digits.

Definition 2

N is called a residue code with check base b if and only if:
N is a separable code.
$N = \{X: X = IC,$ given that C is the binary representation of the residue, modulo b, of I considered as an integer number$\}$.

Definition 3

N is called a Low Cost Residue (LCR) code if and only if N is a residue code with check base $b = 2^p - 1, p \geq 2$. p is the number of check bits.

In the separate residue codes the number N is coded as a pair $(N, |N|_b)$, where $|N|_b$ is the check symbol of the number N. ($|N|_b$ is defined as the least non-negative integer congruent to N modulo b, where b is the base of the residue code.) The ability of this code to check arithmetic operations can be easily proved [9], as shown in Theorem 1.

Theorem 1 (adapted from [9]). Let $\{N_i\}$ be a set of numbers with check bits $C_i = |n_i|_b$. The residue of the sum is $|\Sigma n_i|_b$ and the residue of the product is $|\Pi n_i|_b$. Then

(1) $|\Sigma n_i|_b = |\Sigma|n_i|_b|_b$
   i.e. the sum of the check bits modulo b is equal to the modulus of the sum of the numbers n_i;
(2) $|\Pi n_i|_b = |\Pi|n_i|_b|_b$
   i.e. the product of the check bits modulo b is equal to the modulus of the product of the numbers n_i.

PROOF. Assume that $n_i = a_i b_i + r_i$ and $0 \leq r_i < b$
(1) Then

$$|\Sigma n_i|_b = |\Sigma(a_i b + r_i)|_b$$
$$= |\Sigma a_i b|_b^\dagger + |\Sigma r_i|_b$$
$$= |\Sigma r_i|_b = |\Sigma|(n_i)|_b|_b$$

(2) Then

$$|\Pi n_i|_b = |\Pi(a_i b + r_i)|_b$$
$$= |\Pi a_i b|_b^\dagger + |\Pi r_i|_b$$
$$= |\Pi r_i|_b = |\Pi|(n_i)|_b|_b.$$

These very elementary proofs for addition and multiplication can be extended to other arithmetic operations [6] wuch as subtraction and
†Since the modulus of $(ab) \bmod b = 0$ (a integer) complementation. Table 9.1

Table 9.1 Common arithmetic operations

Operation	Output	Output residue
Add	$N_1 + N_2$	$\left\| \left\| N_1 \right\|_b + \left\| N_2 \right\|_b \right\|_b$
Subtract	$N_1 - N_2$	$\left\| \left\| N_1 \right\|_b - \left\| N_2 \right\|_b \right\|_b$
Complement	$-N_1$	$\left\| - \left\| N_1 \right\|_b \right\|_b$
Multiply	$N_1 * N_2$	$\left\| \left\| N_1 \right\|_b * \left\| N_2 \right\|_b \right\|_b$

shows the more commonly used arithmetic operations with their corresponding residue outputs.

Checking arithmetic operations using residue codes is relatively straightforward, as the residue codes are closed under operations such as addition, subtraction, etc. Checking logical operations is not so simple, since residue codes are not closed under bitwise logical operations, such as AND, OR and Exclusive-OR. However, by using the simple relationship between arithmetic and logical operations stated in Theorem 2 [6] below it is possible to use the same residue codes to check logical operations.

Theorem 2. Let N_1 and N_2 be two n-bit binary vectors. Then

$$N_1 + N_2 = (N_1 . N_2) + (N_1 \vee N_2)$$

and

$$N_1 \vee N_2 = (N_1 + N_2) - 2(N_1 . N_2)$$

where the operations '.', '$\vee$' and '$\oplus$' are used to denote the bit by bit logical AND, OR and Exclusive-OR respectively.

This theorem can be proved by showing that the relationship holds for one bit and then extending it to all n bits.

Using these relationships, an expression can be derived for the output residue of the '$\vee$' and '$\oplus$' operations as shown below:

$$\left| N_1 \vee N_2 \right|_b = \left\| \left| N_1 \right|_b + \left| N_2 \right|_b - \left| N_1 . N_2 \right|_b \right\|_b$$

and

$$\left| N_1 \oplus N_2 \right|_b = \left\| \left| N_1 \right|_b + \left| N_2 \right|_b - 2 \left| N_1 . N_2 \right|_b \right\|_b.$$

Table 9.2 gives the output residues for the logical operations.

There are three main factors which affect the choice of the base for the residue code:

(1) Generally the complexity of the checking hardware increases sharply as the base increases;

Table 9.2 Common logical operations

Operation	Output	Output residue
AND	$N_1 . N_2$	$\|N_1 . N_2\|_b$
OR	$N_1 \lor N_2$	$\|\|N_1\|_b + \|N_2\|_b - \|N_1 . N_2\|_b\|_b$
Exclusive-OR	$N_1 \oplus N_2$	$\|\|N_1\|_b + \|N_2\|_b - 2\|N_1 . N_2\|_b\|_b$

(2) As the base increases more bits are needed to represent the residue;

(3) The Error-Detection Ability (EDA) (see Section 9.5.1) of the residue increases only slowly as the base increases.

Therefore, in order to keep the extra hardware required as small as possible (i.e. fewer bits in the residue) a base 3 LCR code ($b = 3, p = 2$) is used. This also has the advantage that a simplification can be made to the residue equations for checking logical operations, as shown below. Let

$$N_3' = |N_1 . N_2|_3;$$

then

$$|N_1 \lor N_2|_3 = |N_1 + N_2 + \bar{N_3'}|_3$$

$$|N_1 \oplus N_2|_3 = |N_1 + N_2 + N_3'|_3$$

The hardware required to implement all of these operations will be discussed in the next section.

9.3 HARDWARE REQUIRED TO IMPLEMENT A RESIDUE CHECKING SCHEME

To allow the production of VLSI designs which use the mod 3 residue code as a means of checking their own operation, it is necessary to provide certain basic building blocks. The blocks required are as follows:

(1) *Mod 3 adder.* This block is required in the addition of residues and the generation of residues from the information bits. This can be performed by a normal two-bit carry lookahead adder with the carry-out looped back to the carry-in [3]. The operation could be performed by a ripple carry adder; however, the use of carry lookahead speeds up the addition operation (this will be shown to be an important attribute). The truth table for the operation of the adder is shown in Fig. 9.1(a). The table shows that the output of the adder is not as expected for a mod 3 residue, since some outputs which should be 00 are in fact 11. This anomaly can be resolved by the use of a comparator which can cope with the two

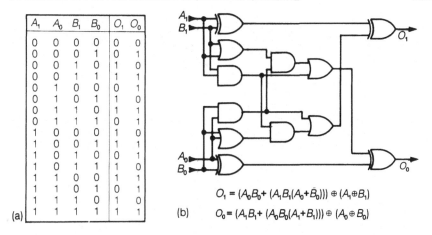

A_1	A_0	B_1	B_0	O_1	O_0
0	0	0	0	0	0
0	0	0	1	0	1
0	0	1	0	1	0
0	0	1	1	1	1
0	1	0	0	0	1
0	1	0	1	1	0
0	1	1	0	1	1
0	1	1	1	0	1
1	0	0	0	1	0
1	0	0	1	1	1
1	0	1	0	0	1
1	0	1	1	1	0
1	1	0	0	1	1
1	1	0	1	0	1
1	1	1	0	1	0
1	1	1	1	1	1

(a)

(b)

$$O_1 = (A_0B_0 + (A_1B_1(A_0 + \bar{B}_0))) \oplus (A_1 \oplus B_1)$$

$$O_0 = (A_1B_1 + (A_0B_0(A_1 + B_1))) \oplus (A_0 \oplus B_0)$$

Fig. 9.1 Mod 3 adder block: (a) truth table; (b) logic diagram.

representations. The use of the 11 residue representation has many advantages which will be discussed in the design of the comparator and in the checking of Programmable Logic Arrays (PLAs) [10] (Section 9.4.3). The logic diagram for this block is shown in Fig. 9.1(b).

As the mod 3 adder was to be used in the design of self-checking circuits it was simulated to determine a minimum test pattern set for single stuck-at faults. The simulation also took into account the actual layout. The minimum test set was found to consist of only six patterns {0000, 0011, 0111, 1100, 1110, 1111} for the postulated fault class.

(2) *Residue generator.* The generation of the residue bits from the information bits is required when a comparison must be made between the expected and actual residues. Fortunately this operation can be performed easily for a low cost residue code ($b = 2^p - 1$) [4], another reason for the choice of this code. The residue can be generated by the use of a tree of mod 3 adders as shown in Fig. 9.2. As the residue generator is a tree structure it is desirable that each internal block operates as quickly as possible in order to improve the overall speed of the residue generation.

(3) *Mod 3 comparator.* At some stage in a self-checking design it is necessary to determine if two residues are equal; this operation is performed by the comparator. The comparator is therefore the most important block in the design, since it must be capable of checking not only the operation of the circuit for which it is responsible but also for its own internal faults. In fact the block must be designed in a totally self-checking manner. This task is made easier if the set of input patterns it is likely to receive during normal operation is as large as possible; this objective can be achieved by the use of the 11 value as a representation of the zero residue. Therefore the two

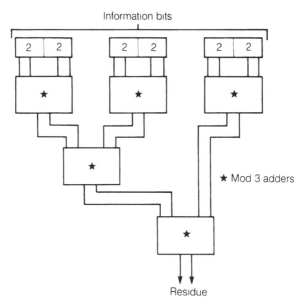

Fig. 9.2 Residue generation tree.

pairs of inputs to the device can now receive the required four input
patterns {00, 01, 10, 11} instead of just three, making internal faults easier
to detect, which makes it easier to fulfil the self-testing property of
self-checking circuits.

The truth table for this device is shown in Fig. 9.3(a). As can be seen the
output is in the form of a 1-out-of-2 code. The output values 01 and 10
result from a correct comparison and 00 and 11 result if a fault has
occurred. The logic diagram is shown in Fig. 9.3(b). As with the mod 3
adder the device was simulated for single stuck-at faults. It was found that
all the postulated faults were in fact detectable using the set of normal
input patterns; however, this would not have been the case if the 11 input
had not been available.

(4) *Two-rail encoder* [11]. On some chips more than one comparator may be
present; therefore in order to use as few output pins as possible in taking
the error information off chip all the 1-out-of-2 code lines can be
compressed into a single 1-out-of-2 code line by two-rail encoder from the
comparators etc. The truth table and logic diagram are presented in Fig.
9.4(a) and 9.4(b).

(5) *Mod 3 multiplier*. As discussed in the general theory of residue codes it is
possible to predict the residue for the multiplication of two *n*-bit numbers;
hence it is necessary to build a multiplier for mod 3 residue numbers. The

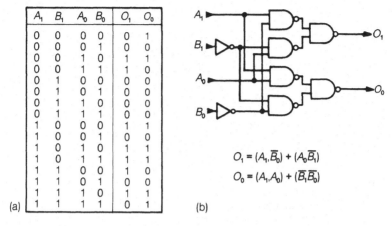

A_1	B_1	A_0	B_0	O_1	O_0
0	0	0	0	0	1
0	0	0	1	0	0
0	0	1	0	1	1
0	0	1	1	1	0
0	1	0	0	0	0
0	1	0	1	0	0
0	1	1	0	0	0
0	1	1	1	0	0
1	0	0	0	1	1
1	0	0	1	0	0
1	0	1	0	1	1
1	0	1	1	1	1
1	1	0	0	1	0
1	1	0	1	0	0
1	1	1	0	1	1
1	1	1	1	0	1

$$O_1 = (A_1 . \overline{B_0}) + (A_0 \overline{B_1})$$
$$O_0 = (A_1 . A_0) + (\overline{B_1} \overline{B_0})$$

(a)   (b)

Fig. 9.3 Mod 3 comparator: (a) truth table; (b) logic diagram.

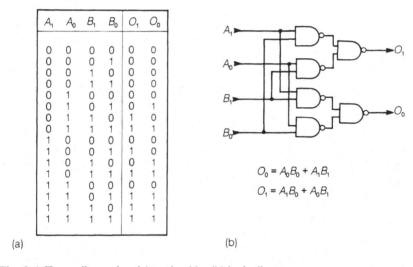

A_1	A_0	B_1	B_0	O_1	O_0
0	0	0	0	0	0
0	0	0	1	0	0
0	0	1	0	0	0
0	0	1	1	0	0
0	1	0	0	0	0
0	1	0	1	0	1
0	1	1	0	1	0
0	1	1	1	1	1
1	0	0	0	0	0
1	0	0	1	1	0
1	0	1	0	0	1
1	0	1	1	1	1
1	1	0	0	0	0
1	1	0	1	1	1
1	1	1	0	1	1
1	1	1	1	1	1

$$O_0 = A_0 B_0 + A_1 B_1$$
$$O_1 = A_1 B_0 + A_0 B_1$$

(a)   (b)

Fig. 9.4 Two-rail encoder: (a) truth table; (b) logic diagram.

truth table for this block is identical to that of the comparator shown in Fig. 9.4(a).

These blocks are the minimum requirement for the design of self-checking circuits using residue codes. Each one could be produced in advance in a cell based VLSI design system. The next sections will now highlight areas where these blocks may be utilized and the problems that are encountered in their use.

9.4 APPLICATIONS OF RESIDUE CODING

To demonstrate the applicability of the residue coding technique to a wide range of functions and to determine how much extra checking hardware (ie silicon area) was required the following circuits were chosen as examples:

(1) $N \times N$-bit multiplier [12]
(2) N-bit ALU
(3) N-bit register (memory elements)
(4) PLA structures [10]

9.4.1 Multiplier

As multiplication is an arithmetic operation it is possible to derive a relationship between the input and output of the device, as demonstrated in the previous section. The block diagram in Fig. 9.5 shows an example layout which can be used to check the multiplication operation. In the diagram A and B are the inputs and C is the product $A * B$. The residues of A, B and C are given by R_a, R_b and R_c respectively, $R_c = |R_a * R_b|_b$. Once the output residue has been generated it is compared with the residue generated from the actual multiplication operation itself.

The checking hardware was applied to an $N \times N$-bit multiplier [12], which is constructed from two basic building blocks, a 2×2-bit multiplier, and a full adder. The use of such a simple arrangement allows the extension of the multiplier from its minimum 4×4-bit multiplication configuration to a 32×32-bit configuration. The ability to extend the design allows the comparison of checking logic and design logic sizes to be made over a wide range of input bit lengths, as discussed in Section 9.5. As the checking logic is independent of the implementation of the function, it can be incorporated into any device which performs a multiplication operation.

9.4.2 ALU

The main problem with applying residue checking to ALU designs is that they can perform both logical and arithmetic operations. Consequently, extra provision has to be made to check the logical operations. Figure 9.6 shows an example of how the checking logic could be applied to an ALU. The main differences between incorporating residue checking into an ALU and a multiplier, which can only perform an arithmetic operation, are:

(1) The inclusion of logic to produce the mod 3 residue of the 'AND' operation ($|A.B|_b$). This is used in the generation of residues for other logical operations such as 'OR' and 'Exclusive-OR';
(2) The selection circuit to choose the residues to be used in the final generated residue. This is necessary because the ALU performs many different operations. Table 9.7 shows an example of the residues required in

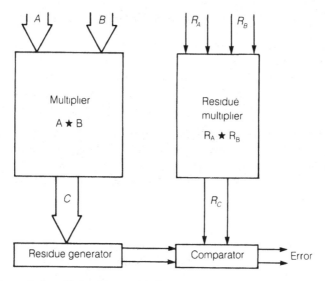

Fig. 9.5 Layout of a self-checking multiplier.

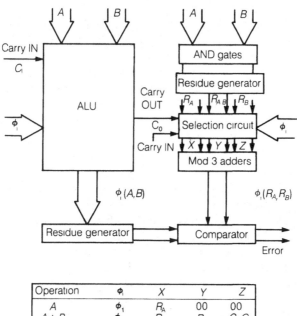

Operation	ϕ_i	X	Y	Z
A	ϕ_1	R_A	00	00
A + B	ϕ_2	R_A	R_B	$C_0\,C_i$
A OR B	ϕ_3	R_A	R_B	$\overline{R}_{AB}$
A $\oplus$ B	ϕ_4	R_A	R_B	$\overline{R}_{AB}$

Fig. 9.6 Layout for a self-checking ALU.

generating the output residues. A more thorough treatment of the application of this technique can be found later when the design of the ALU and checker hardware is discussed in detail. The need for the carry-in and carry-out in the selector circuit will be discussed in Section 9.8.1. The output of the ALU can be checked in a similar way to the multiplier.

The ALU to which this technique was applied was expandable from 4 bits to 32 bits per input allowing the checking logic to be applied to a wide range of input bit lengths for comparison purposes. With slight modifications to the selector circuit (to allow for different functions) the design shown can be applied to any ALU.

9.4.3 PLA

The residue checking technique was applied to a range of PLA structures of various complexity: for example, the traffic-light controller described in Mead and Conway [10], a DMA controller and an electronic combination-lock controller. The problem with trying to produce a standard technique to check PLAs is that PLAs do not perform standard functions, whilst ALUs and multipliers always perform basically the same operations, the internal architecture being irrelevant to the checking problem. Consequently in arithmetic operations the checker is essentially independent of the hardware to be checked, and therefore since the two systems (checker and arithmetic hardware) are separate, unchecked common hardware does not exist. The aim in checking a PLA is to provide checking independent of the hardware providing the outputs, without the overheads involved in direct duplication. Mod 3 checking is applied to the three PLAs mentioned by two methods. Fig. 9.7 gives a block diagram for each method used.

Method 1

This technique simply adds the two extra residue bits onto the side of the OR plane before minimization of the product terms. However, because the system being checked now includes the check bits (i.e. checker hardware) it is possible that certain faults will be undetectable. For example, if a product term failed by becoming stuck at 0, then the checker would also fail to detect this because the output would be all zeros, including the residue, giving a valid codeword. To overcome this problem extra false product terms can be added to the AND plane. The false product terms are those terms which are associated with an all-zero output and are usually ignored in PLA designs. The false product terms are associated with a residue of 11; thus if a true product term (i.e. a term which produces at least one true output) were to fail as previously described, its output could be treated as invalid, since an all-zero output should produce a 11 residue, instead of 00. Most errors in the AND plane can be detected by this method. Errors in the OR plane usually produce single bit output errors [13] which can be detected mod 3.

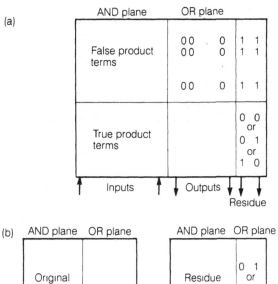

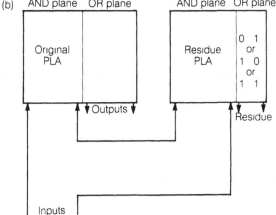

Fig. 9.7 Example of PLA layouts:
(a) Method 1
  Output = 0 valid residue 11
  Output ≠ 0 valid residues 00 or 10 or 01.

(b) Method 2
  Output = 0 valid residue 00
  Output ≠ 0 valid residues 11 or 01 or 10.

Method 2

The second technique involves the addition of a second PLA, the residue PLA, to the original PLA. The residue PLA has exactly the same inputs as the PLA being checked; however, it only has two outputs, which are the residues of the outputs of the original PLA. The AND plane of the residue PLA can be a minimized version of the original PLA. This method has several advantages over the first method:

(1) Since an error is very unlikely to occur in both PLAs simultaneously, identical product terms to those used in the original PLA can be used in the residue PLA. This means that it is unnecessary to generate the false product terms, which require a large area overhead;
(2) As the residue PLA is independent of the PLA being checked, addition of the checker hardware to the system does not affect the checked PLA; consequently the intended operating characteristics will not be affected.

This checking method has similar error-detection characteristics to method 1, as errors are highly unlikely to occur simultaneously in both PLAs, and the detection characteristics are mainly due to the properties of the code.

In both the cases presented the residue generation circuit must be capable of checking the all-zero output. The two methods above present slightly different problems. The first method requires that an all-zero output be checked against a 11 residue output. The second method allows an all-zero output with a corresponding 00 residue, and here, if a 00 residue is normally produced by a valid non-zero output it would not be possible to detect a faulty product term; for this reason a 00 residue should be avoided, if possible, for non-zero outputs. There are two possible ways of achieving this:

(1) Extend the OR plane by one bit, to allow only 01 or 10 residues, and increase the checker size;
(2) Use a 11 output for these terms instead, as already discussed in connection with the residue generator.

Method 2 is preferable for two reasons: it does not affect the original PLA and it fits in well with the residue generation and checker hardware already proposed. Method (b) for the implementation of the 00 residue in the PLAs is also to be preferred since it will easily fit in with the scheme proposed. However, in the next section methods 1 and 2 will be used to show how the size of the PLA can be expected to increase with their use. It should be noted that if the mod 3 adders discussed in Section 9.3 are used then only the all-zero value will generate a 00 residue. In terms of a PLA this will mean that the 00 residue should never appear in the residue PLA (method 2) and will be erroneous if it is present since this will indicate a zero value in the original PLA which should not normally be present after minimization.

9.4.4 Memory elements

As a memory element is only a storage device (i.e. it performs no function between the input and output), it is only necessary to store the residue associated with a particular bit sequence. This requires the addition of two extra blocks of storage (for mod 3 checking) to the word storage. The checking of memory elements will be discussed further in connection with the design of the data path circuit.

9.5 PROPERTIES OF THE RESIDUE CODES

There are two important questions which govern the choice of test method in a particular application:

(1) The cost in terms of silicon area required to implement the chosen test method when compared to the design without the test structures;
(2) The time penalties imposed on the design by any parts of the test hardware.

The extra area required by this checking technique was therefore calculated in order to compare it with other methods; the results produced will be discussed in this section. The increase in area resulting from the additional mod 3 residue checking hardware in the multiplier and ALU structures is shown graphically in Fig. 9.8, where only extra area taken by the test hardware, and not the interconnection of the hardware, has been considered.

The multiplier shows the best performance as the number of bits in the design increases. This is because the residue prediction hardware size stays almost constant, although the checker circuits do increase in size. The ALU

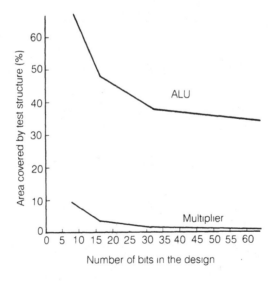

Fig. 9.8 Mod 3 residue check.

Table 9.3 Size increases in PLA structures

PLA name	Number of inputs	Number of outputs	Number of product terms	Method 1 Extra Size increase (%)	Outputs	Product terms	Method 2 Residue PLA Extra area required (%)	Inputs	Outputs	Product terms
Traffic light controller	5	8	10	8	2	0	73	5	2	9
DMA controller	12	12	26	27	2	8	67	12	2	22
Electronic combinational lock control	15	6	18	213	2	50	93	15	2	20

does not show such a good performance, partly because of the increase in the 'AND' residue prediction hardware as the number of bits grows.

Table 9.3 shows the results obtained from the three types of PLA to which mod 3 residue checking was added. Generally, it was found that the more overlap between product terms (i.e. the fewer inputs to the product terms) the better the minimization of the false product terms and the residue PLA AND plane. The residue PLA method, although requiring more silicon area, was generally easier to generate by software means and produced a more constant size increase, allowing size increases for other residue PLAs to be predicted. The checking hardware associated with the PLAs was not included in any of the estimates, but could be expected to increase the size by an extra 10–20%.

It is difficult to produce estimates for the size of the checking logic required by memory elements, since they could be used in conjunction with other devices, e.g. ALUs, which are capable of performing the check operation themselves, or alternatively the memory elements could be used in a larger block, such as an RAM, in which case checking could be applied to the block as a whole, allowing a complete RAM chip to be checked by one piece of checking hardware. For a 16-bit register, an increase of about 14% can be expected in the register size to store the residue, ignoring the checker overhead.

It is interesting to compare the effectiveness of the mod 3 residue checking scheme with the single-bit parity checking scheme. This comparison was made for the multiplier and ALU designs. The single-bit parity check can be applied to arithmetic and logical operations in exactly the same manner as a mod 3 residue code. Table 9.4 gives a breakdown of the operations that can be performed using the parity information. Proofs of some of these relationships can be found in Lewin [14] and Garner [15]. The parity bits of the input data A and B are represented by P_a and P_b respectively. The only major difference

Table 9.4 Parity check operations

Operation	Parity operation
A	P_A
$A + B$	$P_A \oplus P_B \oplus P_C$
$A . B$	$P_{A.B}$
$A \oplus B$	$P_A \oplus P_B$
$A \vee B$	$P_A \oplus P_B \oplus P_{A.B}$
$A \times B$	$(P_A . P_B) \oplus P_C$
$\overline{A}$	$\left[\begin{array}{l} P_A \text{ even no. bits} \\ \overline{P}_A \text{ odd no. bits} \end{array}\right.$

P_A = Parity bit for A
P_B = Parity bit for B
P_C = Parity bit for the carries

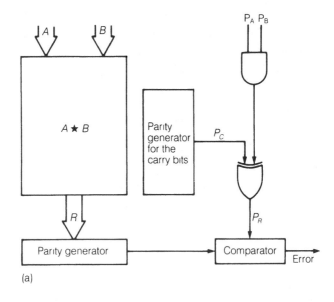

(a)

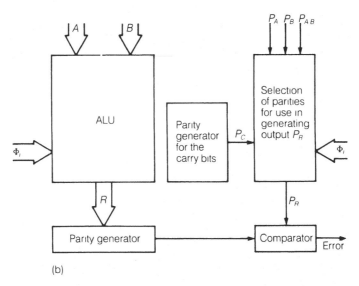

(b)

Fig. 9.9 Parity checking schemes: (a) parity check for a multiplier; (b) parity check for an ALU.

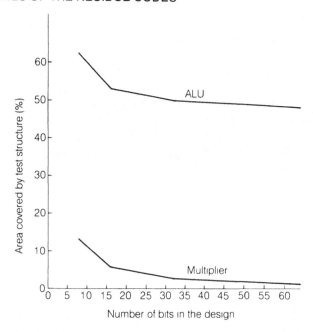

Fig. 9.10 Parity check results.

between the single-bit parity check and the mod 3 residue check is the need to include the parity of the carries in any addition-type operations; this also includes the multiplication operation. One similarity between the two checking schemes is the need to generate the parity of the AND operation independently.

The checkers and comparators for the parity codes can be built from Exclusive-OR gates; a two-rail coded output can be generated from these circuits [16] if desired. Figure 9.9(a) and (b) shows the hardware necessary to perform the parity check on the multiplier and ALU respectively. Figure 9.10 shows the results of adding the single-bit parity check to the two devices. The extra area required is mainly due to the need to generate the parity of the carries. In fact the area required for the parity scheme on the multiplier is almost twice that of the mod 3 checking scheme. Although the single-bit parity check gives a very poor performance it is shown later that the Hamming code may have potential in the design of error-detecting/correcting hardware for VLSI chips.

9.5.1 Error-detection ability

It may initially be thought that a mod 3 code can detect only 66% of all errors, but this is not so. All single-bit errors, which are the errors most likely to occur, can be detected by a mod 3 residue code, since it has a Hamming distance of 2. Multiple-bit errors are less likely to occur [17] and therefore less likely to affect the results. Also, in adders and therefore ALUs, it is possible to alter the structure [18] in such a way as to eliminate errors which can occur mod 3; this technique will be demonstrated in Section 9.6. Some idea of the error-detection ability (EDA) [19] of mod 3 codes, taking into account the probability of multiple-bit errors, can be obtained as follows:

Assume that the single-bit error rate is P_s (i.e. P_s is the probability that a single-bit error will occur); then the average probability that a multiple-bit error will occur in an n-bit number is

$$P_{MB} = \left[\sum_{i=2}^{n} \frac{n! \cdot P_s^i}{i!(n-1)!} \right] \bigg/ 2^n$$

This can be rearranged to give

$$P_{MB} = \sum_{i=2}^{n} \frac{C_i^n P_s^i}{2^n}.$$

If the base of the residue is R, then $1/R$ of these faults will be undetectable errors. Therefore the probability of an undetectable error is:

$$P_{UD} = \frac{P_{MB}}{R} = \sum_{i=2}^{n} \frac{C_i^n P_s^i}{2^n R}$$

Therefore the probability of detecting the error is given by

$$P_{DE} = 1 - P_{UD}$$

$$P_{DE} = 1 - \left[\sum_{i=2}^{n} \frac{C_i^n P_s^i}{2^n R} \right].$$

in this case $R = 3$. The graph in Fig. 9.11 shows the EDA for $n = 12$.

As a check on the abilities of the mod 3 code to detect faults, a small-scale simulation was made on the 4×4-bit multiplier, discussed earlier. In the simulation the most probable single-bit errors and a range of multiple-bit errors were introduced. The overall probability that an error would be detected was found to be 99.8% for the mod 3 residue code and only 68% for the single-bit parity code. Most of the faults were introduced into the adder part of the circuit, since this is the point at which faults that occur may cause burst errors, especially if the carry is affected; several output bits could then be altered at the same time.

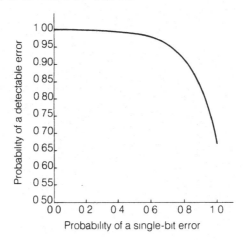

Fig. 9.11 EDA for $n = 12$.

9.5.2 Miscellaneous operations

This section will discuss a few problems that arise in the application of this mod 3 residue technique to certain arithmetic operations and their possible solutions. The main problems arise because of the use of two-bit adders with end-around carry to generate the residue. This produces a 11 residue for input data divisible by 3, except for the zero datum. Consequently it is, on occasions, necessary to take corrective action to restore the residues to consistent values. The conditions that necessitate the corrective action are the cases when the residue produced is 00 and the actual information bits are not zero and when the residue is 11 and the information bits are zero.

Another problem arises because of the residue codes themselves. When performing modulo 2^n-arithmetic, as in a normal binary adder, it is possible to produce a carry-out in the $(n + 1)$th bit position. The residue generated however will in fact cover all the $(n + 1)$ bits. Therefore the residue has to be corrected so that it only applies to the n bits of the normal word.

Possible solutions to these problems will be discussed in connection with the ones complement and twos complement number system. The rotate and logical shift operations will also be discussed.

9.5.3 Ones complement number system

Mod 3 checking can be applied easily to this number system since the carry-out bit is usually added back into the number using an end-around carry.

Ones complement negation simply involves performing the NOT operation on every bit in the number. This operation can also be performed on the

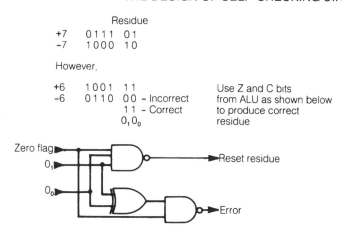

```
                    Residue
        +7    0111  01
        -7    1000  10

        However,

        +6    1001  11                      Use Z and C bits
        -6    0110  00 - Incorrect          from ALU as shown below
                    11 - Correct            to produce correct
                   0₁0₀                     residue
```

Fig. 9.12 Ones complement negation.

```
        +6    0110  11              +6    0110  11
        +1    0001  01              +4    0100  01
        +7    0111  01      (+10)   -5    1010  01

                               Overflowed. Residue still valid

        -3    1100  11
        -4    1011  10
              0111
              1--→1←— End-around carry
        -7    1000  10
```

Fig. 9.13 Ones complement addition.

residue to produce the residue of the negated number. However, because of the form of the residue encoding implemented, corrective action must be taken whenever a number with residue 11 is negated, if the negation operation does not produce the all-zeros output. Alternatively, since the ones complement system has two representations of 0, only the all-ones case could be used, which would mean that a residue of 11 would never have to be inverted. A few examples of this operation are presented in Fig. 9.12, a circuit which can be used to correct the residue is also shown.

The addition of ones complement numbers is accomplished in the normal way. The residues are simply added together using the mod 3 adder discussed earlier. The final residue is the residue of the sum produced by the addition operation. No further corrective action is required. Figure 9.13 shows a few examples of this operation.

```
 +7    0111  01              -4    1011  10
+(-4)  1011  10            +(-1)  1110  10
       ----                       ----
       0010                       1001
       1---►1                     1---►1
 ---   ----  --             ---   ----  --
 +3    0011  11             -5    1010  01
```

Fig. 9.14 Ones complement subtraction.

Subtraction can be performed as normal. The normal sequence of events is to negate the subtrahend and then to add this number to the minuend. The residues can therefore undergo the same sort of operation, since they will be negated and then added. The additions can be performed by a mod 3 adder. Figure 9.14 shows a few examples of this operation.

9.5.4 Twos complement number system

The application of mod 3 checking to the twos complement number system is not as simple as the description given for the ones complement number system. The main problem with the twos complement number system is the generation of a carry out from the most significant bit; unlike the ones complement system, the carry is not incorporated back into the number. Therefore in a real system information (the carry bit) will be lost, and the residue will no longer be correct, as it applies to the complete word including the carry bit. There are a number of possible solutions to this problem. The simplest solution is to regenerate the residue from the data whenever a carry occurs. However, this will break down the independence of the two operations and may make the system unreliable.

The technique chosen to solve this problem relies on information that is usually already available in the adder, that is the carry bit itself and the zero flag. The corrected residue can be obtained by subtracting the carry bit from the generated residue. The result produced will be correct providing the answer is not zero. When the result is zero the status of the zero bit must be examined. If the zero bit is true, then the residue may or may not be correct because of the dual representation used for the zero residue. If a zero residue should have resulted, then the final residue could be either 00 or 11, but it will not be 01 or 10. Therefore if the zero flag is true and the resulting residue is 01 or 10 an error has occurred and can be flagged. Otherwise the residue can be cleared to zero for the zero datum, providing the necessary corrective action. Of course any corrective action required for the carry bit should be performed first. Therefore the zero flag bit is checked. The carry bit is also checked, since if it is faulty it causes an error of $\pm 2^n$ in the residue [5], which is easily detected by the mod 3 residue code. The error is detected when the comparison is made between the residue and the residue generated from the information bits. Figure 9.15 illustrates the technique.

R_A
R_B
A
B
Residue adder
Carry in
Adder
Carry out
Carry in
Zero
Residue adder
Residue correction
Error
C
R_C

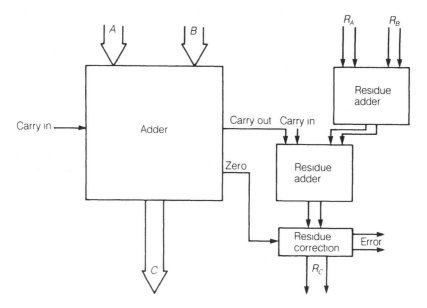

Fig. 9.15 Correcting the residue using the carry bit.

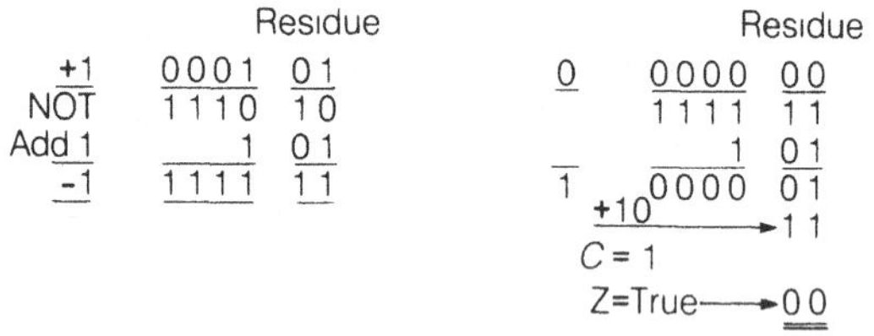

Residue
+1 0001 01
NOT 1110 10
Add 1 1 01
-1 1111 11
Residue
0 0000 00
1111 11
1 01
1 0000 01
+10 11
C = 1
Z=True 00

Fig. 9.16 Twos complement negation.

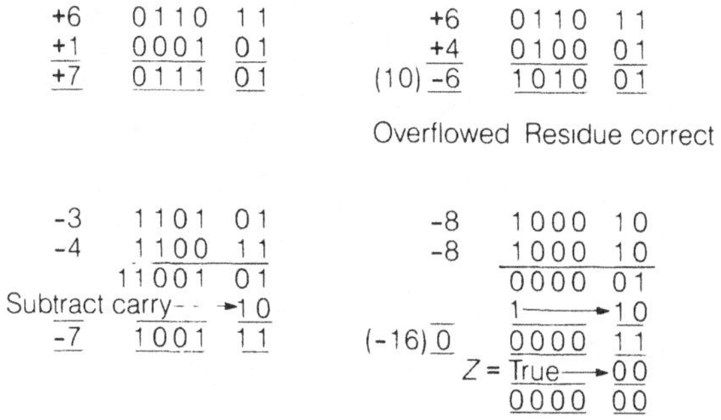

+6 0110 11
+1 0001 01
+7 0111 01
+6 0110 11
+4 0100 01
(10) -6 1010 01
Overflowed Residue correct
-3 1101 01
-4 1100 11
11001 01
Subtract carry 10
-7 1001 11
-8 1000 10
-8 1000 10
0000 01
1 10
(-16) 0 0000 11
Z = True 00
0000 00

Fig. 9.17 Twos complement addition.

The method used to negate twos complement numbers is to invert all the bits in the number and then add 1 to the result. This operation can also be performed on the residue. The residue is first complemented and a constant value of 01 is added. Figure 9.16 illustrates this method for the corrective action discussed above.

The addition operation is illustrated in Fig. 9.17; addition is performed in the normal way. The two residues of the numbers are added and the corrective action taken. The subtraction operation is illustrated in Fig. 9.18.

```
+7    0 1 1 1   0 1          +7    0 1 1 1   0 1
+-4   1 1 0 0   1 1          +-8   1 0 0 0   1 0
      0 0 1 1   0 1          -1    1 1 1 1   1 1
          1 ──────▶ 1 0
 +3   0 0 1 1   1 1
```

Fig. 9.18 Twos complement subtraction

The operations described for the two number systems are the most basic available. Therefore further operations could be performed by using these basic techniques. Consequently these more advanced operations will also be checked satisfactorily, since they use operations which are themselves checked.

9.5.5 Rotate operation

The left and right rotates are illustrated in Fig. 9.19. Since all the bits are preserved in this operation, the residue will remain constant and can simply be rotated left or right along with the number. Single-bit shifts are shown in the example.

9.5.6 Logical shifts

Logical shifts are more difficult to check because of the loss of bits as the shift operations take place. Figure 9.20(a) shows the logical shift operation. The circuit in Fig. 9.20(b) demonstrates how the loss of these bits can be dealt with; the system uses the same corrective action as the twos complement number system. The left shift operation simply involves rotating the residue left then subtracting the carry-out from this value. The right shift is very similar in that the residue is rotated right and then any carry-out is added to the residue. It is also necessary to clear the residue if the result of a shift produces a zero result; again this can be checked as before.

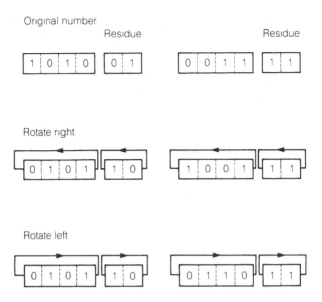

Fig. 9.19 Examples of the rotate operation: the rotate operations do not need any corrective action on their residues.

9.6 THE DESIGN OF A SELF-CHECKING DATA PATH USING CODING TECHNIQUES

As an example of how residue codes can be used in practice, the data path for a CPU will be designed to incorporate the coding technique. It is intended that the data path will be a programmable cell in a larger cell library, and will be used to control digital signal processing chips, where it may be embedded within the overall design. The ability of the CPU to check its own operation will consequently be of prime importance, as few of its pins will be controllable directly at the periphery of the chip. Applications are not limited to this particular field, as the CPU has a general purpose architecture and could be used in designs requiring a programmable, self-checking, unit as one of the blocks.

9.7 DATA PATH ARCHITECTURE

A block diagram of the data path is shown in Fig. 9.21. This shows two sections not concerned with testing – the ALU and the data registers. The ALU and data registers along with the control PLA/ROM will now be discussed.

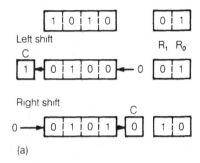

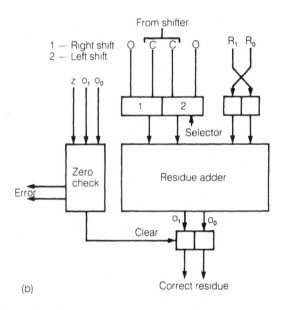

Fig. 9.20 Logical shift operations: (C denotes Carry bit).

9.7.1 Data path registers

The data path registers are used as the local store for all operations to be performed by the ALU. Since the data path will be programmed by the designer the architecture was based on a common commercially available CPU so as to make the programming task easier, consequently this determined the number of data registers and how they are used. In the design the registers are eight bits wide; however, it is possible to increase this to any number of bits. The eight-bit wide data path represents the worst case

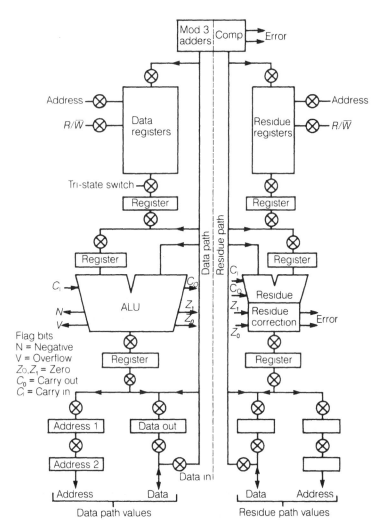

Fig. 9.21 Block diagram of data path.

situation for the checking hardware overhead. There are eight registers in the data path circuit, which perform the following operations:

(1) Program counter;
(2) Stack pointer;
(3) Six general purpose registers.

9.7.2 Data path ALU

The data path ALU is modelled upon the ALU described by Mead and Conway [10]; however, there are several modifications which improve speed, reliability and permit mod 3 residues checking to be incorporated.

The Manchester carry chain [20] used by Mead and Conway is well suited for implementation by saturated symmetrical bipolar transistors; however, it is not very efficient when MOSFET transistors are used, even when

Table 9.5 List of micro-operations performed by the ALU

Arithmetic operations	Mnemonic	Operation
Add A to B	ADD	$A + B$
Add A to B with carry	ADC	$A + B + C_{in}$
Subtract B from A	SUB	$A - B$
Subtract B from A with carry	SBC	$A - B - C_{in}$
Increment A with carry	ICCA	$A + C_{in}$
Increment A	INCA	$A + 1$
Identity A	IDA	A
Increment B with carry	ICCB	$B + C_{in}$
Increment B	INCB	$B + 1$
Identity B	IDB	B
Complement A	COMA	$\overline{A}$
Negate A	NEGA	$\overline{A} + 1$
Complement B	COMB	$\overline{B}$
Negate B	NEGB	$\overline{B} + 1$
Decrement A with carry	DCCA	$A - C_{in}$
Decrement A	DECA	$A - 1$
Decrement B with carry	DCCB	$B - C_{in}$
Decrement B	DECB	$B - 1$
Literal 1	LT1	1
Clear	CLR	0
Negate A and add to B	NAD	$-A + B$
Negate A and add to B with carry	NAC	$-A + B - C_{in}$
Logical operations		
AND A,B	ANDL	$A \cdot B$
OR A,B	ORL	$A \vee B$
Exclusive-OR A, B	EORL	$A \oplus B$

precharging is employed. Consequently the carry path of the ALU was re-designed to use a regular four-bit carry lookahead block with carry ripple between blocks. This redesign allowed precharging to be removed from all internal sections of the ALU, but the functional block technique, as used in the Mead and Conway ALU, is retained.

The operations performed by the ALU have also been reduced to conform with the proposed use of the data path. Table 9.5 gives a list of these operations with their corresponding mnemonics. This list shows a reduced instruction set which will cover most requirements. AND gates and mod 3 adders have been incorporated to generate the mod 3 residue of the bitwise logical AND of the two inputs; this is used in the generation of the residues for the ORL and EORL operations and is required by the residue prediction hardware. Figure 9.22 shows the structure of a single four-bit slice of the ALU.

9.7.3 Data path control

This section will describe, briefly, the approach to be taken in the control of the data path. The data path will have two levels of control. One level will be concerned with controlling the ALU for all its possible operations and will be provided in the full design. Only six bits are needed to control the ALU if this decoding scheme is used. The next level of control will be the microcode PLA (or ROM) which will control the whole data path. The intention is to allow the designer to either build his own controller PLA or use a predesigned microcode PLA which will support the instruction set shown in Table 9.6; obviously this is only one possible implementation. It is apparent from Table 9.6 that there are no right shift or arithmetic shift operations. This omission is intentional and has been made to keep the design smaller and therefore more reliable. These operations can be performed by combinations of other available operations, although there is some speed penalty. The shift operations can be checked using the residue code if hardware is required to perform these operations.

9.8 CHECKING HARDWARE

The main blocks that are required to implement the mod 3 checking scheme, as discussed in Section 9.3, comprise a mod 3 residue generation tree, a mod 3 comparator and a two-rail encoder.

9.8.1 ALU residue prediction

The hardware used to predict the residues for the ALU is the major piece of checking hardware shown in Fig. 9.21. The prediction of the residues for the

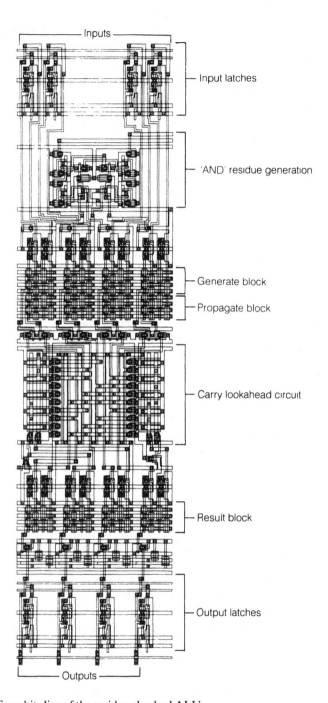

Fig. 9.22 Four-bit slice of the residue checked ALU.

Table 9.6 Instruction set supported by data path

Mnemonic	Description
ADD	Add two values
ADC	Add two values with carry
ANDF	Logical AND of two values
CLRR	SET value to zero
COMR	Complement value
NEGR	Negate value
DECR	Decrement value
EORF	Exclusive OR two values
INCR	Increment value
MOVL	Read value into data registers
ORF	Logical OR of two values
MOVS	Store value from data registers
SUB	Subtract one value from another
SBC	Subtract with carry
SWAP	Swap two data registers
CMPR	Compare two values
JMP	Jump to an address
JMP CC	Jump to an address on condition
CC = N, Z, V, C.	
TEST	Test for minus
BIT	Bit test
PUSH	Push value onto stack
POP	Pop value from stack
LSL	Logical shift left
CCF	Clear carry flag
SCF	Set carry flag

ALU was discussed in Section 9.4.2. The output residue of the ALU can be predicted using the residues shown in Table 9.7.

As discussed previously the method used to generate the residues may cause inconsistencies in some circumstances. Consequently it is necessary to take corrective action at some point. This is achieved by the use of the residue correction circuit which uses the zero detect bits from the ALU output to determine when to correct the residue. The output of the ALU is checked for zero using two NOR gates; one of the outputs is inverted so that the two bits form a two-rail code. These two bits are then applied to the input of the residue correction circuit along with the residue values to be corrected. The main corrective action is taken when the residue is 00 and the zero flag is false and when the residue is 11 and the zero flag is true: in both cases the residue is simply inverted. This is not as much of a drawback as it may first appear since this effectively means that the zero flag bits are checked against the residue and are also checked for being a valid 1-out-of-2 codeword. If there are any inconsistencies an error can be flagged, again as a two-rail code.

Table 9.7 Residues required to predict output residue for the ALU

Mnemonic	Residue number 1	2	3
ADD,ADC	R_A	R_B	$C_O C_I$
SUB,SBC	R_A	$\bar{R}_B$	$C_O C_I$
ICCA,INCA,IDA	R_A	0 0	$C_O C_I$
ICCB,INCB,IDB	0 0	R_B	$C_O C_I$
COMA,NEGA	$\bar{R}_A$	0 0	$C_O C_I$
COMB,NEGB	0 0	$\bar{R}_B$	$C_O C_I$
DCCA,DECA	R_A	0 0	$C_O C_I$
DCCB,DECB	0 0	R_B	$C_O C_I$
LT1,CLR	0 0	0 0	$C_O C_I$
NAD,NAC	$\bar{R}_A$	R_B	$C_O C_I$
AND	0 0	0 0	$R_{A.B}$
OR	R_A	R_B	$\bar{R}_{A.B}$
EOR	R_A	R_B	$R_{A.B}$

R_A = Residue of the A input
R_B = Residue of the B input
$C_O C_I$ = Carry out C_O and Carry in C_I
$R_{A.B}$ = Residue generated from a duplicate AND operation
The output residue is formed by adding the three residues together in a mod 3 adder.

Table 9.7 also shows the need to include the carry-in and the carry-out in the calculation of the residues. The need for the carry-in is obvious: since it is an input to the ALU it must also be an input to the ALU residue prediction circuit. The carry-out from the ALU must be included because of the problem, discussed previously, of lost information whenever a carry occurs. The carry-out can cause an error of only $\pm 2^n$, which can be detected by the mod 3 residue code; consequently the carry bit is also checked for correct operation. Unfortunately it is not possible to check the negative (N) or overflow (V) flag by interaction with the residue prediction hardware.

9.8.2 Checking the data registers

Memory systems are normally checked and errors corrected by the use of Hamming [7] SEC/DED codes. This is inappropriate in this case since an 8- or 16-bit register would require more than the two bits necessary for the mod 3 residue error-detecting code. Also as the residue code is used to check the ALU, and other operations, it would be necessary to convert the Hamming code into a residue code to preserve the independence of the system; this is expensive in terms of both silicon area and time. Although residue coding for the data registers is not ideal it is desirable in the environment in which it will be used.

There are two possible methods available for checking the data registers:

(1) Two residue codes could be incorporated into the data registers, giving four extra bits. One set of residue bits would be used to check that the data stored in the registers are correct. The second residue would be used to check that the address selection circuit was functioning correctly;

(2) Two independent register sets could be implemented. One set of registers would be used to hold the data and the other to hold the residues. Thus addressing errors should be detected by a mismatch between the residue and the data register values, since the probability of both registers being in error simultaneously with the same fault is very low.

The second method is preferred since it will preserve, as much as possible, the independence of the residue prediction data path and the actual data path.

9.8.3 Controller checking

The controller is checked, simply, by duplicating the low-level decoding for the ALU and residue prediction circuit, i.e. they are both driven independently, and by the use of a residue PLA for the 'microprogram' PLA. The use of duplication at the lower level reduces the number of signal lines that need to be routed around the data path, thereby reducing the size of the data path.

The use of the residue PLA technique for the microprogram control allows the specification of the different control PLAs without a great deal of effort on the part of the designer to incorporate the residue PLA.

9.8.4 Placement of checkers

In order to maximize the detection ability of the code and keep the overhead requirements for the technique to a minimum it is necessary to place the code checkers at strategic points in a circuit; guidelines for their placement have been given by Sellers *et al.* [18] and Wakerly [11]. Furthermore, Wakerly gives a concise definition for the placement of checkers in a circuit in terms of the 'detection lossless' property of a circuit which can be used to keep the overall circuit fault secure. This property is best illustrated by a few examples. Consider a functional block capable of performing operations on codewords in an arithmetic code, such as the residue code. There are various possible output results depending on how the block behaves in the presence of non-codewords on its inputs. For example a circuit is said to be single-error detection lossless if only one of its inputs can be in error, producing a non-codeword output; if more than one codeword is incorrect then the output of the block may be a codeword, therefore the input error will go undetected. A block is said to be zero-error detection lossless if no incorrect inputs can be tolerated, i.e. any incorrect input may produce an erroneous codeword output. Alternatively a block may be n-error detection lossless if any one of its n inputs can be non-codewords. With these conditions Wakerly proposes placing checkers in a circuit so that it remains fault secure as follows. All the inputs to a zero-error detection lossless block must be checked. If an n-input block is m-error detection lossless then $(n - m)$ inputs must be checked. Also all loops in the circuit must be checked.

In the data path the ALU acts as an adder/subtractor and consequently is single-error detection lossless with regard to these operations. However, it also acts as a logical unit performing AND, OR and Exclusive-OR operations which are zero-error detection lossless. Therefore both inputs to the adder should be checked, the second condition being the dominant one. The data registers are simply multiplexers and behave as n-error detection lossless multiplexers; therefore it is simply necessary to check the output from the registers. Since the data path has a bus structure already available only one checker is attached to the bus. This effectively checks every word that is moved around the bus, thereby satisfying all the constraints. It is even possible to check input and output data. With only a single checker performing all the check operations the bus bandwidth could be limited. Although the placement of the checker is very area efficient it limits processor speed; other arrangements could be used if higher performance were required.

It was proposed in Section 9.4 that only the outputs of a block should be checked, but it has been shown here that inputs also need to be checked. Consequently, if blocks checked in this manner are placed together the inputs will in fact be checked from the outputs of the preceding block. However, the first block in any chain which is fed from the primary inputs of a chip will not be checked at the input stage, thus checkers should be fitted to the primary inputs of all chips to fulfil the checking requirements.

9.9 DATA PATH IMPLEMENTATION

This section will deal with the actual data path implementation using the previously discussed cells. Figure 9.23 shows the full eight-bit layout of the ALU, including the necessary generation of the mod 3 AND residue. Figure 9.24 shows the complete layout of the self-testing data path. The interconnection of the cells was performed by the BEDE autolayout program [21]. This data path now represents a cell that could be used in any future designs.

After the data path cell had been designed it was necessary to determine how effective the residue code was at detecting faults within the circuit. Stuck-at-fault analysis, although ineffective for providing full coverage of bridging and other classes of faults, is still a useful guide in determining the overall testability of a design. Until a greater understanding of other types of faults is obtained, it will continue to dominate the testing scene. However, as

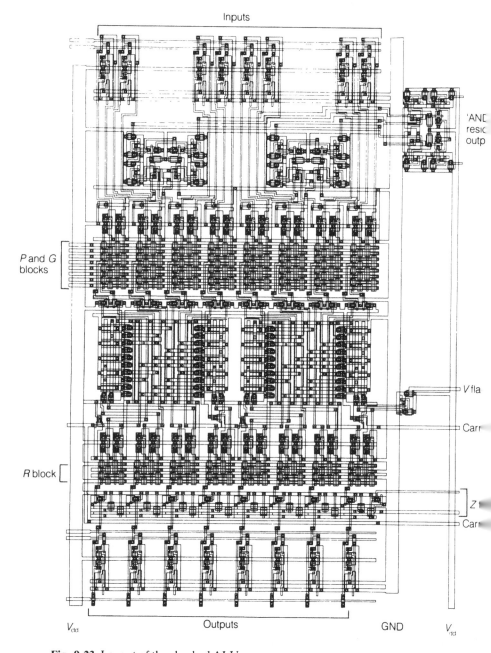

Inputs

'AND
resic
outp

P and G
blocks

R block

V fla

Carr

Z

Carr

V_{dd} Outputs   GND V_{dd}

Fig. 9.23 Layout of the checked ALU.

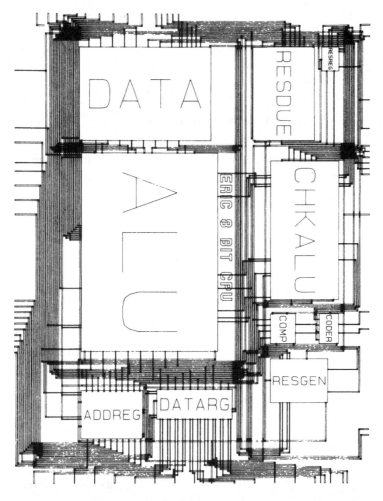

Fig. 9.24 Checked data path cell. ALU, Data path ALU (Fig. 9.23); DATA, Data registers; RESIDUE, Residue registers; CHKALU, Check ALU; RESGEN, Residue generator; COMP, comparator (Fig. 9.3); CODER, Two-rail encoder (Fig. 9.4); ADDREG, Address register (output); DATARG, Data registers (bidirectional); RESREG, Temporary residue register.

CMOS devices become more prevalent, stuck-open faults must be considered. The ability to alter the structure of a CMOS layout so that stuck-open faults, should they occur, can be modelled as classical stuck-at type faults is a useful design method [22]. This leads to the conclusion that a set of design rules should be adopted which not only considers the minimum separation of the signal lines but also considers the layout of the design to prevent hard-to-model (and hard-to-test) faults from occurring [23]. This approach was taken with the design of the data path, although slightly easier to achieve with the

NMOS technology that was used. This allowed only stuck-at type faults to be considered when attempting to verify that the mod 3 checking hardware would perform its task. This type of approach could of course be taken with CMOS designs, with similar results. The results of the simulation are discussed in Section 9.10.

9.9.1 Cost of implementing the checking hardware

The area overhead incurred by incorporating residue checking into the data path circuit is outlined below. The circuit is implemented using $6\,\mu\text{m}$ 5 V NMOS technology.

Data register

Area occupied by:

$$8 \times 8\text{-bit data register} = 1.3 \times 10^6 \ (\mu\text{m})^2$$

$$8 \times 2\text{-bit residue register} = 6.2 \times 10^5 \ (\mu\text{m})^2$$

$$\text{Percentage extra area} = \underline{47\%}$$

ALU

Area occupied by:

$$8\text{-bit ALU with carry lookahead} = 3.0 \times 10^6 \ (\mu\text{m})^2$$

$$\text{Residue prediction hardware} = 1.2 \times 10^6 \ (\mu\text{m})^2$$

$$\text{Percentage extra area} = \underline{40\%}$$

Data path cell

Total area occupied by:

$$
\begin{aligned}
\text{Data path} &= \text{area of ALU} + \text{area of data registers}\\
&= 3.1 \times 10^6 + 1.3 \times 10^6\\
&= 4.3 \times 10^6 \ (\mu\text{m})^2
\end{aligned}
$$

$$
\begin{aligned}
\text{Checking hardware} &= \text{area of residue prediction hardware}\\
&\quad + \text{area of residue registers}\\
&\quad + \text{area of checker and comparator}\\
&= 6.2 \times 10^5 + 1.2 \times 10^6 + 8 \times 10^4 + 2.3 \times 10^5\\
&= 2.1 \times 10^6 \ (\mu\text{m})
\end{aligned}
$$

Percentage extra area overall $= \underline{49\%}$

The degradation in performance of the system is minimal; the major drawback is the necessity to wait for the carry and zero flags to be generated by the ALU before the residue prediction hardware can complete its operation. This problem increases with the number of bits in the design.

It has been reported that the extra hardware required to test a 32-bit CMOS processor chip using scan path is estimated to be between 25% and 40%, increasing to 70% when extra control logic etc. is included. In the case of this data path, because of the large number of registers already present in the design if scan paths or BILBOs are to be used the increase in size may only be of the order of 20%, mainly due to the inclusion of multiplexer blocks in the register for the scan path test method and Exclusive-OR gates with feedback paths if a BILBO method is to be used. However, scan path and BILBO techniques are intended to be employed as a single test of the design either before or after packaging; sometimes they may be used to assist diagnosis of a faulty component. However, residue codes are not only intended to provide ease of testing initially but also to continuously monitor the operation of the chip once it is in service. Residue codes have a further advantage over BILBO techniques in that it removes the need to generate the good signature which can take a large amount of computer time. Residue codes avoid this in that they essentially provide the designer with a preset signature against which a comparison can be made, thereby reducing the simulation time.

9.10 TESTING THE MOD 3 RESIDUE DATA PATH

The data path can be tested by several methods:

(1) The easiest approach is to simply exercise all the blocks in the data path from the microcode PLA/ROM. This may contain a section of code which performs some suitable functional test while a test pin is held true;

(2) A reduced form of scan path may be used where patterns are shifted in but need not be shifted out, since the error-detecting hardware is able to produce a GO/NO-GO signal, reducing the time required for testing. This approach adds little extra area to the design because of the number of registers already available.

The first method is probably the better as it alleviates test costs in terms of specialized automatic test equipment needed to test such devices.

The ALU is the most complex part of the design and as such was subjected to the most rigorous investigations to determine how well the mod 3 checking hardware performed. The addition/subtraction functions are the most hardware intensive operations performed by the ALU in that all parts of the ALU are exercised, including the carry chain. Consequently the addition function was taken as a typical example on which to perform most of the investigations. This involved the use of a special case of the Boolean difference, due to Langdon and Tang [24] to show that all single stuck-at errors within a four-bit slice of the ALU were detectable using a mod 3 residue; if they are detectable in a four-bit ALU slice they will be detectable in any number of connected four-bit slices, since ripple carry is used between each block of bits. This proved by an analytical method that it was possible to detect all single errors using the mod 3 approach. To confirm this result the four-bit ALU section was simulated for all possible single stuck-at faults in the carry path. It was found that all the faults were detectable using a mod 3 residue code, as predicted.

The next stage was to determine the number of test patterns required. This was found to be the $n + 5$ patterns shown in Table 9.8, this sequence of patterns would be repeated for all the possible ALU operations in order to detect faults in the controller. This sequence of tests also gives the residue prediction ALU a reasonable set of inputs cycling through sequences of residues on each residue input.

Since each individual block is checked in its operation and these blocks are linked together to form a data path which is self-checking, the data path can be used as a block capable of checking its own operation. This permits the data path to be used as a self-checking block in a higher level of the hierarchy. Consequently if this test strategy is followed throughout the design the complete system can be made self-checking.

The main objective in using the residue code technique is to provide a means of detecting errors and reporting them as they occur. However, it is desirable in a computer system that degradation takes place gradually and not abruptly as can occur with error-detecting codes. Stand-by units can be used to take the place of a failed or failing unit thereby enabling repairs to be carried out before a total failure occurs; there are many such systems which use this principle to give fault tolerant computing to the user. However, it is desirable that the chip itself continues to function correctly in the presence of the fault. One possible way of achieving this is to use an error-detecting/correcting code; errors could then be detected and corrected up to a certain limit. This approach is discussed in the next section.

Table 9.8: Test patterns for the ALU

C_{in}	A_3	A_2	A_1	A_0	B_3	B_2	B_1	B_0	R_A	R_B	Remarks
0	0	0	0	0	1	1	1	1	00	11	1 pattern, tests P, G & R stages
1	0	0	0	0	1	1	1	1	00	11	1 pattern, tests carry chain
X	0	0	0	1	1	1	1	1	01	11	⎫
X	0	0	1	0	1	1	1	0	10	10	⎬ Test carry lookahead
X	0	1	0	0	1	1	0	0	01	11	1 pattern/bit
X	1	0	0	0	1	0	0	0	10	10	⎭
X	1	1	1	1	0	0	0	0	11	00	1 pattern, carry stages
0	0	0	0	0	0	0	0	0	00	00	⎱ Exclusive-OR faults in
0	1	1	1	1	1	1	1	1	11	11	⎰ final stages

9.11 ERROR-DETECTING/CORRECTING CODES IN THE DESIGN OF THE DATA PATH

One well-known code that has single error correcting and double error detecting (SEC/DED) properties and also requires very simple checkers is the Hamming code [7]. An eight-bit word, as used in the data path, would require five check bits to give a total of 13 bits overall and complete SEC/ DED properties. The bit positions over which the check bits are to be calculated can be obtained in the same way as described in [7]. To optimize the check bits' performance bit positions 0, 1, 2, 4 and 8 will be used. The example below illustrates the technique, where P_n = check bits, B_n = data bits.

	B_7	B_6	B_5	B_4	P_8	B_3	B_2	B_1	P_4	B_0	P_2	P_1	P_0
Bit positions	12	11	10	9	8	7	6	5	4	3	2	1	0
	1	1	1	1	1	0	0	0	0	0	0	0	0
	1	0	0	0	0	1	1	1	1	0	0	0	0
	0	1	1	0	0	1	1	0	0	1	1	0	0
	0	1	0	1	0	1	0	1	0	1	0	1	0
Overall check	1	1	1	1	1	1	1	1	1	1	1	1	1

If the check bits are now calculated the following table is produced:

$$P_0 = B_7 \oplus B_6 \oplus B_5 \oplus B_4 \oplus B_3 \oplus B_2 \oplus B_1 \oplus B_0 \oplus P_8 \oplus P_4 \oplus P_2 \oplus P_1 \quad \text{even parity}$$

$$P_8 = B_7 \oplus B_6 \oplus B_5 \oplus B_4 \qquad\qquad\qquad\qquad\qquad\qquad \text{even parity}$$

$$P_4 = B_7 \oplus \qquad\qquad B_3 \oplus B_2 \oplus B_1 \qquad\qquad\qquad \text{even parity}$$

$$P_2 = \qquad B_6 \oplus B_5 \oplus \qquad B_3 \oplus B_2 \oplus \qquad B_0 \qquad \text{odd parity}$$

$$P_1 = \qquad B_6 \oplus \qquad B_4 \oplus B_3 \oplus \qquad B_1 \oplus B_0 \qquad \text{odd parity}$$

This gives the values of the check bits. The next stage is to determine their characteristics when used in logical and arithmetic operations, as with the mod 3 residue code. For the purpose of this exercise it is better to think of each check bit as being like a single parity bit which has either odd or even parity. Section 8.2.2 describes the Hamming code in more detail.

9.11.1 Logical operations

In residue codes the equations used to derive the relationship between the check bits and data bits are shown below. These equations can also be used for Hamming codes in a modified form:

$$\text{OR} \qquad |N_1 \vee N_2|_b = ||N_1|_b + |N_2|_b - |N_1 . N_2|_b|_b$$

$$\text{EXOR} \qquad |N_1 \oplus N_2|_b = ||N_1|_b + |N_2|_b - 2|N_1 . N_2|_b|_b$$

In this case the base $b = 2$ and the $|N_1|$ can be replaced by the more familiar P_1 to represent the single bit parity. Again it is necessary to generate the parity bits for the logical AND operation between the two data words. As the parity system is base 2 it is possible to replace addition and subtraction by the Exclusive-OR operation. Also since only single bits of information are being considered, multiplication by 2, in the binary system, is the same as a left shift with a 0 being placed in the lowest bit position; this effectively means that the multiplication operation is unnecessary in the EXOR equation. Therefore rewriting the above two equations based on these simplifications gives

$$\text{OR} \qquad A \vee B = P_a \oplus P_b \oplus P_{a.b}$$

and

$$\text{EXOR} \quad A \oplus B = P_a \oplus P_b$$

these operations have to be performed on every check bit.

9.11.2 Arithmetic operations

Two arithmetic operations will be considered – addition and bitwise complementing – since these operations will allow the implementation of the data path instruction set.

The addition operation is checked in a similar manner to that of the residue codes, provided check bits have been generated from the carries [14]. If this information is available, then the check bits for the addition operation can be generated from the following relationship:

$$P_r = P_a \oplus P_b \oplus P_c$$

Where P_r is the check bit of the result and P_a, P_b and P_c are the check bits for the two words A and B, to be added, and the carry information; this operation is performed over all five check bits in the code word. The example below illustrates the technique.

Example 1.

The addition of two 8-bit words including the code bits:

	C	B_7	B_6	B_5	B_4	B_3	B_2	B_1	B_0	P_8	P_4	P_2	P_1	P_0
A		1	1	1	1	0	0	0	0	0	1	0	0	1
B		1	1	0	0	1	0	0	0	0	0	0	0	1
C	1	1	0	0	0	0	0	0	0	1	1	0	0	1
R	1	1	0	1	1	1	0	0	0	1	0	0	0	1

$\longleftarrow$ Bits covered $\longrightarrow$ $\qquad$ $\longleftarrow$ Check bits $\longrightarrow$

The loss of information from the top bit does not matter, therefore twos complement operations do not require any special consideration for the carry bit. The ones complement operations will produce correct results provided that any end-around carry is included in the generation of the carry check bits.

The next important operation to consider is complementation. This operation is required, usually, as a preliminary part of a subtraction operation. To perform the complement operation on the check bits it is necessary to determine whether or not the check bits are generated from an odd or even number of data bits. If an odd number of data bits are complemented then the check bit will also have to be inverted to remain consistent, but for an even number of data bits only the data bits themselves need to inverted – the check bit can remain the same. This is a result of the way in which each check bit is produced.

In the scheme proposed some of the check bits are made up of an even number of bits (P_8, P_4 and P_0) and the others (P_1 and P_2) an odd number. This means that it is only necessary to complement P_1 and P_2 when performing this operation on the rest of the data. An example is given below.

Example 2

The complement operation:

B_7	B_6	B_5	B_4	B_3	B_2	B_1	B_0		P_8	P_4	P_2	P_1	P_0
1	0	1	1	1	0	1	1		1	1	1	0	1
0	1	0	0	0	1	0	0		1	1	0	1	1

This is satisfactory for ones complement negation, but for twos complement numbers it is necessary to add 1 to the resultant complemented number, consequently it is necessary to include the check bits for the added 1 and the resulting carries to produce the correct check bits.

The shift operations are very difficult to check using this method because of the movement of bits within the word, and they are probably better checked by duplication of the shift hardware. However, it must be remembered that a single-bit left shift can be performed by adding the data to itself.

The next section will discuss the application of this technique to the data path described in connection with the residue codes.

9.11.3 Data path implementation

The data path implementation is identical to that used for the residue code except that the form of the checkers etc. must be changed. The ALU performs the same functions and Table 9.9 shows the required Hamming bits to

Table 9.9 Hamming code bits required to predict ALU output bits

| Mnemonic | Bit number | | |
	X	Y	Z
ADD,ADC	P_A	P_B	P_C
SUB,SBC	P_A	$\overline{P_B}$	P_C
ICCA,INCA,IDA	P_A	0	P_C
ICCB,INCB,IDB	0	P_B	P_C
COMA,NEGA	$\overline{P_A}$	0	P_C
COMB,NEGB	0	$\overline{P_B}$	P_C
DCCA,DECA	P_A	LTM	P_C
DCCB,DECB	LTM	P_B	P_C
LT1,CLR	0	0	P_C
NAD,NAC	$\overline{P_A}$	P_B	P_C
AND	0	0	$P_{A\,B}$
OR	P_A	P_B	$P_{A\,B}$
Exclusive-OR	P_A	P_B	0

P_A = parity bits for the A input
P_B = parity bits for the B input
P_C = parity bits generated from the carrier
$P_{A.B}$ = parity bits generated for the AND operation
LTM = parity bits for the constant value 1
The output parity is generated by adding the three parities (X, Y and Z)
in an Exclusive-OR gate

perform each function. The checkers and generators are constructed from Exclusive-OR gates. It is possible to generate a syndrome from the Hamming code which indicates the number of the bit that is in error when a fault occurs; consequently to correct this single-bit error it is necessary to use a decoder. The decoder simply uses the error syndrome as its input, the number of outputs being equal to the number of data bits and check bits. Each output is Exclusive-ORed with its corresponding data or check bit, as shown in the original code generation scheme. If an error occurs the error syndrome is decoded to the bit in error, the Exclusive-OR gate input goes high and inverts the incorrect bit as it passes through the gate. If the decoder is faulty and does not correct the proper bit, then a double error occurs which is detectable but not correctable using this scheme. Therefore single-bit faults can be corrected, and subsequently reported to the error control chip; this could allow maintenance to occur before the system stops completely.

Figure 9.25 illustrates a schematic layout for the Hamming code checked data path. Its layout is very similar to that of the residue coded data path except for the decoder correction circuit. The check ALU is slightly smaller, as are the code generation circuits; however, the extra parity generation circuits for the ALU can increase the size. However, in order to compare it with the previous design a few estimates of the size of the proposed blocks

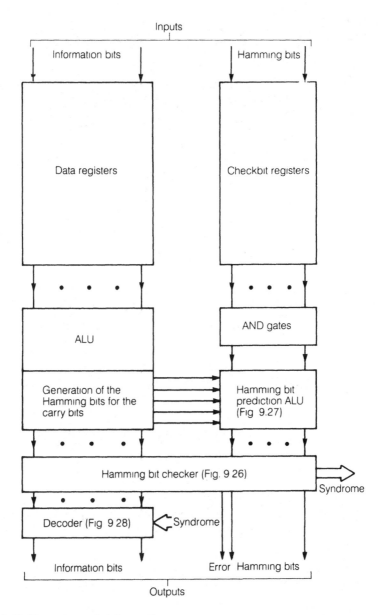

Fig. 9.25 Hamming coded data path.

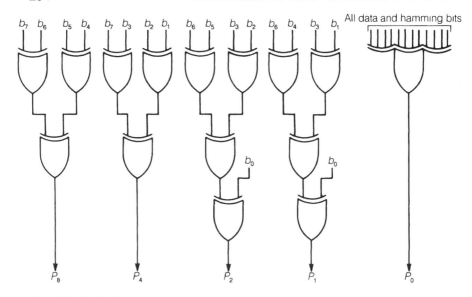

Fig. 9.26 Code bit generators.

were made based on the sizes of the blocks already designed for the residue data path. Figures 9.26 to 9.28 show the internal structures of the cells shown in Fig. 9.25.

Data registers

Area occupied by:

$$8 \times 8\text{-bit data register} = 1.3 \times 10^6 \ (\mu m)^2$$

$$8 \times 5\text{-bit check bit register} = 8.0 \times 10^5 \ (\mu m)^2$$

$$\text{Percentage extra area} = \underline{61\%}$$

These sizes are easily obtained from the dimensions of the RAM used in the residue data path. The data register is duplicated, but five-bits are required by this code.

ALU

Area occupied by:

$$8\text{-bit ALU with carry lookahead} = 3.1 \times 10^6 \ (\mu m)^2$$

$$\text{Check ALU} + \text{carry generation etc.} = 1.7 \times 10^6 \ (\mu m)^2$$

$$\text{Percentage extra area (approx.)} = \underline{55\%}$$

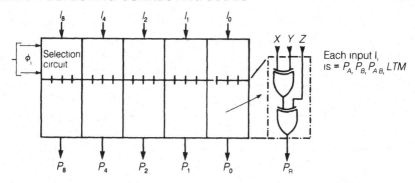

Fig. 9.27 ALU check cell.

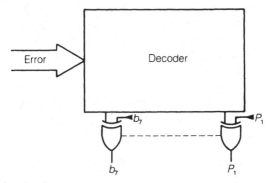

Fig. 9.28 Decoder circuit.

The values are based on estimates for the size of the single-bit parity prediction ALU discussed in Section 9.5. The carry circuit for the Hamming coded data path must be replicated in order to generate the Hamming bits for the carry operation securely.

Data path cell

Total area occupied by:

$$\text{Data path} = \text{area of ALU} + \text{area of data registers}$$
$$= 3.1 \times 10^6 + 1.3 \times 10^6$$
$$= 4.4 \times 10^6 \ (\mu m)^2$$

$$\text{Checking hardware} = \text{area of check bit prediction hardware}$$
$$+ \text{area of checker}$$
$$+ \text{area of decoder, etc.}$$
$$= 1.7 \times 10^6 + 8.0 \times 10^5 + 1.6 \times 10^5 + 4.4 \times 10^5$$
$$= 3.1 \times 10^6 \ (\mu m)^2$$

Percentage extra area overall = <u>70%</u>

The estimate of 70% increase in area required to implement this technique does not take into account any increase in area that may arise due to the construction of each subsystem from its constituent blocks, upon which this estimate is based. If the same operation is performed for the residue data path the overall area increase is about 35%. Therefore since the actual value was 49% the area lost due to the interconnection within blocks is about 37% of the total area. This then gives a figure of approximately 96% for the area occupied by the test hardware for the Hamming coded data path, if the same level of increase occurs in the area as with the residue coded data path. This figure is very encouraging since duplication is likely to increase the area by greater than 100% and only offer the same performance as the error-detecting codes. It must also be remembered that as the number of information bits increases the relative size of the Hamming code checking hardware will decrease, since the number of Hamming bits required increases only logarithmically. Therefore at higher bit densities the overall area increase will be even less. The only drawback the Hamming code scheme has is its inability to perform the checking of shift operations.

It is also quite feasible to use the Hamming code to check the operation of PLAs in a similar manner to the method used for the residue code. Instead of the second PLA producing a residue code it could be used to produce a Hamming code output.

9.12 CONCLUSIONS

This chapter has discussed methods of applying the mod 3 low cost residue code to a range of VLSI devices. Basic hardware requirements, such as the mod 3 adder, have been presented. Also techniques for solving some of the problems associated with the ones complement and twos complement number system have been demonstrated.

The mod 3 residue self-checking method has been presented allowing the use of a unified test philosophy throughout the design of a VLSI chip. The extra hardware required by this technique when compared with LSSD and BILBO techniques is offset by several advantages:

(1) There is continued monitoring of the chip for hard errors once in service;
(2) Most temporary faults are detected as they occur;
(3) A series of similarly checked chips can be linked to a central 'checking chip', allowing the indication of errors when they occur and more importantly an indication of which chip caused the error;
(4) There is no need to generate the system response to a large number of input vectors in order to predict the check signature, resulting in probable reduction of design costs as well as testing times;

(5) The coding scheme presented is very easy to understand and simple to apply. This may be an asset when convincing a designer to incorporate testability techniques. The incorporation of residue checking in a cell based system can be made transparent to the user.

The ability of these codes to detect and report temporary faults as they occur is probably their most significant advantage. As has already been demonstrated in Chapter 2 the number of temporary errors is likely to increase as the scale of integration becomes even greater and component sizes decrease. The importance of continuous checking at little extra hardware cost is therefore a factor which will weigh more heavily in future VLSI systems, and residue codes have been shown to be a cost effective way of providing such checking.

The previous sections have presented methods for introducing fault tolerance into a VLSI design at small cost, when compared to duplication and triple modular redundancy techniques. The action taken when these types of system indicate an error is very much dependent on the environment in which they are operating. The methods used need to determine the type of fault that caused the error, i.e. permanent or temporary.

In the case of the error-detecting code, if an error is detected it is necessary to halt the system and use rollback techniques in either hardware or software, to restore the system to a state prior to the occurrence of the error. The operation can then be re-run, and if the operation now completes successfully then the error was temporary in nature. However, if the operation again fails at the same point then the fault is permanent. If standby chips are available these can be employed in the replacement of the faulty device; alternatively the system can be shut down to await repair.

If an error-detecting/correcting code is used, then the error will be corrected immediately. Thus the only way of differentiating between permanent and temporary errors is to be able to record the source of the fault: if the same fault occurs persistently then it is a permanent fault, but if it does not happen again within a preset time frame then it must have been a temporary fault. In the case of a permanent fault it is necessary to schedule a repair as soon as possible so that multiple errors do not overwhelm the correcting hardware.

In both cases the errors should be logged by a central checking chip with the capability of pinpointing the error to a single chip or circuit on a chip. This greatly assists in the repair of the failed system. It can also assist in the removal of errors within a design or reliability studies on the design, and it would certainly help to reduce the mean time to repair. These possibilities are available with NMR techniques, but the use of coding techniques may have several advantages:

(1) The residue codes require much less hardware than duplication in almost all cases; similarly the SEC/DED Hamming code appears to require less

area than the equivalent TMR techniques. Usually the residue coding scheme appears to be very simple to construct and implement;

(2) The use of TMR and duplication techniques require the design of reliable voter circuits on which the reliability of the design depends. Also with these techniques only crucial blocks are usually checked with inter-circuit/block communications normally being left unchecked. This is not so with the coding techniques presented since the code words are used throughout the design so that data highways are also checked.

The coding techniques presented here seem to offer the necessary savings in silicon area over duplication and TMR to make them cost effective for implementation in VLSI. These techniques are extremely useful for detecting temporary faults which, as was seen in Chapter 2, will almost certainly dominate in future VLSI designs even if extra physical design rules are employed.

Since extra active area is being used to implement these designs the yield of the device [25] will almost certainly drop. However, it may also drop because the scheme might detect faults that were not considered in the original design phase. If an error-detecting/correcting scheme is employed then it is possible to obtain three classes of chips after initial testing. There is a set of chips that are functional and working to specification and the usual set of non-functional chips. A third set of chips now also exists that may function due to the presence of the coding scheme correcting the errors caused by the internal faults, but they would fail if another error occurred. It is therefore possible for the yield of useful devices using this type of coding scheme to be higher than expected. The third class of chips may be used in less demanding applications where their error-correcting abilities are not required. This may be useful in Wafer Scale Integration (WSI).

9.13 REFERENCES

1. Eichelberger, E. B. and Williams, T. W. (1977) A logic design structure for LSI testability. *Proc. 14th Design Automation Conference*, 462–8.
2. Konemann, B., Mucha, J. and Zwiehoff, G. (1979) Built-in logic block observation techniques. *Cherry Hill Test Conference*, IEEE 79CH1509-90, 37–41.
3. Ashjaee, M. J. and Reddy, S. M. (1976) On totally self-checking checkers for separable codes. *Proc. 6th Fault Tolerant Computing Symposium*, 151–6.
4. Rao, T. R. N. (1974) *Error Coding for Arithmetic Processors*, Academic Press.
5. Avizienis, A. (1971) Arithmetic error codes: Cost and effectiveness studies for applications in digital system design. *IEEE Trans. Computers*, **C-20**(11), 1322–31.
6. Monteiro, P. and Rao, T. R. N. (1972) A residue checker for arithmetic and logical operations. *Proc. 2nd Fault Tolerant Computing Symposium*, 8–13.
7. Hamming, R. W. (1950) Error detection and correcting codes. *Bell Tech. J.*, April, 147–60.
8. Garner, H. L. (1959) The residue number system. *IRE Trans. Electronic Computers*, **EC-8**, 140–7.

9. Breuer, M. A. and Friedman, A. D. (1977) *Diagnosis and Reliable Design of Digital Systems*, Pitman.

10. Mead, C. and Conway, L. (1980) *Introduction to VLSI Systems*, Addison-Wesley.

11. Wakerly, J. F. (1979) *Error Detecting Codes, Self Checking Circuits and Applications*, North-Holland.

12. Yung, H. C. and Allen, C. R. (1984) Part I – VLSI implementation of a hierarchical multiplier. *Proc. IEE*, **131**(2), Pt G, 56–60.

13. Wang, S. L. and Avizienis, A. (1979) The design of totally self checking circuits using PLAs. *Proc. 9th Fault Tolerant Computing Symposium*, 173–89.

14. Lewin, D. (1980) *Theory and Design of Digital Computer Systems*, Nelson.

15. Garner, H. L. (1958) Generalised parity checking. *IRE Trans. Electronic Computers*, September, **EC-7**, 207–13.

16. Khakbaz, J. and McCluskey, E. J. (1984) Self testing embedded parity checkers. *IEEE Trans. Computers*, **C-33**(8), 753–6.

17. Prasad, B. A. (1982) Model for VLSI test optimisation using the interrelationships of fault coverage, yield and test quality. *Proc. 12th Fault Tolerant Computing Symposium*, 173–89.

18. Sellers, F. F., Hsiao, M. and Bearnson, L. W. (1968) *Error Detecting Logic for Digital Computers*. McGraw-Hill.

19. Fujiwara, E., Mutoh, N. and Matsuoka, K. (1984) A self testing group parity prediction checker and its use for built-in testing. *IEEE Trans. Computers*, **C-33**(6), 578–83.

20. Kilburn, T., Edwards, D. B. G. and Aspinall, D. (1960) A parallel arithmetic unit using a saturated transistor fast carry circuit. *Proc. IEE*, Pt B, **107**, 573–84.

21. Liesenberg, H. K. E (1985) *A Layout Module for a Silicon Compiler*, PhD Thesis, Newcastle University.

22. Murray, A. F. and Denyer, P. B. (1985) Testability and self test in NMOS and CMOS VLSI. *Proc. IEE*, **132**(3), Pt G, 93–104.

23. Galiay, J., Crouzet, Y. and Vergniault, M. (1979) Physical versus logical fault models in MOS LSI circuits, impact on their testability. *Proc. 9th Fault Tolerant Computing Symposium*, 195–202.

24. Langdon, G. G. and Tang, C. K. (1970) Concurrent error detection for group lookahead binary adders. *IBM J. Research and Development*, September, 567–73.

25. Sze, S. M. (1983) *VLSI Technology*. McGraw-Hill.

10
PROGRAMMABLE LOGIC ARRAY AND RANDOM ACCESS MEMORY TESTING

10.1 INTRODUCTION TO PROGRAMMABLE LOGIC ARRAY (PLA) TESTING

The Programmable Logic Array (PLA) [1] is used extensively in the design of VLSI circuits because of its extremely regular structure and ease of construction. In general, PLAs are used as finite state machines, instruction decoders, or when a regular array of random combinational logic is required. The versatility of the PLA is further enhanced by the range of CAD tools that are available to generate the customization layers of a typical MOS PLA. These CAD tools include minimization programs that are capable of automatically minimizing a Boolean function for implementation as a PLA [2]–[4] or optimally assigning states to produce finite state machines [5], [6].

The use of PLAs in VLSI circuits is likely to increase since their use will enable designs of large complexity to be implemented easily and quickly. However, this will bring with it a need for suitable test methods to verify the design once it has been fabricated. Therefore in the first half of this chapter PLA testing techniques will be discussed. In general the available test techniques attempt to take advantage of the PLA's regular structure in order to make the production of test diagnostic patterns straightforward. Also since the PLA structure is so regular faults other than the simple stuck-at fault can be taken into account; therefore the fault models considered allow a more realistic fault coverage than is generally available with other design techniques. Three different PLA test methods will be considered:

1. test pattern generation methods which use the personality matrix of the PLA
2. built-in test methods which are used to augment the PLA and produce a simplified test set or a totally function independent test set
3. concurrent error-detection techniques that allow faults within the PLA to be detected while it is in the field as well as simplifying the initial test pattern application.

In Section 10.5 the use of PLAs to produce Totally Self-Checking (TSC) circuits will be discussed. The TSC circuits are more easily implemented as PLAs since the fault models used are very comprehensive; therefore the TSC behaviour of a circuit can be easily proved.

10.2 PLA STRUCTURE AND FAULT MODELS

Typically the PLA structure implements a two-level AND–OR circuit in the form of an AND array and OR array. The AND array is driven by either single input decoders producing true and complement outputs or double-bit decoders [7] which can produce the four possible combinations of two inputs. A typical PLA and decoder structures are shown in Fig. 10.1. The outputs from the decoders are called *bit lines*, and are used to select *product term lines* in the AND array. The output is then formed by the logical OR of the active product term lines. The PLA can be constructed from bipolar, NMOS or CMOS technology. Figure 10.2(a) shows the structure of a NOR–NOR NMOS PLA, with single-bit input decoders. If there are n inputs then there will be $2n$ bit lines. The PLA in Fig. 10.2(a) has four product term lines and two outputs, i.e. $m = 4$ and $p = 2$. The intersection of a product term line with a bit line or output line is called a crosspoint. To program the PLA it is simply a matter of placing transistors at the desired crosspoints, as shown in Fig. 10.2(a). In the case of the NMOS PLA the AND and OR planes are programmed as NOR gates. Implementation of any sum of products Boolean function is performed by selecting the correct places for the crosspoint transistors. The product terms therefore represent implicants in the sum of products Boolean function. This programming step can be performed automatically by CAD tools; hence the designer need not be concerned with the synthesis of the required function but only its specification.

A shorthand form is normally used to represent the positions of the programming transistors within the AND and OR array; this is called the personalization matrix. In the AND array part of the matrix a 1(0) shows the existence of a transistor on the true (false) bit line. An X or 'don't care' indicates no programming transistor is present. In the OR array a 1(0) indicates that the implicant is connected (not connected) to the output. The personalization matrix for the NOR–NOR PLA in Fig. 10.2(a) is shown in Fig. 10.2(b). This is sometimes referred to as cubical notation.

Depending upon the function implemented by the PLA it is possible for the AND and OR arrays to contain very few crosspoint transistors (i.e. the arrays are very sparse); if this situation arises then an optimization technique called folding [8] can be used to reduce the amount of silicon area needed to implement the PLA. The folding technique simply involves splitting or cutting input or product term lines so that more than one row or column can

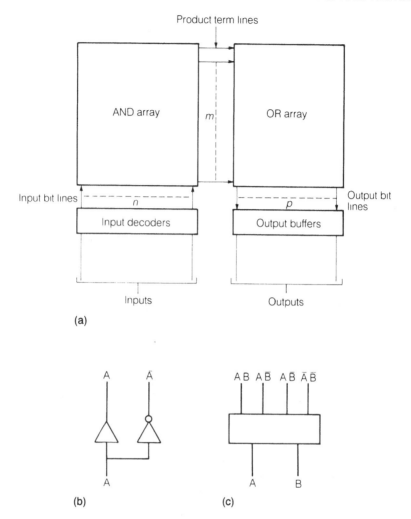

Fig. 10.1 Construction of a general PLA architecture: (a) normal PLA structure; (b) single-bit decoder; (c) two-bit decoder.

occupy the same space i.e. row or column folding takes place. However, as well as the normal faults found in unfolded PLAs it is necessary to consider a special fault that only occurs in folded PLAs. This fault occurs when the row or columns that have been split in order to realize the folded PLA are shorted together at the point where they have been cut. If this fault is considered, then it is necessary to decide which logic value will dominate when the shorted lines are driven to different states. Usually either logic 0 or 1 is considered to

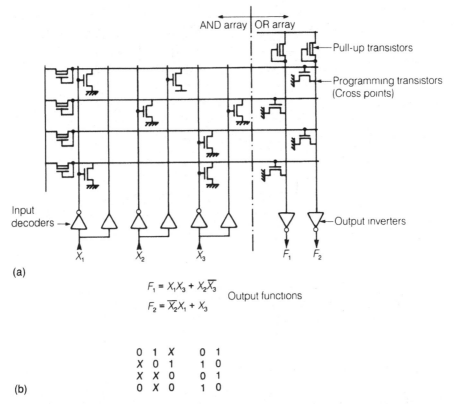

(a)

$$F_1 = X_1X_3 + X_2\overline{X_3}$$
$$F_2 = \overline{X_2}X_1 + X_3$$

Output functions

(b)

0	1	X		0	1
X	0	1		1	0
X	X	0		0	1
0	X	0		1	0

Fig. 10.2 An NMOS Technology PLA: (a) structure of a typical NMOS NOR–NOR PLA; (b) personality matrix for the above PLA.

dominate for all cases, which means that test generation can be greatly simplified. Any test scheme must therefore be capable of testing folded PLAs.

As discussed in Chapter 2 the stuck-at fault model is widely accepted for the testing of combinational logic circuits. With PLAs it is possible to introduce another fault model, the crosspoint defect. If a transistor is not present at the intersection of a row and column in the AND or OR array when it should be, this is termed a *missing crosspoint fault*. However, if a transistor occurs where it should not be this is termed an *extra crosspoint fault*. The crosspoint faults can lead to four classes of defect with regard to the AND and OR arrays [9]. If a crosspoint is missing in the AND array the implicant concerned grows (*growth fault*) since it is no longer dependent on that input. An extra crosspoint in the AND array causes a *shrinkage fault* since an implicant is now dependent on an extra input. A missing crosspoint in the OR array causes a *disappearance fault* since the output of the array is no longer dependent on that

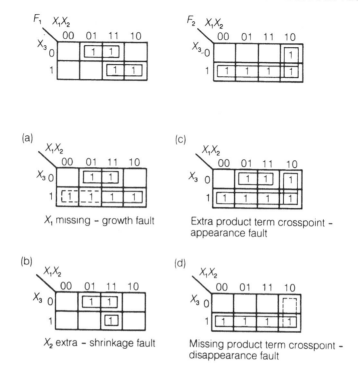

Fig. 10.3 Crosspoint defects in PLA structures.

implicant. Finally, an *appearance fault* occurs when an extra crosspoint in the OR array causes an implicant to appear in the output function. The four failure modes are illustrated in Fig. 10.3 using the *K*-maps for the NMOS PLA shown in Fig. 10.2(a). Smith [9] shows that a test set that detects these four categories of faults also detects the stuck-at 0/1 faults on the input inverters, AND plane inputs and outputs, OR plane inputs and outputs and output inverters. These conditions remain true providing the PLA implementation is not redundant.

Ramanatha and Biswas [10] introduce a fifth fault called the *vanish fault*. This type of fault occurs when an extra crosspoint appears on a bit line of a product term that already has a crosspoint on the inverse of that bit line. In this case the product line will always be zero as at any one time, in a single-bit decoder environment, one of the bit lines is always active. Therefore the product line effectively vanishes from the PLA. This fault is easy to detect provided that the PLA has no redundant product term lines which can be active at the same time as the faulty product line. Generally PLAs should be designed to be non-redundant, so this fault is always detectable.

Tamir and Sequin [11] present an even more general fault model for PLA faults. As well as the normal stuck-at faults, a weak 0/1 fault is introduced. This type of fault could occur due to a shift in threshold voltage levels or floating gate inputs causing the output of a gate to lie between the values of logic 0 and 1. If this fault occurs on a gate with a fan-out of more than one then it is possible for a driven gate to interpret a logic 1(0) value as logic 0(1), in which case the fault is described as a weak 1(0) fault. It is further assumed that a gate will always interpret a weak 0/1 fault in the same way. In a PLA a device that always interprets a logic 1 as 0 (i.e. a weak 1 fault) is equivalent to a missing crosspoint. A total of 12 different faults are considered by Tamir and Sequin in the design of their PLA structure. The types of faults considered range from weak 0/1 faults on input, product and output lines, shorts between these lines, extra crosspoints and breaks in the lines. If a short fault occurs then it is assumed that one logic value will be dominant. If each 'pull-down' transistor is capable of pulling down two normal load devices in NMOS or discharging twice the normal capacitance in CMOS, then this criterion can be met since the output will be forced to logic 0 if one of the shorted lines is logic 0.

A further useful result regarding the ability of fault sets designed to cover single faults and their ability to cover multiple faults is given by Agrawal [12], [13]. It is already well established that in any non-redundant fan-out-free network a single stuck-at fault set will cover multiple faults of size 2 or 3. This can be extended to 98% of multiple faults of size 8. The result is obtained by considering how the growth, shrinkage, appearance and disappearance faults interact. (In the paper these faults are described as contact-0 or contact-1 faults.)

10.3 PLA TEST PATTERN GENERATION TECHNIQUES

The techniques to be described in this section do not require any modifications to the structure of the PLA. Most of the techniques use the personality matrix of the PLA to be tested as the starting point for the test generation procedure. As with normal combinational circuits it may be thought possible that random or pseudo-random patterns could be used for testing, but without modifying the structure of the PLA (see later) these types of test are generally ineffective. The main reason for their ineffectiveness is due to the possibly high fanin of the AND gates, i.e. a large number of programmed crosspoints. Therefore, the probability of a random pattern detecting a missing crosspoint is extremely small for a large number of inputs; if the number of inputs is n then the probability is approximately $1/2^n$, i.e. extremely small.

Test generation techniques that are specifically targeted at PLA structures are far superior to more traditional test algorithms, such as the D-algorithm [14], as they are capable of taking account of the more specialized fault models that can be applied to PLAs. Several techniques [9], [15]-[18] have been proposed to perform this operation. All of the algorithms operate in a similar manner, that is a pattern is applied in order to detect the fault and sensitize the output to the fault condition. The test patterns for a particular fault are readily generated from the personality matrix of the PLA. To illustrate the techniques the algorithms used by Smith [9] and Eichelberger and Lindbloom [16] will be briefly outlined.

The algorithm proposed by Smith [9] consists of the following steps. First a pattern must be applied to the inputs so that the output of the faulty and fault-free PLA are different. This can be achieved using the sharp operation '$\#$'. The sharp operation is a binary operation between two cubes, used to produce the vertices in one cube that are not in the other cube or list of cubes, that is the operation will attempt to produce inputs to a PLA that only activate the product line with the fault. For example if a growth fault has occurred in an implicant P_t, then one of the crosspoints is missing and a new implicant P_t' is formed. Therefore to obtain the input pattern to detect this fault the operation $P_t' P_t$ is performed. The pattern produced is simply the single cube representing P_t with the position corresponding to the fault complemented. If a shrinkage fault has occurred then the 'don't care' at the position of the fault is changed to 1 if a true bit line gained an extra crosspoint transistor or 0 in the case of a false bit line fault. In the case of appearance or disappearance faults the input P_t remains the same; it is simply the output that is affected.

Once the test input pattern has been generated it is necessary to propagate its effect to an output, i.e. a path sensitization step must occur. This can be achieved by choosing an output which is connected, via a crosspoint in the OR array, to the implicant under test. However, there may be other implicants that use this output. Therefore the final test input pattern must be selected from the list of vertices generated by performing the sharp operation between the test input pattern (cube) and a list of implicants (cubes) also used in the formation of the output chosen to propagate the fault information, so that only the implicant under observation will control the output. If no such pattern can be found, then another output column must be selected to propagate the fault and the sharp operation performed with the implicants that are ORed with that output. This procedure continues until all such output columns are exhausted, in which case no test can exist for the implicant in question. After the input pattern has been selected an attempt is made to determine what other faults are detected by this test pattern. If other faults are detected then they are removed from the list of possible faults, thereby reducing the size of the overall test set. It has also been demonstrated that the order in which the faults are considered can have a bearing on the total number of test patterns produced. It is therefore recommended that shrinkage and growth faults are considered first, followed by the disappearance and appearance faults. Normally, if this method is used then a large proportion of the appearance and disappearance faults are detected by the shrinkage and growth fault test sets; consequently fewer test patterns are required. Although only single fault conditions are considered it is possible to extend the algorithm to multiple faults.

The algorithm proposed by Eichelberger and Lindbloom [16], called PLA/ TG, only considers test patterns for the used crosspoints in the AND–OR array of the PLA; although it can be extended to unused crosspoints. The first part of the test pattern generation algorithm involves the production of tests to exercise each product term line by turning it on. This is accomplished by specifying a pattern which activates all the crosspoints on the product lines. If a crosspoint does not exist then a don't care value is written into that position in the pattern. These test patterns simply test for the product line stuck-at-0 Once all the patterns have been generated, by inspection of the personality matrix, a second set of patterns is generated that will test each bit line for being stuck-at-1. To do this it is simply a matter of inverting, in turn, each of the bits in the first test set. As this operation proceeds for each new pattern generated an attempt is made to join it with a test pattern from the first test set, in an operation called subsumption, thereby reducing the total number of test patterns. Once this operation is complete a 'random assignment' of 0 or 1 is made to any don't care terms that remain in the test set; this stage attempts to sensitize a path to the output. The next step is to determine whether or not these test patterns detect the required faults. This can be done by checking to see if only the product term line under test is ORed with the output and no other product term lines are active. If this condition is true then the fault is detected, alternatively the pattern is removed from the test set. At the end of this checking stage, if undetected faults still exist, then the original tests for these faults are used in a further subsumption and random assignment operation. The test pattern generated is again checked as before. These operations will continue until each additional pass detects no further faults or until a preset limit on the number of iterations is reached. In the case of a PLA using a two-bit decoder, if the tenth pass through the algorithm is reached then the initial patterns are regenerated. A further enhancement can be made for any PLA; if no new faults are detected on a pass then the subsumption operation is not performed on the subsequent pass.

10.4 STRUCTURAL PLA TESTABILITY TECHNIQUES

The previous section has described the various algorithms that can be used to generate tests for unmodified PLAs. The main drawback with test generation techniques is that the patterns produced are totally dependent on the function of the PLA; also situations may arise when test patterns cannot be generated for certain faults. A further problem occurs if the PLA is embedded within other logic (this happens in VLSI circuits), rendering it impossible to apply the test patterns in the form required. Therefore the quest for techniques that modify the PLA in some way to enhance the ability to generate test patterns is an active area of research. Also techniques which add extra hardware to the overall PLA structure, shown in Fig. 10.1, are finding favour since they allow patterns to be applied locally to the PLA and in some cases the patterns used are independent of the function implemented by the PLA. These techniques can also increase the variety of faults covered.

The first group of test techniques to be discussed only alters the personalization matrix of the PLA. That is, either extra input or output columns are added to the PLA in order to enhance the testability of the design. One of the first such techniques to generate a fully testable PLA was proposed by Dong and McCluskey [19]. This scheme comprises four different steps. The first step involves reducing the cover between product terms; that is if a product term for one output function is active every time a product term in another output function is active, then this condition is removed by producing two product terms that do not cover each other. For example, consider a four-input PLA with inputs A, B, C and D; if one output is associated with the product term AB, this will cover another product term ABD, assigned to another output. These two terms can be split into ABD and ABD' i.e. ABD now drives both outputs and ABD' just the initial one. Therefore it is now impossible for any single crosspoint or stuck-at fault in the OR array on the output of ABD to be covered by AB. Any redundant terms are also removed after this operation. The next step is to reduce the amount of partial overlap between product terms on the same output. For example if AB and ACD occur as product terms for one of the outputs, then this can be changed to AB and $AB'CD$ removing the overlap. This allows single crosspoint faults or stuck-at faults to be detected in the AND array. These two steps are repeated until all such occurrences are eliminated from the minimized PLA. The final step is to add extra outputs connected to some of those product terms that have been used in the first two steps. This makes sure that the operation of particular product terms is not obscured by other product terms, i.e. the extra output is used as an additional observation point in the PLA. If this PLA is implemented, then every single stuck-at fault or crosspoint defect in the PLA can be tested; also since the procedure can be applied with the minimization phase the final result is nearly minimal and yet fully testable.

A similar technique is proposed by Min [20] to produce an Easy Test Generation PLA (ETG PLA). This scheme essentially arranges the product terms into pseudo-non-concurrent relation groups. (A non-concurrent PLA is one in which only a single product term is active for any input pattern.) A relation group is simply a collection of product terms which have similar sequences in the personality matrix. Each particular relation subgroup is then made non-concurrent by adding extra product terms within the subgroups where necessary. Only the subgroups are non-concurrent; therefore product terms in other subgroups may be active at the same time, hence the term pseudo-non-concurrent. Since the PLA is almost non-concurrent the overlap between output terms is not great. However, to be able to detect all faults it is necessary to add extra outputs in the OR array to make the PLA exclusive. A PLA is exclusive if the outputs of the product terms when taken as pairs differ in at least two bits. This condition is achieved in ETG by using [$M/2$]-out-of M codes as extra bits on the outputs, M being the extra number of check-bits. Furthermore, each relation subgroup may require a different set of code bits in order to distinguish it from other subgroups. The test sequence is easily generated for this PLA and designed to detect all stuck-at faults, bridging faults and missing or extra crosspoint defects. Since the PLA is pseudo-non-concurrent the extra area overhead required is not as great as making the PLA totally non-concurrent.

Instead of adding extra observable outputs an alternative solution is to use extra control inputs on the AND array of the PLA. This allows individual product terms to be controlled with ease. Ramanatha and Biswas [10] and Bozorgui-Nesbat and McCluskey [21] have both proposed schemes to achieve this objective. The main theme behind the Ramanatha and Biswas scheme is the production of tests that are capable of detecting otherwise undetectable faults within the PLA's structure. This is accomplished by making the PLA crosspoint irredundant, using extra inputs connected to specific product terms. The suggested algorithm calculates both the number of extra inputs required and the positions of the programming crosspoints needed for each product term. Programming transistors are only allowed on the false bit lines. Therefore during normal operation these extra bits can be set to logic 1 and during test application they can take on the logic values of 0 or 1. It is possible to calculate an upper bound on the number of extra inputs required, depending on how many product terms have undetectable crosspoint faults.

The scheme proposed by Bozorgui-Nesbat and McCluskey [21] attempts to extend the AND plane with extra inputs so that only one product line and its associated circuitry are activated at any one time. The test pattern or 'select set' is then simply generated from the personality matrix of the PLA. For each product term the range of possible input patterns that can activate the line is produced. A main test pattern for each product line is then chosen which is at least Hamming distance 2 away from all the other members of the select sets of other product lines. An 'auxiliary test pattern' set is then generated for each main test pattern by simply inverting each bit of the main test pattern in turn. Therefore if there were n inputs then the overall test set for a product term would contain $n + 1$ patterns, i.e. a main test pattern and n auxiliary patterns. In order to generate the main test pattern initially there are two possible methods. The first is simply to select by inspection patterns that are a Hamming distance of 2 from other main test sets. The second method is to add extra inputs to the PLA so that the personality matrix is changed, simplifying the task of generating the main test pattern. Unfortunately the assignment of the programming crosspoints to the extra inputs to achieve this objective is an NP-complete problem. However, a simple heuristic is suggested which can be used to overcome this problem. The heuristic involves adding extra inputs to the PLA. Initially all the crosspoints on the extra lines are assumed to be on the complement bit lines. The effect of moving the position of each crosspoint in turn on the Hamming distance of the main test set is then calculated. As long as the net improvement is positive this sequence of events will continue, otherwise it stops. Once this objective has been achieved the generation of the test patterns is straightforward. Changing a single bit in the main test pattern to produce the auxiliary patterns, for a product line, will not affect any other product line as it is now at least a Hamming distance of 2 away. As each pattern is applied to the inputs the output produced is verified for correctness. After the test sequences have been applied the extra inputs are no longer required; they must therefore be disabled for normal operation. However, circuitry has to be included in the PLA to disable any extra input lines that were used to ease test generation since both bit lines (true and complement) might have been used for programming; the simple expedient of applying a 1 to the inputs, as in the last test method, cannot be used. This test method is capable of detecting all multiple stuck-at faults, extra crosspoints and missing crosspoints. The method compares favourably in terms of area overhead and test pattern length when judged against the methods [33], [36] to be described in the next sections.

Using an unmodified PLA it is, in general, impractical to apply random patterns for test purposes since they are ineffective. However, Eichelberger and Lindbloom [22] have proposed a scheme for modifying a PLA so that random patterns can be used as tests. The necessary alterations to the PLA are shown in Fig. 10.4. The inputs R_n represent random pattern test points. The test switch is used to select between test or normal operation; under test this switch is set to logic 1. In this case the inputs R_0 and R_1 are used to select one of the four segments of the AND array; the inputs to the other three sections are forced to logic 1. The inputs to the fourth section are controlled by the normal inputs. Using this technique the probability of generating an AND gate test is greatly enhanced since the number of inputs is reduced. The probability of generating a test can be increased further by the use of the product term select circuit controlled by inputs R_2, R_3, R_4 and R_5; this circuit has the effect of allowing only one product term to be selected at any one time, thereby eliminating the problem of sensitizing an output to the fault being tested. In tests with an unmodified PLA it was found that the best deterministic test pattern generator could detect only 94.3% of the non-redundant faults and with 10000 random patterns only 82.3% of faults were covered. However, when the PLA was modified only 4300 random patterns were

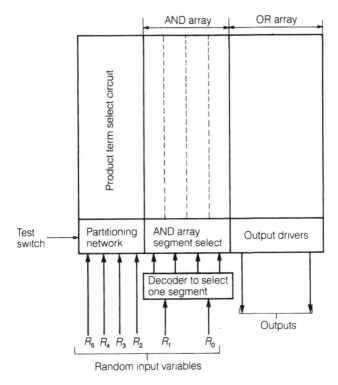

Fig. 10.4 Eichelberger and Lindbloom [22]: random pattern testable PLA.

required to detect 98.0% of the faults. The undetectable crosspoint faults have been made testable by using the product term selection circuit.

10.4.1 Hardware augmentation techniques

The techniques described in the previous section effectively alter the personalization matrix of the PLA in order to enhance the testability. No extra hardware, except a few more columns in the AND or OR array, is required beyond that already available. The techniques to be discussed in this section, however, endeavour to increase the testability of a PLA by adding extra functional blocks to the fundamental structure as shown in Fig. 10.1. The first two techniques that use this approach to testing were proposed by Hong and Ostapko (FITPLA) [23] and Fujiwara *et al*. [24], [25] in 1980. The basic hardware requirements of both techniques are shown in Figs 10.5 and 10.6 respectively. As can be seen from the two figures both of these techniques have certain features in common. For instance shift registers are used by both methods to allow the selection of individual product term lines as

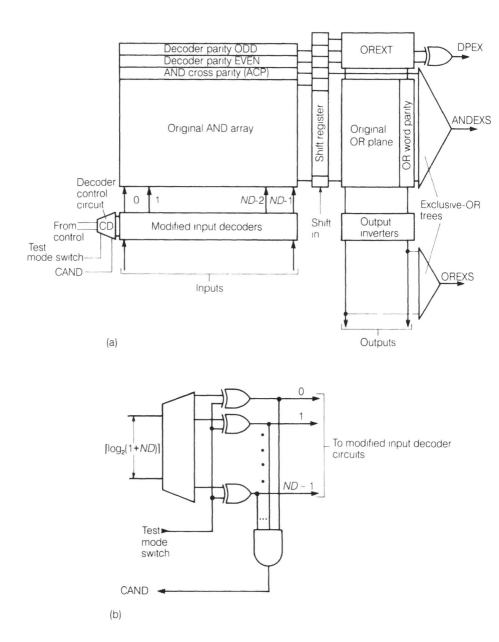

Fig. 10.5 FITPLA test scheme [23]: (a) overall test scheme; (b) decoder control circuit.

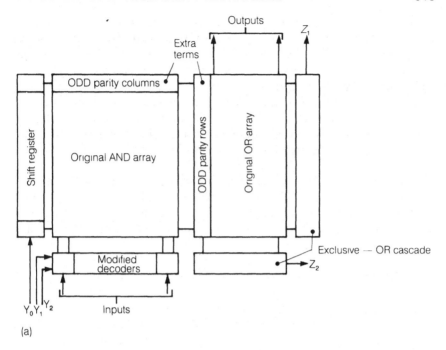

(a)

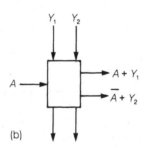

(b)

Fig. 10.6 Fujiwara *et al.* [24], [25]: PLA test technique: (a) overall test scheme; (b) modified decoder circuit.

required. Similarly Exclusive-OR gates are used by both methods as parity checkers on the AND and OR planes. To use the parity method of checking, the number of used crosspoints on the input bit lines and product lines in the OR array must be odd, which is achieved by using extra columns in the AND (called AND cross parity in FITPLA) and OR planes so that extra crosspoint devices can be added where required. Using the parity technique it is possible to check for single missing or extra crosspoint faults, since if either of these conditions occurs the parity will become even and the parity check will fail.

Also the input decoders are modified so that individual bit lines can be selected. The PLA structure proposed by Hong and Ostapko has the additional feature of a parity check on each individual bit line. This is obtained by connecting all the even bit lines to the decoder even parity row in the AND array and all the odd bit lines to the decoder odd parity row. Therefore during the test operation the combined output from these two rows should be 01 or 10. If an output of 00 or 11 is produced then an error has occurred within the bit line decoders; this can be checked using the DPEX output. However, the two methods use slightly different sequences during the application of the tests.

In the case of FITPLA the test patterns are applied as follows. Initially the AND plane is tested by individually selecting each bit line. To select a single bit line the Test Mode Switch (TMS) is set to 0, the number of the decoder to be selected is then applied to the decoder control circuit (CD) and the input of that decoder is cycled through all of its possible input values, selecting all the bit lines attached to the decoder. This sequence will test the crosspoints within the AND array, since the parity will be checked by the ANDEXS Exclusive-OR tree using the And Cross Parity (ACP). An incorrect parity value would indicate either a missing or extra crosspoint. The OR plane is tested in a similar manner: a single one bit is placed into the shift register; if NOR technology is used these bits must be complemented, and all the bit lines are deselected by placing the value ND, i.e. a value equal to the number of input decoders, on the control decoder and setting $TMS = 0$. As the single bit is shifted through the register each of the OR plane lines is tested for correct parity, using the OR word parity line and the OREXS Exclusive-OR tree. In order to test the Exclusive-OR trees it is necessary to apply a special test pattern sequence. In the case of the ANDEXS tree this is achieved by the single 1 passing through the shift register followed by a special pattern of 11 being shifted through the register. This applies the sequence 01, 10, 00 and 11 to the parity tree, which is a sufficient test. The OREXS tree has to be tested using the OREXT bits in the OR array. Within this region of the OR plane the necessary sequence of patterns to test the OREXS is stored; therefore as the 1 bit is shifted along, this portion of the OR plane is activated and the correct test sequence will be applied to the OREXS tree, the all-zeros pattern being applied when the single 1 is in the ACP position. The final tests make sure that the control decoder will not affect the normal operation of the PLA. In this case TMS is set to 1 and the sequence $ND, 0, 1, \ldots, ND - 1$ is applied to the control decoder. The outputs are checked by monitoring the output of CAND. Since the all-ones pattern followed by a single zero on each line is applied to CAND by this input sequence. If all tests are successful the system is returned to normal mode by setting $TMS = 1$ and the input of the decoder to ND. Using this technique all single crosspoint defects are tested.

The testing sequence used by Fujiwara *et al.* for an AND–OR PLA is slightly different since the testing of an Exclusive-OR gate cascade is less complicated. The initial part of this test sequence is used to detect crosspoint defects or stuck-at faults within the OR array. To perform this operation the input bit lines are disabled by setting $Y_1 = 1$ and $Y_2 = 0$ and all decoder inputs to zero. At this point the shift register contains all zeros and Z_1 and Z_2 will also be zero therefore they will be tested for being s-a-1; the following tests on the AND and OR array are now performed. A single one bit is shifted into the shift register, which activates each product term line in turn and hence each line in the OR array. The output parity of the OR array is checked by observing the output Z_2; the output Z_1 is also observed to detect stuck-at faults within the AND array. The control lines Y_1 and Y_2 and inputs to the decoder are then inverted and the whole sequence repeated, again observing Z_1 and Z_2. This test sequence as well as detecting crosspoint defects and stuck-at faults in the OR array can also detect stuck at 1/0 on each of the product terms. The testing sequence is now switched to the AND plane. In order to perform this operation the shift register is set to all ones. The decoder control lines are then set to $Y_1 = 0$ and $Y_2 = 1$ with all the inputs to the decoders set to 1 except for a single zero input. Therefore a true bit line is set to 0 and all other bit lines are set to 1, which activates all the crosspoints to which the bit line is attached; since this should be an odd number of product lines in a fault-free PLA a correct parity bit will be generated. However, a faulty PLA will produce an incorrect parity bit, this could be caused by a crosspoint fault (extra or missing) or a stuck line. Therefore line Z_1 can be checked for correct parity in the AND plane, as this single zero bit is applied to each decoder in turn. Once all true bit lines have been checked the input patterns are inverted and the test is repeated to check all the false bit lines. This sequence of test patterns will check all crosspoint faults and stuck-at-1 faults in the AND array. It is claimed that all single stuck-at faults and crosspoints faults are detected by this method. In between these two 'prime' test sequences a one bit is shifted into the shift register so that it becomes all ones. This sequence helps check the operation of the shift register and also sets the shift register to the right values for the next test.

The main drawbacks with these two methods of hardware augmented test is that the Exclusive-OR cascades can introduce large delays when in test mode. Also the replacement of a cascade by a tree structure, although speeding up the parity check operation, has inferior multiple fault detection properties. However, both techniques do allow a function independent test to be performed. That is, the test sequences and responses do not depend on the function implemented by the PLA. It is also true to say that the total number of patterns is linearly dependent on the number of inputs and product terms. However, the overhead required by the Exclusive-OR gates can be excessive; consequently the test schemes which follow attempt to eliminate these structures altogether. One such scheme proposed by Khakbaz [26] has an

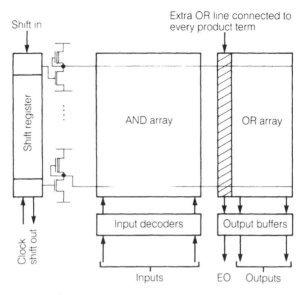

Fig. 10.7 Khakbaz [26] PLA test technique.

extremely low overhead and high fault coverage. An outline of this method is shown in Fig. 10.7. The only extra pieces of hardware required by this technique are a shift register to control the product term lines and an extra column in the OR plane which has a crosspoint on every product term line. The test sequence for this method is also straightforward, although the input patterns and the output results do have to be determined from the personality matrix of the PLA to be tested. The test pattern application can be broken down into four phases. The first phase tests the shift register to make sure it can shift data in and out correctly and the extra output (EO) to check that it is not stuck at 0 or 1. Stage two of the test sequence checks for the existence of devices on the outputs of the shift register by deselecting (i.e. placing all ones in the shift register) all the product lines and then attempting to activate each product line in turn by applying an appropriate pattern at the inputs of the PLA. If all the devices on the shift register are present, then the output EO should remain at zero. The third stage involves controlling one product line at a time by placing a zero in the shift register bit corresponding to the product line required and a one in all the rest. The product line in question is then activated from the inputs: if it is correct EO should go to one. The bits of the input pattern to select the product term are now inverted in turn. If the position that is inverted is a don't care, then EO will remain at one, but if the position has a programmed crosspoint, then the output EO will go to zero. Therefore all the crosspoint defects in the AND plane will be checked. The final stage simply involves selecting each product term in turn and reading the corresponding output produced and checking it against that expected from the function implemented by the PLA. Although this technique does require some knowledge of the function of the PLA the area overhead is extremely small. This technique is capable of detecting all single and multiple stuck-at faults and all crosspoint defects or indeed any combination of these faults. This technique is very easily applied to folded PLAs.

Another technique which produces a testable PLA with low area overhead has been proposed by Reddy and Ha [27]. In this scheme the PLA is modified so that a single controlling pass transistor is incorporated onto the true bit line of the single-bit decoder. These pass transistors are then controlled by a control line C. Also if the complement bit line is not connected to any product lines, then it is deleted and no pass transistor is placed on the true bit line. Similarly a true bit line is deleted if it does not connect with any of the product lines. By the use of a pass transistor a dynamic latch is formed on the input of the true bit line. Therefore if the control line C equals one then the decoder behaves normally, however if C equals zero then the previous value of the input is stored on the input of the true bit line. If the decoder input then changes, only the complement bit line will be affected. In two operations it is possible to set both the true and complement bit lines to the same value for test purposes. The test patterns for the augmented PLA can either be derived from the implicants of the function or from the implicants and their complement terms which are sometimes produced as a by-product of the minimization procedure. Since a single test may require two patterns to set up the particular bit lines, judicious ordering of the test sequence can help reduce the overall length of the test. However, this method of test ordering would depend on how long the charge was retained on the node of the dynamic latch. As this time is usually large compared with the test application time any stored state would last for several hundred test patterns. The faults that this technique is capable of detecting are multiple stuck-at, crosspoint and bridging faults. In tests carried out with 56 different PLAs the average number of tests required was found to be 488 reducing to 377 if the order of application was considered. This was achieved with only a 2% increase in area required to implement the PLA. This technique can also be applied to folded PLAs.

10.4.2 Self-testing PLAs

The current trend in VLSI testing is to produce circuits that are not only easily testable but are also capable of performing the tests themselves, usually on-chip; this eliminates the problem of applying specific test patterns to a PLA when it is embedded within a complex design. One technique called BILBO [28] has been proposed to perform this operation for combinational circuits and this section will look at the techniques that are available to produce self-testing PLAs.

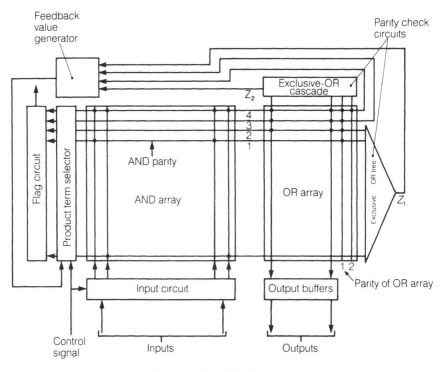

Fig. 10.8 Yajima and Aramaki (ATPLA) [29]: PLA test technique.

One of the first self-testing schemes was the Autonomously Testable PLA (ATPLA) proposed by Yajima and Aramaki [29] in 1981. The main components required for this technique are shown in Fig. 10.8 in relation to the normal PLA hardware. In addition to extra hardware it is also necessary to add four extra product term lines to the AND array and two extra output columns to the OR array. In each case one of the lines is used to make the number of used crosspoints in the AND and OR arrays odd. The parity line in the AND array is only connected to the OR parity line. Two of the four extra AND array lines are only connected to every crosspoint in the OR array except the second OR array line, which is connected to the fourth extra AND line. This fourth line is connected to all the crosspoints in both arrays. To perform the test the product term selector is set to all ones (this is simply a shift register as discussed in the previous sections) and a zero is put into the first bit of the input circuit, all other bits being set to one. This again is a shift register which controls all the bit lines; the control signal is then set to test mode. The four signal values produced by the lines shown in Fig. 10.8 are fed back to the *feedback value generator* which produces the next bit for the product term selector register; if the PLA is fault free this will be a 1. A logic 1 is also applied to the input of the 'input circuit', so a single zero occurs on each bit line as the register is clocked. When the zero is shifted out of the input circuit, it becomes all ones, it will now remain in this state throughout the test; the feedback value generator produces zero, because of the arrangement of its inputs, to input into the product term selector register. The new sequence tests the AND array. This situation continues until the product term selector register is all zeros, then because of the arrangement of the feedback lines into the feedback value generator a single one is produced and shifted into the product term selector register; this bit is now shifted along followed by zeros from the feedback value generator. If at any time during the test sequence a fault is detected, then one of the feedback lines will be wrong and an incorrect bit will be fed into the product term selector register modifying the test patterns applied to the PLA and compressing the fault information. The test, however, guarantees to keep this information until the end of the test only if single bit errors occur; however, if multiple errors occur then faults could be masked due to the way in which test data is stored in the product term select register. At the end of the test sequence either the values in the product term selector register can be shifted out for checking or alternatively the encoded data from the product term selector register stored in the flag circuit can be observed; in either case any errors can be detected. All single crosspoint defects are tested using this test method. With the proposed scheme the patterns applied and the outputs generated are independent of the function implemented by the PLA.

A testing technique based on the use of BILBO [28] registers has been suggested by Daehn and Mucha [30]. The test patterns for an n-input NOR gate are shown in Fig. 10.9(a); also shown is a non-linear feedback shift register (NLFSR) capable of generating this pattern sequence. In order to detect bridging faults between bit lines in a two-bit decoder the patterns shown in Fig. 10.9(b) are required; again these can be generated by an NLFSR. These two structures form the basis of the proposed testing scheme. The extra piece of hardware required is a multiple input signature analysis register [28]; this is simply a linear feedback shift register which is capable of receiving a series of parallel inputs and compressing them into a final unique signature. Figure 10.9(c) shows the overall test structure. The BILBO structures are simply a combination of the generator NLFSR and the signature LFSR. In order to test the two NOR planes of the PLA the following sequence is used. BILBO 1 generates the patterns for NOR plane 1 and the results are compressed into a signature by BILBO 2. Once all the patterns have been generated, the signature is shifted out of BILBO 2 for comparison to the pre-calculated result. The scene of testing is then shifted to NOR plane 2 where BILBO 2 now generates the inputs to test the plane and BILBO 3 compresses the result. After the test sequence is finished the signature is again shifted out for comparison. If the outputs of the PLA are connected to the inputs of another

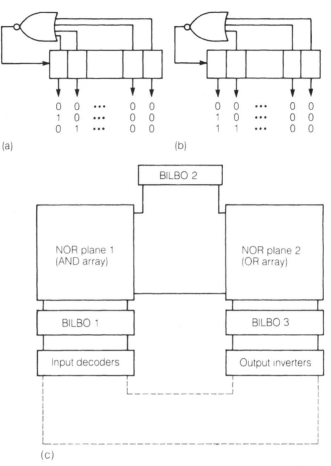

Fig. 10.9 Daehn and Mucha [30]: self-testing PLA scheme: (a) non-linear feedback shift register for the NOR planes: (b) non-linear feedback shift register for the two-bit partitioning network; (c) typical layout for the test scheme.

PLA or if the outputs can be connected back to the inputs of the PLA, then BILBO 3 can be used to generate patterns to check the output buffers and input decoders. The result of this test will be collected by BILBO 1; the signature is again evaluated by shifting off-chip. BILBO 3 can use the NLFSR scheme shown in Fig. 10.8(b) to test the inputs of two-bit decoders. In general the test patterns applied are not dependent on the personalization of the PLA. However, the signatures generated in the BILBOs are dependent and simulations must be performed in order to calculate their value. This test scheme is capable of detecting all crosspoint defects and bridging faults. The number of patterns required is linearly dependent on the number of inputs, product terms and outputs.

Another scheme which uses a test strategy similar to that of Daehn and Mucha has been proposed by Grassl and Pfleiderer [31] and is illustrated in Fig. 10.10. This scheme uses two shift registers to activate the crosspoints. One shift register is incorporated into the input lines and the other is attached to the product term lines by a selector circuit. The selector circuit is used to control two product lines from only one shift register bit. This method of controlling the product term lines has been used so that the pitch of the shift registers output can be brought into line with that of the product terms. The outputs of the OR plane are connected to the inputs of a multiple input

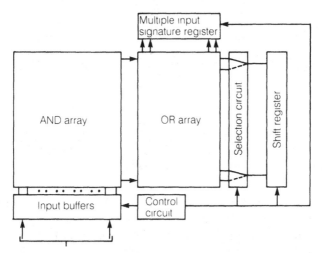

Fig. 10.10 Grassl and Pfleiderer [31]: self-test PLA scheme.

signature analysis register so that the output responses from the test sequence can be compressed into a more manageable form. Initially the test sequence itself is simple, a single bit line is activated, then using the selection circuit and shift register on the product lines each product line is activated in turn. Another bit line is selected and the whole procedure of product line activation is repeated. This sequence of events carries on until all input bit lines have been tested. To complete the test the signature produced in the MISR is compared to the signature for a good PLA. This test scheme has been applied successfully to the testing of an eight-bit ALU-folded PLA. The extra area required by the test hardware was only 44% of the total PLA area. The scheme is capable of detecting all shorts between adjacent lines as well as stuck-at-1/0 faults and missing or extra crosspoint defects; also the testing hardware is checked.

Hassan and McCluskey [32] have suggested the use of a linear feedback shift register and MISRs to test a PLA. The test scheme is illustrated in Fig.

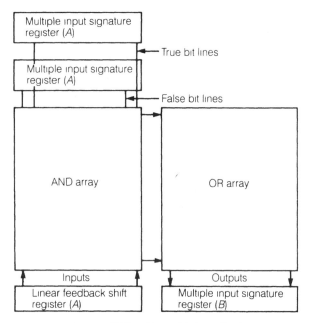

Fig. 10.11 Hassan and McCluskey [32]: self-test PLA method.

10.11. Each of the LFSRs marked A in the circuit has the same characteristic polynomial; this is very important if the scheme is to function correctly. If a PLA has a greater number of outputs than inputs then the MISR marked B in Fig. 10.11 can be constructed from several type A MISRs or some of the outputs can be compacted using exclusive-OR gates before feeding into the MISR. These techniques allow a trade off between the fault coverage and area overhead to be made. The operation of the test is as follows. The input LFSR is cycled through all of its possible output patterns, except the all-zero case. The MISRs compress the outputs produced into signatures. When the test is over the signatures generated are compared with good signatures; if they match then the PLA is 'fault free'. This scheme is capable of detecting all multiple bit line stuck-at faults, as well as most stuck-at faults in the AND plane and on the product lines, and all output stuck-at faults. When this scheme was applied to an example PLA an increase in area of only 23% was recorded. The main drawback, however, with this technique is that as the number of inputs increases the length of time required to exhaustively test the PLA increases exponentially, but the scheme is applicable to PLAs with a large number of product terms and only a few inputs.

One particularly elegant solution to the problem of PLA testing originally proposed by Fujiwara [33] in 1984 and extended by Treuer *et al*. [34], [35] is illustrated in Fig. 10.12 along with the test patterns required. The test scheme

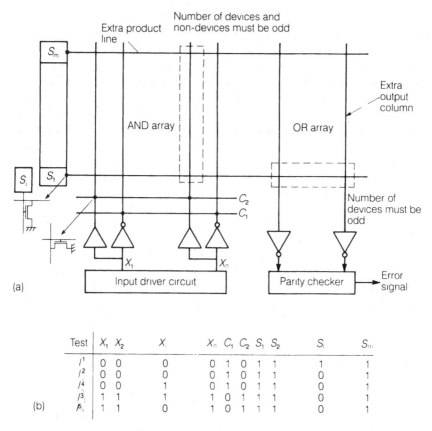

Fig. 10.12 Treuer *et al.* [34], [35]: test scheme: (a) PLA structure; (b) test sequence used on the PLA.

uses extra control lines C_1 and C_2 to allow each individual bit line to be controlled from the normal inputs. A shift register is also added so that individual product term lines can be selected. The number of devices and non-devices on the bit lines is made odd by the use of extra product lines. If the number of product lines is odd then only one extra product line is needed; otherwise it is necessary to add two extra lines. The number of crosspoints on the output lines is also made odd by using an extra output column. Hardware is also required to drive the inputs $X_1, \ldots, X_n$ through all the required function independent input sequences. A parity check circuit, using exclusive-NOR gates, is added to the outputs. The test patterns in Fig. 10.12(b) have the following properties. The first pattern I^1 disables all the product terms and produces a zero parity output in a fault-free PLA; this pattern is used to check for stuck-at faults. The two pattern sets I_j^2 and I_j^3 are used to test the crosspoints in the OR array. In a fault-free PLA these patterns will activate the m product lines in turn, producing an output parity bit of one in each case, since the outputs produce odd parity. Using the patterns an output vector of m ones will be produced. The patterns I_{ij}^4 and I_{ij}^5 test the AND plane. One bit line is enabled at a time while all the product term lines are selected in turn. If a crosspoint is present on the product term line, then it will be pulled down to zero and the output will be a zero parity bit, since all the outputs will be zero. If no crosspoint is present then the product line will stay high and the output will generate a one parity bit, since the output will be an odd number of ones. However, because the number of crosspoints and non-crosspoints on the product lines is odd then there will be an odd number of 0 and 1 parity bits produced. Using these parity bits it is possible to check the function of the PLA in such a way that it does not depend on the PLA's personality. Each new output parity bit is exclusive-ORed with a bit representing the accumulated parity of all the previous test vectors; this bit is called the cumulative parity. If the cumulative parity bit is checked on only $2n + 2m + 1$ occasions it is possible to check the operation of the PLA. As can be seen the input sequence and output responses are independent of the PLA on which the scheme has been implemented. Using this scheme a PLA can test itself for all single crosspoint stuck-at faults, bridging faults and almost all multiple faults. This scheme has been implemented on many PLAs, producing an increase in area of as little as 15% in some cases for NMOS PLAs.

A test scheme proposed by Saluja *et al.* [36], [37] and illustrated in Fig. 10.13, aims to detect all multiple stuck-at faults and is totally independent of the function implemented by the PLA. This scheme uses a shift register and OR gates to control the bit lines in the AND array. Another shift register is used to perform a similar function on the product term lines. An extra product term is added to the AND (Z_1) array and connects to all the bit lines. An extra column is also added to the OR array (Z_2) and is connected to all the product term lines. The test pattern used basically consists of shifting ones and zeros into the shift registers, in a similar manner to the previous example, with $C = 0$ and then $C = 1$. This scheme can be used in a self-testing mode by simply adding an on-chip generator for the test patterns and a multiple-input signature register to compress the PLA outputs, including Z_1 and Z_2. If the MISR is used, then the output signature is dependent on the function programmed into the PLA.

The final scheme to be discussed under the heading of self-testing PLAs is proposed by Hua *et al.* [38] for testing Finite State Machine PLAs (FSMBT), and is illustrated in Fig. 10.14. This scheme uses parity checking on the AND and OR plane by adding extra lines to both arrays. In the case of the AND array the contacts on the extra product line are arranged to make the number of non-contacts on the bit lines odd. However, the contacts on the extra output column of the OR array are arranged so that the output is odd or even

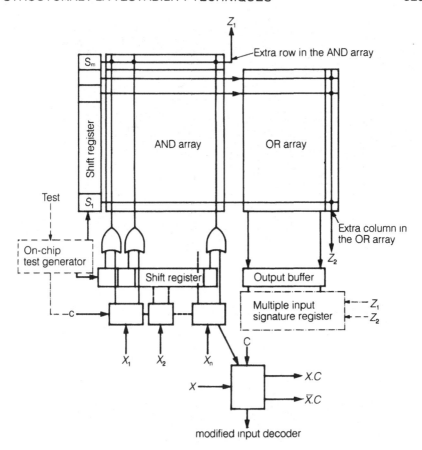

Fig. 10.13 Saluja *et al.* [37]: self-test method.

parity depending upon how many columns were present in the OR plane originally. Two further lines are connected to the AND plane: one connects to all the product term lines and feeds into the test pattern generator; the other connects to all the input lines and feeds into the augmented decoder, as well as acting as a test output (E_1). The test pattern generator and augmented decoder are simply shift registers controlled by the lines T_1 and T_2. The patterns used in the registers are those normally associated with NOR gate testing and shown in Fig. 10.9(a). If $T_1 = T_2 = 0$ this is normal machine operation. $T_1 = 1$ and $T_2 = 0$ are used when testing the AND plane. $T_1 = 0$ and $T_2 = 1$ are used when testing the OR plane. The test response can be checked by monitoring lines E_1, E_2 and E_3. The scheme covers shorts between adjacent lines, stuck-at-0/1 faults and missing or extra crosspoints.

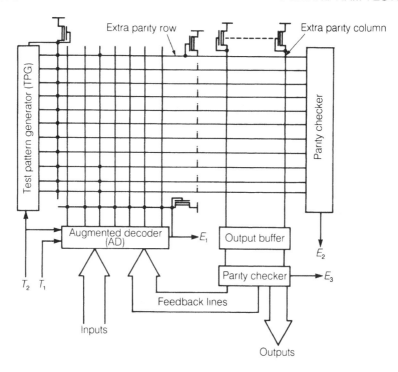

Fig. 10.14 FSMBT (adapted from [38]).

10.5 CONCURRENT ERROR DETECTION IN PLAS

So far all the schemes described for the testing of PLAs have been aimed at testing the circuit when it is off-line. That is either to prove the circuit works after fabrication or for the periodic checking of the device once it is in service. However, neither of these approaches can detect transient or ageing faults when they occur and before they affect the operation of the device. In order to achieve this objective of enhanced testing it is necessary to detect the errors while the device is operating. These concurrent error-detection techniques usually involve using some form of encoding on the outputs of the PLA to detect the faults as they occur. For example a codeword output from the PLA signifies correct operation and a non-codeword output indicates that a fault has occurred. One particularly efficient method of constructing a PLA of this type is based on Totally Self-Checking (TSC) circuits [39]. The definitions concerning this type of circuit have already been given in Chapter 8. There are many schemes [11], [40], [41] which build the PLA directly as a TSC circuit and then use TSC checker circuits on the outputs and inputs of the PLA to

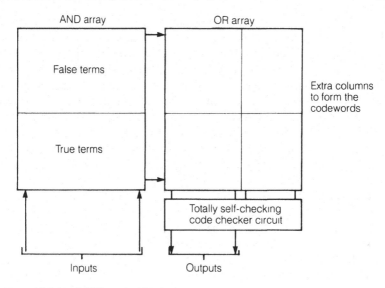

Fig. 10.15 Mak *et al*. [43]: coded PLA.

determine if the system is operating correctly. The main drawback with these techniques is that to make sure no undetectable faults occur in the checker circuits all input combinations must be applied to the circuits, and this is usually difficult for most PLAs. However, Wang and Avizienis [40], who pioneered this work in 1979, do propose a systematic method of constructing a TSC PLA. Generally, in order to produce a TSC circuit the PLA must be non-concurrent, that is only one product line should be active at a time. This restriction is not too great since many VLSI PLAs with a large number of product terms may naturally be non-concurrent. Also it may be possible to partition the output lines into different code sets in order to be able to apply all possible input combinations to the TSC checker circuit used on that output group, thus more than one TSC may be employed on checking the outputs. The faults covered by this scheme are single stuck-at faults, crosspoint defects and shorts between adjacent lines.

A broader definition of TSC is given by the Strongly Fault Secure (SFS) circuits proposed by Smith and Metze [42] which can cover multiple faults. For this reason the Mak *et al*. [43] scheme, illustrated in Fig. 10.15, has adopted the SFS definition for the design of their PLAs with concurrent error detection. They also use the fact that any errors in the OR plane of a PLA will produce undirectional errors (i.e. all 1(0) outputs will be changed into 0(1) outputs by the fault); this allows m-out-of-n codes [39], Berger codes [44] or modified Berger codes (defined in [43]) to be used to encode the output of the PLA. A further restriction is placed on the PLA in that all normal inputs to the PLA must activate at least one product term line in the AND array. To achieve this objective the false product terms normally excluded from the design of any PLA must be taken into account. Thus the size of the PLA will normally be larger than expected. Of course the number of outputs in the OR plane must also be expanded to allow for the extra bits required by the encoded output. Mak *et al.* present a set of rules which allow for the systematic generation of SFS PLAs incorporating one of the three codes. Increases in PLA area of the order of 40% can be expected when using this technique. Again all the normal fault types are detected by this method. Since it is very difficult to be able to guarantee the generation of all possible inputs to adequately check the checker circuit on the outputs of the PLA Mak *et al.* suggest two solutions. The first requires the design of a new checker circuit for each SFS PLA produced. The checker circuit is designed in such a way that all the possible patterns that can be produced by the PLA will exercise the checker adequately. The second solution involves adding registers to the OR array which contain the extra code patterns necessary to completely check the checker's circuits. During normal operation the PLA AND plane would be periodically disabled and these registers activated to flush out any latent faults that may have occurred during the operation of the PLA. This scheme allows any TSC check circuits to be used in detecting faults in the output of the PLA; however, the size of the OR array will be increased.

Khakbaz and McCluskey [45] have proposed a concurrent error-detection scheme for PLAs (illustrated in Fig. 10.16) which does not involve more than minor structural alterations to the PLA. However, the PLA used must be large, non-concurrent, use single-bit input decoders and be implemented in NOR–NOR logic. A further advantage of this technique is that it can be used to initially test the PLA once it has been fabricated; therefore the extra area overhead due to the checking circuits is effectively being used for two operations. The PLA is checked by three different coding schemes. TSC two-rail checkers are used to check the input bit lines. A TSC 1-out-of-m code checker is connected to the product lines. Since the PLA is non-concurrent only one product line should be active for each input pattern; therefore any faults that cause a deviation from the norm will be easily detected. The output of the PLA is also encoded in some code, for example single-bit parity; a checking circuit is then used to check the output of the PLA. When this technique was added to a PLA with 25 inputs, 40 outputs and 300 product terms only a 37% increase in the circuitry required was recorded; the usual fault set is detectable using this technique.

The two techniques discussed above both require that only one product term must be active for any input. The technique proposed by Chen *et al.* [46], as well as being applicable to ROMs and non-concurrent PLAs, can be extended to concurrent PLAs. The scheme proposed for the checking of concurrent PLAs is illustrated in Fig. 10.17. In this technique three extra

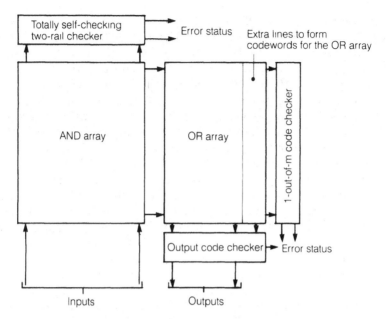

Fig. 10.16 Khakbaz and McCluskey [45] concurrent error-detection method.

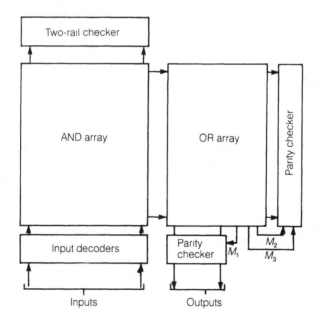

Fig. 10.17 Chen *et al.* [46]: concurrent error-detection scheme for concurrent PLAs.

outputs are added to the OR plane. One bit, M_1, represents the odd parity of the OR plane outputs. The bits M_2 and M_3 represent the odd and even parity, respectively, of the product lines of the original circuit. That is, these bits are used as a parity check on the number of active product lines in the AND array. The generation of M_1, M_2 and M_3 occasionally require the addition of extra product terms to the AND plane. In an example by Chen *et al.* with 20 inputs, 100 product terms and 30 outputs, 15 extra product terms were needed and the overall increase in hardware required was only 34%. The types of fault covered by this technique include single stuck-at-1/0 faults, missing or extra crosspoints and shorts between adjacent and orthogonal lines.

This part of the chapter has discussed the types of test methods that have been used to detect faults in PLAs. In general the techniques range from 'traditional' test pattern generation methods to more sophisticated methods which modify the PLA so that a universal test input and response can be obtained. Obviously the choice of method will depend on the constraints of the circuit. For example if the inputs and outputs of the PLA are easily accessible, then test generation techniques might be the preferred method; however, a PLA that is deeply embedded within a design may require extra hardware in order to eliminate the problems of test pattern application and test response evaluation. Within the domain of VLSI, PLAs are likely to become embedded within the overall design; therefore techniques aimed at allowing the built-in self-testing of PLAs will become dominant, the extra area required for their implementation being the price that must be paid for the easy testing of devices.

10.6 INTRODUCTION TO RANDOM ACCESS MEMORY TESTING

As the complexity of VLSI circuits grow, the need to include fast local Random Access Memory (RAM) on the chip increases. Also the reduction in device geometries will bring with it higher density RAMs on a single chip: already one-megabit RAMs are being produced commercially, with four-megabit RAMs projected in the near future. Therefore the requirement is for tests which are capable of detecting faults in both embedded and single-chip memories. However, this will be slightly easier for single-chip RAMs since all of the signal pins are accessible to the test equipment. Nevertheless in the case of embedded RAMs test algorithms which can efficiently detect RAM faults and also be implemented as self-testing structures are required in a VLSI environment. The following sections will describe some of the techniques and methods available to test RAM structures. An additional requirement is for RAMs which are capable of tolerating temporary faults, as discussed in Chapter 2.

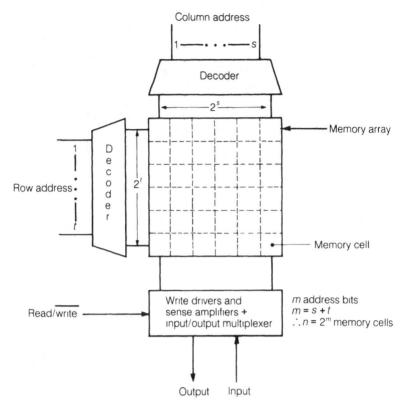

Fig. 10.18 Typical memory array architecture.

10.7 RAM STRUCTURES AND FAULT MODELS

Figure 10.18 illustrates the typical functional elements of a simple RAM. If the RAM has an n-bit capacity then there will be m address bits, where $m = \lceil \log_2 n \rceil$. Usually the m-bit address is partitioned into two bit fields of s bits and t bits, controlling 2^s columns and 2^t rows. The t bits select one of the 2^t rows, the s bits then select one of the 2^s columns and a signal line in the output multiplexer, the intersection of the row and column is the selected cell. If a read operation is in progress then data will be read out of the selected cell; if a write operation is being performed then the data on the input is written into the selected cell.

There are two possible types of RAM cell, static and dynamic. This terminology generally relates to the method used to store the information in the bit cell. The two forms are shown in Figs 10.19(a) and (b) respectively. The static

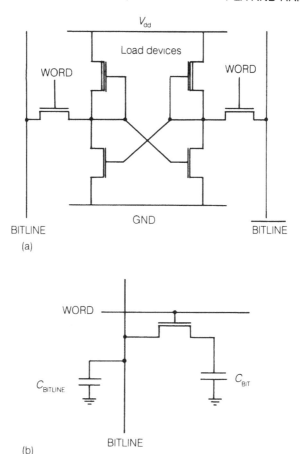

Fig. 10.19 Typical RAM cell structures: (a) static NMOS bit cell; (b) dynamic MOS cell.

form of ram cell uses cross-coupled inverters to store the state of the bit. To write into the cell the data value is placed on the BITLINE and $\overline{data}$ is placed on $\overline{BITLINE}$. The WORD line is then driven true and the inverter will flip to the required state. To read the cell the BITLINEs are precharged. The word line is again driven true; one of the BITLINEs will then be discharged by the pull-down transistor of the ON inverter producing the output state via the sense amplifiers. This type of cell requires six transistors to perform its operation. Therefore in order to place more devices on a chip a cell with either smaller geometry transistors or a smaller number of transistors is required. A reduction in the number of transistors required can be achieved by using the dynamic cell shown in Fig. 10.19(b). This type of cell relies on the storage of charge on an MOS capacitor. However, because of the problems of charge sharing etc. the design of this type of cell is more complicated than that of a static cell. Despite these problems, most high-density RAMs currently use this form of construction. Unfortunately, this type of cell does have the drawback that in order to maintain the stored state of the device, it is necessary to 'refresh' the cell data periodically so that the charge does not leak away. This operation simply involves rewriting the data into the cell in order to restore the charge. The refresh operation can be performed very simply on a row-by-row basis. That is each row is selected in turn and the refresh performed on that row, no column address being required. The sense amplifiers in this case are used to detect the charge stored on the MOS bit capacitance and restore the logic level to its correct value before producing the output data or rewriting the data during a refresh operation.

In common with all testing strategies it is preferable to have some fault models upon which to base the test algorithm. Since the structure of an RAM is so regular the fault models can be more comprehensive; also certain other peculiar effects specific to RAM architectures can be considered. Basically there are three areas of an RAM which need to be considered when testing [47]–[49]: these are the memory cell array, decoder logic and read/write logic. Faults can occur for numerous reasons, some of which have been described in Chapter 2; also capacitive coupling faults between adjacent bit lines and memory cells can cause problems. Therefore the following faults need to be considered [48]–[50] when testing RAMs.

Memory cell array

This is the portion of the memory where the data is stored, either in static or dynamic latches:

1. One or more bit cells are stuck at 0 or 1;
2. Some memory cells fail to undergo 0 to 1 or 1 to 0 transitions;
3. Unwanted coupling occurs between one or more pairs of physically adjacent cells. That is, a transition of 0 to 1 or 1 to 0 in one cell, say cell A, causes a transition in a neighbouring cell, B. This effect may also depend on the state of other adjacent cells. However, it is the case that a transition in cell B under similar conditions may not have an effect on cell A. This is sometimes referred to as a pattern sensitive fault;
4. Bridging fault occurs between adjacent cells producing a logical AND or OR of the data values depending on the dominant logic values;
5. Dynamic memory cell fails to hold its charge for the stated time (sometimes called 'sleeping sickness') [50];
6. The precharge operation fails to occur properly, producing incorrect values in a memory cell after a read operation.

Decoder logic

The function of a decoder is to select one cell within the memory array. However, the following faults may occur:

1. The decoder may fail to select the addressed cell and could access non-addressed cells possibly due to stuck-at faults;
2. Stuck-at faults in the decoder logic causing it to access non-addressed cells as well as the addressed cells.

According to Thatte and Abraham [48] these faults can be associated with the memory array. If multiple cells are accessed then this can be considered as a coupling fault between adjacent cells. In the case of a non-addressed memory cell this can be considered a stuck-at 0/1 fault depending upon the convention assumed in storing the data.

Read/write logic

The faults considered when testing read/write logic are:

1. Stuck-at faults in the sense-amplifier or write driver logic. These faults can be considered as multiple stuck-at faults in the memory;
2. Bridging faults or capacitive coupling faults between the bit lines. This fault can be considered equivalent to coupling faults between adjacent memory cells;
3. In a write/read sequence data may not be produced at the correct time;
4. Sense amplifier saturation can cause an incorrect output if a long sequence of the same data is read followed by a change to the opposite value.

This list of RAM faults is quite comprehensive and tries to cover most of the typical faults that can be encountered in attempting to derive a test strategy for any RAM circuit. Therefore to prove a RAM is functional the following conditions in any test algorithm are necessary:

1. Each cell must be capable of storing a logic 0 and 1. Also each cell should be able to undergo 0 to 1 and 1 to 0 transitions;
2. For every pair of cells A and B, cell A should undergo 0 to 1 and 1 to 0 transitions with cell B containing both logic 0 and then logic 1, without the contents of B being disturbed. The roles of A and B should be reversed and the operation repeated;
3. Dynamic RAM cells must be capable of storing data for a minimum time without requiring a refresh;
4. The addressing circuits should be tested for correct access to every memory cell.

In the next section algorithms aimed at achieving these objectives will be discussed.

10.8 TEST PATTERN GENERATION ALGORITHMS FOR RAMS

In order to test an RAM it is necessary to find an optimum sequence of test patterns that will cover the types of faults discussed in the previous section. The efficiency with which a test algorithm operates is usually related to the number of read and write operations required. If an RAM has n bits then an algorithm may have a complexity of order $O(n^2)$, that is it must perform Kn^2 read/write operations in performing the test, where K is some constant value. Throughout this discussion the complexity of the test will be used as an indication of the efficiency of the algorithm. This measure becomes very important as the number of bits in the RAM grows, since the difference in test time for an $O(n^2)$ and an $O(n \log_2 n)$ algorithm can be quite large. Therefore since RAMs are usually high volume devices, test algorithms with the lowest complexity and highest fault coverage are very desirable.

The following algorithms have been used to test RAM structures [50], [51]:

1. *The solid pattern*: This pattern consists of all ones or zeros in the RAM. This can be used as the starting point for other test sequence or to test the total worst case power dissipation of the RAM. Complexity $O(n)$.
2. *Checkerboard pattern*. A pattern is written into the RAM so that neighbouring cells have opposite data values. The pattern is illustrated in Fig. 10.20(a). This pattern will test for shorts between adjacent cells. Complexity $O(n)$.
3. *Column (row) bars* (Fig. 10.20(b) and (c)): A one is placed in every odd column (row) and a zero in all even columns (rows). The values can then be read and checked. The columns (rows) are inverted and the read operation repeated. Shorts between adjacent columns (rows) are detected by this sequence. In the case of dynamic RAMs the column bar also tests the refresh timing. Complexity $O(n)$.
4. *Parity pattern*: The value written into the memory cell is based on the parity of the row and column addresses of that cell. Using this test sequence the decoder circuits and memory cell are functionally tested. Complexity $O(n)$.
5. *Marching patterns*: A solid pattern of zeros is first placed into the RAM. The contents of the RAM are then read in ascending order. A check is made to determine if the content of the cell is zero, then a one is written into the location. This effectively leaves a sequence of ones in the memory to indicate that the cell has been read. The memory is next read in descending order, the ones being replaced by zeros. At the end of this test the procedure is repeated for an initial starting pattern of solid ones. This test sequence is designed to detect decoder faults, to check that memory cells can undergo zero to one and one to zero transitions and also to detect some interactions between cells. Complexity $O(n)$.
6. *Diagonal test*. The simple form of this test involves writing ones into the

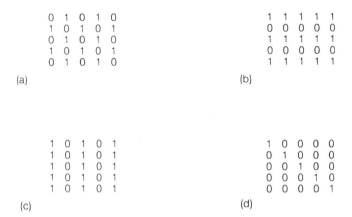

Fig. 10.20 Several patterns used to test RAM devices: (a) checkerboard pattern; (b) row bar patterns; (c) column bar patterns; (d) diagonal test pattern.

memory cells whose row and column addresses are the same. All other cells are set to zero. If the data is now read out in ascending order a long sequence of zeros followed by a single one will be applied to the sense amplifier. Therefore the ability of the sense amplifiers to recover from such a sequence will be tested. The background pattern and diagonal can then be inverted and the test repeated. In addition to testing the sense amplifier, the decoder circuits are also tested. Complexity $O(n)$.

A more complex form of this test involves initializing the memory array as before; however, after the read operation is performed the diagonal is moved to the right by one position. Therefore the shifting diagonal pattern will eventually write a one and zero into every cell in the array. Hence the array will also be tested for stuck-at-0/1 faults. Complexity $O(n \log_2 n)$.

So far all the test sequences described have involved testing the memory cells, decoder circuits and interactions between rows and columns. Faults involving interaction between pairs of memory cells have not been considered; that is, the pattern-sensitive faults. The following tests are aimed at detecting faults of this type.

7. *Walking zeros and ones.* A solid background pattern of all zeros is first written into the array. A single one is then written into the first address and all other locations are checked for a logic zero. The one is then verified and moved onto the next address and the verification procedure repeated. The whole test procedure is continued until the last address is reached. The complement values are then used for the test. This test sequence, as well as detecting pattern-sensitive faults, also checks that each cell can be written with zero or one and that the decoder circuits are functioning correctly.

Furthermore, since every zero cell is read before the one sense amplifier recovery is also tested. Complexity $O(n^2)$.

8. *Galloping zeros and ones (GALPAT)*. This test sequence is identical to the last except that the 1 value is accessed and verified between each read of the zero locations. Since the long sequence of the same value is broken up by reading logic one this scheme cannot be used for testing sense amplifier recovery. Complexity $O(n)$.

In the last two test schemes the method of verifying the zero locations is called ping-ponging because of the way in which the data is accessed. However the full ping-pong as used above is quite expensive in terms of test time. Therefore a reduced form of ping-pong is sometimes used, in which only columns and rows which are physically adjacent to the 1 cell are accessed. This can reduce the complexity of the algorithm to $O(n^{3/2})$. In order to make sure that the logical addresses produced by the test equipment are mapped onto physically adjacent addresses in the memory array it is sometimes necessary to use a topological address scrambler [51] to perform this mapping function. This is necessary since sometimes an RAM may be split internally into columns containing different address blocks, so that one address may occur in a particular block while the next logical address may occur in another block; therefore the two addresses may not be physically adjacent. The topological scrambler simply allows the address fields to be altered in order to create the correct mapping between physical and logical addresses. A typical test sequence for an RAM may involve not only some or all of the above functional tests but also tests to verify that the access and refresh times are within specification, i.e. AC parameters. The refresh time could be verified by writing a specific pattern into the cells and then waiting a predetermined time before reading the data out for checking.

Although the use of $O(n)$ test pattern sequences is more cost effective in terms of testing time the quality of the faults covered by these tests is not as substantial as the $O(n^2)$ test patterns; therefore when deciding upon the tests to use these trade offs have to be taken into account to produce an optimum test solution. One interesting solution to this dilemma has been proposed by Thatte and Abraham [48]. The algorithm proposed involves dividing the memory up into k partitions. The complete memory is written with zeros. One partition then undergoes a zero to one transition. The other partitions are verified; thereafter the first partition undergoes a one to zero transition and the remaining partitions are again verified. This sequence of events is repeated with all the other partitions complemented. The operations are repeated for the other $k - 1$ partitions. By performing these operations each cell is checked for interaction with every cell in the other partitions. If a correct sequence of test patterns is chosen each partition need only undergo two transitions and only $4nk$ patterns are required. Once these operations have been performed, the k partitions are each subdivided into k smaller partitions and the test sequence repeated until the stage is reached where the partition contains only one cell. If $k = 2$ then the total test length is equivalent to $O(n \log_2 n)$. The test procedure obviously detects cells stuck at $0/1$, cells that fail to undergo 0–1–0 or 1–0–1 transitions and cells that are coupled. This algorithm was improved by Nair et al. [49] in order to make the address generation sequence more efficient. To achieve this a bit in the accessed address location is used to determine which half ($k = 2$) of the RAM the cell occupies and therefore which operation should be performed on that cell, i.e. a read or write sequence. The test proceeds by using each bit of the address field in turn on successive iterations of the algorithm. Again the complexity of this scheme is $O(n \log_2 n)$. Another algorithm presented by Suk and Reddy [52] which is based on March tests requires only $O(n)$ operations to detect similar types of faults. Obviously this algorithm reduces the testing time needed for the RAM.

If the types of faults considered are restricted to only stuck-at faults in the RAM, then the test sequences can be further simplified since coupled memory cell pairs do not need to be considered. Knaizuk and Hartmann [53] present such an algorithm for testing a RAM with single or multiple stuck-at $0/1$ faults anywhere within the RAM structure. Although the algorithm only requires $O(n)$ operations ($4n$ to be exact) certain restrictions are placed on the form of the decoder logic, that is the decoder must be implemented in AND and NOT gates without any reconverging fanout paths. However, Nair [54] has modified the original algorithm so that it may operate with any form of decoder. The complexity of the algorithm is still $O(n)$.

The effect of pattern sensitivity on the performance of an RAM is a very important consideration. However, the ping-pong tests do not fully test the pattern sensitivity – they simply test the coupling between pairs of cells; consequently an algorithm is needed to test for pattern-sensitive faults. These types of faults were first discussed by Hayes [55] in 1975 in terms of a sequential machine. When formulating algorithms for this test sequence the cell under consideration, the base cell, is thought of as having adjacent neighbours as shown in Fig. 10.21(a). Where the cells labelled N, W, E and S are the memory cells that are physically adjacent to the base cell. The neighbourhood shown in Fig. 10.21(b) is also considered by some authors [56], [57] when generating test algorithms for pattern sensitive faults. Two types of pattern-sensitive fault have been defined [58]: a static pattern-sensitive fault and a dynamic pattern-sensitive fault. In the case of the static pattern sensitive fault the contents of the base cell are determined by the state of the neighbouring cells N, E, W and S. The dynamic pattern-sensitive fault occurs if the contents of the base cell are affected when the neighbouring cells change in a particular sequence. Obviously these types of faults are open to widely different interpretations: Srini [56], [57] considers only the static type of fault in that it is assumed that the contents of neighbouring cells can affect the value

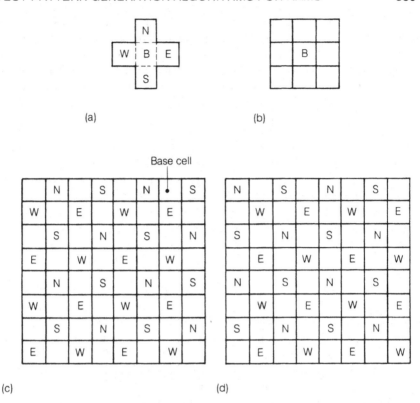

Fig. 10.21 Cell layouts used to test for pattern sensitive faults: (a) type 1 base cell neighbourhood; (b) type 2 base cell neighbourhood; (c) mode 1; (d) mode 2.

of the base cell; Suk and Reddy [59], [60], on the other hand, consider faults where a change in the neighbouring cells can cause a change in the base cell.

To demonstrate how an algorithm to test for pattern-sensitive faults may proceed, the technique discussed by Saluja and Kinoshita [58] will be briefly outlined. The technique proposed uses Hamiltonian paths in a graph to produce test patterns which have the minimum number of write operations. The test algorithm is as follows:

1. Initially the RAM is set to mode 1 (Fig. 10.21(c)). All base cells are set to 0. Each of the neighbouring cells (N, E, W, S) is then written with a binary sequence; there are 16 values from 0000 to 1111. After each write operation the contents of the base cells are read and verified.
2. The above sequence is repeated with the base cells set to 1;
3. The RAM is set to mode 2 (Fig. 10.21(d));
4. Steps 2 and 3 are repeated.

The patterns used in Step 2 to write into the neighbouring cells are chosen so that they correspond to a Hamiltonian path, that is the Hamming distance between each of the patterns is 1. This reduces the number of write operations that need to be performed. If the sequence of patterns is correctly formulated only $41n$ operations are required to perform a test on the memory array.

10.8.1 Testing embedded RAM designs

With advances in VLSI technology the ability of the designer to include memory on the same chip as logic has been enhanced. This mixture of embedded memory and logic has brought with it serious difficulties in testing the memory, since it may not be possible to directly control the address, data and read/write inputs, similarly the data outputs may not be directly observable. Although techniques do exist to enhance the testability of such devices – namely scan path [61] and BILBO [28] – their direct application to memory structures is questionable. The main reason for this difficulty is that memory structures, as discussed in the previous section, can have a wide variety of fault types not just the simple stuck-at 0/1 types. Therefore self-testing techniques have been developed which are specifically aimed at testing embedded memories [62]–[65], [47]. Although very effective algorithms exist – e.g. GALPAT – for testing memories they are difficult to implement in the form of self-testing structures due to their complexity. Therefore, in general, most of the self-testing methods tend to use algorithms of $O(n)$ complexity, including march tests and checkerboard patterns.

One of the primary requirements for testing an embedded memory is to be able to cycle through all the possible addresses within the memory. To achieve this Sun and Wang [63] and Nicolaidis [65] both use linear feedback shift registers (LFSR) attached to the address inputs. LFSRs are used since the hardware requirement is much less than that for a similar size of binary counter; also there are more address transitions using an LFSR, which helps to test for worst case access times. In both methods it is necessary to check for faults within the LFSR since only a simple march sequence is used to test the RAM, so any stuck bits in the counter may not be detected. Sun and Wang use a special realization of the LFSR to overcome this problem, whereas Nicolaidis uses a form of parity prediction linked to the memory structure itself in order to detect faults within the LFSR. An alternative Berger coded check on the LFSR is also suggested by Nicolaidis. The test patterns used on the memory, in both cases, are simple march tests which require the initialization of the memory to a solid pattern of zeros or ones. Nicolaidis employs the march test algorithm suggested by Marinescu [66] with a few slight modifications.

The scheme proposed by Sun and Wang, however, has a few advantages over the Nicolaidis scheme in that the registers used to control the memory

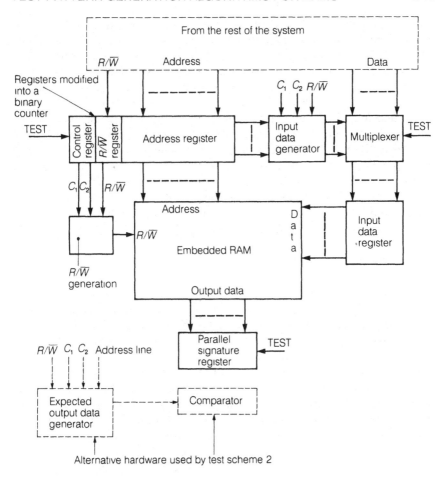

Fig. 10.22 Jain and Stroud embedded test scheme 1 [47].

can be configured into a scan path as well as allowing a single-step mode; special test patterns can also be scanned in and out to test specific memory faults. A further advantage of the single-step mode is that, while applying the marching test pattern during self-test mode, intermediate patterns can be scanned out for comparison off-chip, so the device can be diagnosed further.

Jain and Stroud [47] have presented two schemes for testing embedded memories which have been used on production devices. On the six devices to which the technique was applied area increases of as little as 3–5% were reported. The overall structure of this method is shown in Fig. 10.22. As can be seen from this block diagram the address generation, read/write line and control bits are all generated by a binary counter. In the first scheme proposed (and illustrated in the figure) the address lines and control lines are used to produce the input data for the memory. The test used in this case simply involves writing unique data – namely address or address – into each memory location and then reading it out for verification in ascending then descending order. The output of the memory is fed into a parallel signature analyser, which is disabled during write operations. At the end of the test the signature generated must be compared with a good signature. In the second scheme proposed by Jain and Stroud a checkerboard pattern is used for the test. In this case only one address line is used along with the control bits for generating the input data; also a comparator may be used on the output to verify the data from the RAM instead of a parallel signature analyser. The comparison operation is performed one cycle after the read operation; a parallel signature analyser could be used to compress the data output if required.

10.9 CONCURRENT ERROR DETECTION IN MEMORIES

So far the methods discussed for the testing of memory structures have been aimed at single one-off tests in order to detect permanent failures, such as bit cells stuck at $0/1$, or pattern-sensitive faults. However, the class of techniques to be discussed in this section will look at the methods available to detect/ correct errors caused by transient or alpha particle strikes, i.e. temporary faults (discussed in Chapter 2). Most of the techniques used for this form of detection/correction involve the use of codes. Perhaps the best-known single-error-correcting/double-error-detecting code is the Hamming code [67], [68]. If a word in memory is stored with a few extra bits representing the Hamming check bits for that word, then upon next reading that word, if a single or double error had occurred, i.e. one or two bits had been corrupted, it would be possible to correct the single error and indicate that a double error had occurred. Thus the circuitry external to the memory would either have been unaware of an error (single bit incorrect) or would know that the data from the memory was invalid (double error) and appropriate action could be taken. This technique is illustrated in Fig. 10.23. A further advantage to this scheme is that if a single bit cell has a permanent fault (stuck at $0/1$) the code could be used so that the memory would appear to be fully functional. Obviously if a temporary error occurs in the word containing the fault it would be impossible to correct with a SEC/DED code. However, it does allow an otherwise unusable device to be employed in some less demanding role.

Although Hamming codes are widely used to check memories other codes have been used successfully. It has been suggested by Elkind and Siewiorek [69] that Error Correcting Code (ECC) memories may not be more reliable than their non-redundant counterparts, since if the memory array itself is very reliable then the reliability of the circuitry used to perform the ECC can

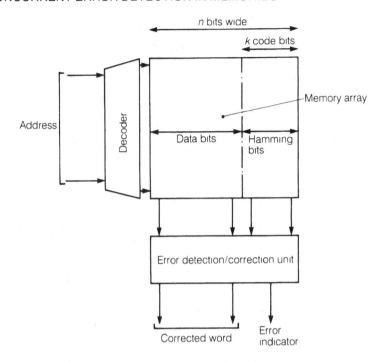

Fig. 10.23 Typical Hamming coded memory architecture.

become the dominant factor. Therefore since Hamming codes require a large gate overhead to perform the SEC/DED operations other code techniques which attempt to reduce this gate overhead have been suggested. Elkind and Siewiorek suggest the use of a block code [70] (i.e. the code is formed over several pieces of data rather than just a single data item as in the Hamming code) in the form of a product code; however, although the extra logic required is reduced implementation can prove difficult.

Two methods which attempt to reduce the redundant area overhead by using product codes are proposed by Osman [71] and Tanner [72]. A product code is illustrated in Fig. 10.24; as can be seen, the code is easily formed. The row parity is produced so that the modulo 2 sum in any row is zero; similarly the modulo 2 sum of any column is also zero and is given by the column parity. Hence, a single-bit error can be detected by the two parity words, since it occurs at the intersection of the two parity bits of the row and column vectors that indicate an error (as illustrated).

The scheme used by Osman to implement the product code is illustrated in Fig. 10.25. Each memory block is an $N \times 1$-bit array. As discussed earlier, when the row address is applied a whole row is accessed, therefore an extra

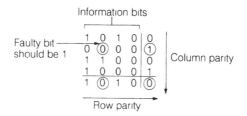

Fig. 10.24 An example of a product code.

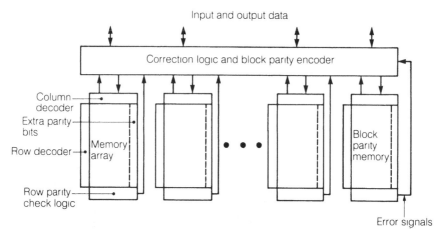

Fig. 10.25 Concurrent error-detection scheme proposed by Osman [71].

parity bit can be associated with the row; the row can then be checked by a parity checker attached to the columns of the memory array. This hardware is used on each of the memory arrays making up the memory word. An extra memory array, called the block parity memory, is also added to the memory word. This holds the parity bits associated with each address of the complete memory word, i.e. it is effectively the column parity. If a word is now accessed and the row parity of a memory array indicates an error and the block parity indicates an error, then the bit generated by the memory array in error simply needs to be inverted to be corrected. If however, the block parity does not indicate an error, then one of the bits in the row must be incorrect, but it is not the bit currently accessed, therefore correction will have to wait until that bit is addressed. If two row parity generators indicate an error or the block parity indicates an error when no rows indicate errors, then a double-bit error has occurred, and this fault cannot be corrected. The scheme proposed by Tanner is slightly different in that the whole memory is structured so that it is a single codeword in the error-correcting code. Using this scheme it is possible to correct all double and some triple errors with only a 3% increase in redundant hardware for a 256K memory. However, any read or write from the memory has to access several bits which make up the codeword. The operation of this scheme simply involves the alteration of the redundant code bits as single bits are written into memory. When data is read it can be verified against the code bits and corrected if necessary. As all the operations are performed internally the actual read/write and correction operations can be made transparent to the user in both of the above schemes.

In order to detect both random and unidirectional errors an extension to the use of product codes is proposed by Pradhan [73]. In this scheme instead of using parity checking for both the columns and rows, a single-bit parity check is used on only the rows. The Berger code is then used to encode the columns.

The above techniques use static information redundancy, that is if the error-correcting capability of the code is overwhelmed by the number of faults occurring, then the memory will fail on the next error. However, Goldberg *et al.* [74] have proposed a scheme in which the memory array containing the fault is switched out in favour of a standby spare unit, so that as devices fail they are replaced. Naturally the supply of new devices will become exhausted, again limiting the number of errors that can be corrected.

An alternative scheme which does not require extra standby units but instead uses address scrambling to achieve a more fault-tolerant system has been proposed by Hsiao and Bossen [75]. This scheme is based on orthogonal Latin squares. A Latin square is an $n \times n$ *array of the digits 0, 1, 2, . . ., n − 1* with each row and column containing permutations of the digits. Two Latin squares are said to be orthogonal if when they are superimposed upon each other, every ordered pair of elements appears only once. Figure 10.26 illustrates the Latin squares that can be generated for $n = 4$. Therefore in order to produce a 4×4-bit memory using this scheme the 4×1-bit memory card shown in Fig. 10.27 can be used. Four of these cards need to be used to produce the desired memory structure. If, initially, the address sequence for each of these cards is as shown in Fig. 10.26, then the memory is fault free, also the memory is encoded in an SEC/DED code. If a double error is detected by the SED/DED then the linear feedback shift register on each of the cards is loaded with 00, 01, 10 and 11 respectively. This will have the effect of skewing the address so that the double bit error is now a single bit error in two locations and can be corrected by the SEC/DED code. Since the memory addresses will be arranged as shown in T_1 (Fig. 10.26) if another double error occurs then a shift pulse is sent to the linear feedback shift registers on three of the memory arrays; this will again rearrange the adresses (T_2). To ensure that no double errors are present a check should be performed on the memory array. If another double error is detected, then

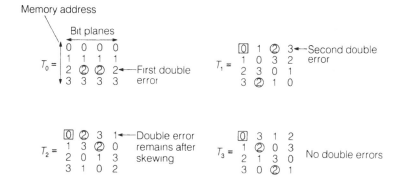

Fig. 10.26 Latin squares – to show address skewing to avoid double errors.

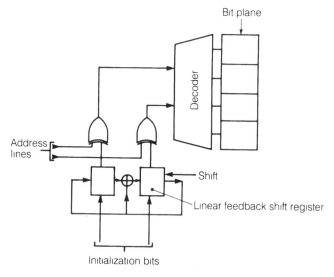

Fig. 10.27 Hardware required by the orthogonal Latin square technique [75]. To produce a 4 × 4-bit array, four of these cards would be required.

another shift pulse can be used to rearrange the addresses yet again to produce the skewed addresses shown in T_3 (Fig. 10.26), thereby allowing the errors to be corrected.

10.10 RANDOM TESTING TECHNIQUES APPLIED TO RAMS

All of the above testing techniques have used deterministic means of generating the test data; however, it is possible to use random pattern testing techniques [76] on memories, although the test sequences generated are usually longer than those of the deterministic pattern generators. However, with regard to testing for coupling between three cells it has been shown [76] that random patterns may have a superior performance in terms of test length. It has also been shown that although the test length may be longer than the optimal deterministic test sets discussed earlier, fewer patterns are required than for GALPAT type tests, and the fault set can be more clearly defined for the random technique.

The above methods illustrate the possible techniques that can be used to test both accessible and embedded RAMs. As the reliability requirements of a system become more stringent techniques that incorporate concurrent error-detection techniques will probably become very important, especially for memories.

10.11 REFERENCES

1. Mead, C. and Conway, L. (1980) *Introduction to VLSI Systems*. Addison-Wesley.
2. Poretta, A., Santomauro, M. and Somenzi, F. (1984) TAU: A fast heuristic logic minimiser. *Proc. ICCAD '84*, November, 206–8.
3. Lim, C. Y. (1982) *A PLA Minimiser Program*. Final year student thesis, Department of Electrical and Electronic Engineering, University of Newcastle upon Tyne.
4. Hong, S. J., Cain, R. G. and Ostapko, D. L. (1974) MINI: A heuristic approach for logic minimization. *IBM J. Research and Development*, **18**(5), 444–58.
5. Demicheli, G., Brayton, R. and Sangiovanni-Vincentelli, A. (1984) KISS: A program for optimal state assignment of finite state machines. *Proc. ICCAD '84*, November, 209–14.
6. Morris, D. T. (1982) *Design, Implementation and Application of an Automated Controller Design System*. Computer Science Department, Leeds University.
7. Fleisher, H. and Maissel, L. I. (1975) An introduction to array logic. *IBM J. Research and Development*, **19**(2), 98–109.
8. Demicheli, G. and Sangiovanni-Vincentelli, A. (1983) Multiple constrained folding of programmable logic arrays: theory and application. *IEEE Trans. Computer-Aided Design*, **CAD-2**(3), 151–66.
9. Smith, J. E. (1979) Detection of faults in programmable logic arrays. *IEEE Trans. Computers*, **C-28**(11), 845–53.
10. Ramanatha, K. S. and Biswas, N. N. (1983) A design for testability of undetectable crosspoint faults in programmable logic arrays. *IEEE Trans. Computers*, **C-32**(6), 551–7.
11. Tamir, Y. and Sequin, C. H. (1984) Design and application of self testing comparators implemented with MOS PLA's. *IEEE Trans. Computers*, **C-33**, 493–506.
12. Agrawal, V. K. (1974) Multiple fault detection in programmable logic arrays. *Proc. 9th Fault Tolerant Computing Symposium*, 227–34.
13. Agrawal, V. K. (1980) Multiple fault detection in programmable logic arrays. *IEEE Trans. Computers*, **C-29**(6), 518–22.
14. Roth, J. P. (1966) Diagnosis of automata failures: A calculus and a method. *IBM J. Research and Development*, **10**, 278–91.

15. Ostapko, D. L. and Hong, S. J. (1978) Fault analysis and test generation for programmable logic arrays (PLA). *Proc. 8th Fault Tolerant Computing Symposium*, 83–89.
16. Eichelberger, E. B. and Lindbloom, E. (1980) A heuristic test-pattern generator for programmable logic arrays. *IBM J. Research and Development*, 24(1), 15–22.
17. Cha, C. W. (1978) A testing strategy for PLA's. *Proc. 15th Design Automation Conference*, June, 326–31.
18. Somenzi, F., Gai, S., Mezzalama, M. and Prinetto, P. (1984) PART: Programmable ARray Testing based on a PARTitioning algorithm. *IEEE Trans. Computer-Aided Design*, **CAD-3**, 142–9.
19. Dong, H. and McCluskey, E. J. (1981) *Design of fully testable programmable logic arrays*, Center for Reliable Computing, Stanford University, Report No. CRC 81-20.
20. Min, Y. (1984) A PLA design for ease of test generation. *Proc. 14th Fault Tolerant Computing Symposium*, 436–42.
21. Bozorgui-Nesbat, S. and McCluskey, E. J. (1986) Lower overhead design for testability of programmable logic arrays. *IEEE Trans. Computers*, **C-35**(4), 379–83.
22. Eichelberger, E. B. and Lindbloom, E. (1983) Random-pattern coverage enhancement and diagnosis for LSSD logic self-test. *IBM J. Research and Development*, **27**(3), 265–72.
23. Hong, S. J. and Ostapko, D. L. (1980) FITPLA: A programmable logic array for function independent testing. *Proc. 10th Fault Tolerant Computing Symposium*, 131–6.
24. Fujiwara, H., Kinoshita, K. and Ozaki, H. (1980) Universal test sets for programmable logic arrays. *Proc. 10th Fault Tolerant Computing Symposium*, 137–42.
25. Fujiwara, H. and Kinoshita, K. (1981) A design of programmable logic arrays with universal tests. *IEEE Trans. Computers*, **C-30**(11), 823–8.
26. Khakbaz, J. (1984) A testable PLA design with low overhead and high fault coverage. *IEEE Trans. Computers*, **C-33**(8), 743–5.
27. Reddy, S. M. and Ha, D. S. (1987) A new approach to the design of testable PLAs. *IEEE Trans. Computers*, **C-36**(2), 201–11.
28. Koenemann, B., Mucha, J. and Zwiehoff, G. (1979) Built-in logic block observation techniques. *Proc. 1979 Test Conference*, 37–41.
29. Yajima, S. and Aramaki, T. (1981) Autonomously testable programmable logic arrays. *Proc. 11th Fault Tolerant Computing Symposium*, 41–3.
30. Daehn, W. and Mucha, J. (1981) A hardware approach to self-testing of large programmable logic arrays. *IEEE Trans. Computers*, **C-30**(11), 829–32.
31. Grassl, G. and Pfleiderer, H. J. (1983) A function independent self test for large programmable logic arrays. *Integration, VLSI Journal*, **1**(1), 71–80.
32. Hassan, S. Z. and McCluskey, E. J. (1983) Testing PLA's using multiple parallel signature analyzers. *Proc. 13th Fault Tolerant Computing Symposium*, 1983, 422–5.
33. Fujiwara, H. (1984) A new PLA design for universal testability. *IEEE Trans. Computers*, **C-33**(8), 745–50.
34. Treuer, R., Fujiwara, H. and Agrawal, V. K. (1985) A low overhead, high coverage built-in self-test PLA design. *Proc. 15th Fault Tolerant Computing Symposium*, 112–17.
35. Treuer, R., Fujiwara, H. and Agrawal, V. K. (1985) Implementing a built in self-test PLA design. *IEEE Design and Test*, April, 37–48.
36. Saluja, K. K., Kinoshita, K. and Fujiwara, H. (1981) A multiple fault testable

design of programmable logic arrays. *Proc. 11th Fault Tolerant Computing Symposium*, 44–6.

37. Saluja, K. K., Kinoshita, K. and Fujiwara, H. (1983) An easily testable design of programmable logic arrays for multiple faults. *IEEE Trans. Computers*, **C-32**(11), 1038–45.

38. Hua, K. A., Jou, J. Y. and Abraham, J. A. (1984) Built-in tests for VLSI finite-state machines. *Proc. 14th Fault Tolerant Computing Symposium*, 292–7.

39. Anderson, D. A. and Metze, G. (1973) Design of totally self-checking check circuits for *m*-out-of-*n* codes. *IEEE Trans. Computers*, **C-22**(3), 263–9.

40. Wang, S. L. and Avizienis, A. (1979) The design of totally self checking circuits using programmable logic arrays. *Proc. 9th Fault Tolerant Computing Symposium*, 173–80.

41. Piestrak, S. J. (1985) PLA implementations of totally self-checking circuits using *m*-out-of-*n* codes. *VLSI in Computers, ICCD '85*, 777–81.

42. Smith, J. E. and Metze, G. (1978) Strongly fault secure logic networks. *IEEE Trans. Computers*, **C-27**(6), 491–9.

43. Mak, G. P., Abraham, J. A. and Davidson, E. S. (1982) The design of PLA's with concurrent error detection. *Proc. 12th Fault Tolerant Computing Symposium*, 303–10.

44. Berger, J. M. (1961) A note on error detection codes for asymmetric channels. *Information Control*, **4**, 68–73.

45. Khakbaz, J. and McCluskey, E. J. (1982) Concurrent error detection and testing for large PLA's. *IEEE Trans. Electron Devices*, **ED-29**(4), 756–64.

46. Chen, C. Y., Fuchs, W. K. and Abraham, J. A. (1985) Efficient concurrent error detection in PLA's and ROM's. *VLSI in Computers, ICCD '85*, 525–9.

47. Jain, S. K. and Stroud, C. E. (1986) Built in self testing of embedded memories. *IEEE Design and Test*, October, 27–37.

48. Thatte, S. M. and Abraham, J. A. (1977) Testing of semiconductor random access memories. *Proc. 7th Fault Tolerant Computing Symposium*, 81–7.

49. Nair, R., Thatte, S. M. and Abraham, J. A. (1978) Efficient algorithms for testing semiconductor random access memories. *IEEE Trans. Computers*, **C-27**(6), 572–6.

50. Breuer, M. A. and Friedman, A. D. (1977) *Diagnosis and Reliable Design of Digital Systems*. Pitman.

51. Healy, J. T. (1981) *Automatic Testing and Evaluation of Digital Integrated Circuits*. Reston Publishing Co.

52. Suk, D. S. and Reddy, S. M. (1981) A march test for functional faults in semiconductor random access memories. *IEEE Trans. Computers*, **C-30**(12), 982–5.

53. Knaizuk, J. and Hartmann, C. R. P. (1977) An optimal algorithm for testing stuck at faults in random access memories. *IEEE Trans. Computers*, **C-26**(11), 1141–4.

54. Nair, R. (1979) Comments on 'An optimal algorithm for testing stuck at faults in random access memories'. *IEEE Trans. Computers*, **C-28**(3), 258–61.

55. Hayes, J. P. (1975) Detection of pattern sensitive faults in random access memories. *IEEE Trans. Computers*, **C-24**(2), 150–7.

56. Srini, V. P. (1978) Fault location in a semiconductor random access memory unit. *IEEE Trans. Computers*, **C-27**(4), 349–58.

57. Srini, V. P. (1977) API tests for RAM chips. *Computer*, July, 32–5.

58. Saluja, K. K. and Kinoshita, K. (1985) Test pattern generation for API faults in RAM. *IEEE Trans. Computers*, **C-34**(3), 284–7.

59. Suk, D. S. and Reddy, S. M. (1980) Test procedures for a class of pattern sensitive faults in semiconductor random access memories. *IEEE Trans. Computers*, **C-29**(6), 419–29.

60. Suk, D. S. and Reddy, S. M. (1979) An algorithm to detect a class of pattern sensitive faults in semiconductor random access memories. *Proc. 9th Fault Tolerant Computing Symposium*, 219–26.
61. Eichelberger, E. B. and Williams, T. A. (1977) A logic design structure for LSI testability. *Proc. 14th Design Automation Conf.*, 462–8.
62. Westcott, D. (1981) The self assist test approach to embedded arrays. *Proc. International Test Conference*, 203–7.
63. Sun, Z. and Wang, L-T. (1984) Self testing Embedded RAMS. *Proc. International Test Conference*, 148–56.
64. Kinoshita, K. and Saluja, K. K. (1984) Built in testing of memories using on chip compact testing scheme. *Proc. International Test Conference*, 271–8.
65. Nicolaidis, M. (1985) An efficient built in self test scheme for functional test of embedded RAMS. *Proc. 15th Fault Tolerant Computing Symposium*, 118–23.
66. Marinescu, M. (1982) Simple and efficient algorithms for functional RAM testing. *Proc. International Test Conference*, 236–9.
67. Hamming, R. W. (1950) Error detection and correcting codes. *Bell Technical Journal*, 2, April, 147–60.
68. Sarrazin, D. B. and Malek, M. (1984) Fault tolerant semiconductor memories. *Computer*, August, 49–56.
69. Elkind, S. A. and Siewiorek, D. P. (1980) Reliability and performance of error-correcting memory and register arrays. *IEEE Trans. Computers*, C-29(10), 920–7.
70. Peterson, W. W. and Weldon, E. J. (1972) *Error Correcting Codes*, 2nd edn, MIT Press, Cambridge, Mass.
71. Osman, F. (1982) Error correction techniques for random access memories. *IEEE J. Solid State Circuits*, SC-17(5), 877–81.
72. Tanner, R. M. (1984) Fault tolerant 256K memory designs. *IEEE Trans. Computers*, C-33(4), 314–22.
73. Pradhan, D. K. (1980) A new class of error correcting/detecting codes for fault tolerant computer applications. *IEEE Trans. Computers*, C-29(6), 471–81.
74. Goldberg, J., Lewitt, K. N. and Wensley, J. H. (1974) An organisation for a highly survivable memory. *IEEE Trans. Computers*, C-23(7), 693–705.
75. Hsiao, M. Y. and Bossen, D. C. (1975) Orthogonal Latin Square configuration for LSI memory yield and reliability enhancement. *IEEE Trans. Computers*, C-24(5), 512–16.
76. Fuentes, A., David, R. and Courtois, B. (1986) Random testing vs deterministic testing of RAM's. *16th Fault Tolerant Computing Symposium*, 266–71.

11
THE APPLICATION OF KNOWLEDGE-BASED SYSTEMS TO THE TESTING OF DIGITAL CIRCUITS

11.1 INTRODUCTION

The use of algorithmic CAD tools in the design of integrated circuits has been well established; however, the design process itself is far from algorithmic and comprises searching a vast solution space for a particular solution which will satisfy given design criteria within certain constraints. As circuits become more complex and design styles and techniques develop, the potential solution space increases and the number of trade-offs to be considered by the designer becomes enormous. Consequently a need has evolved for 'intelligent' CAD tools [1], [2] which will not only provide 'expert' solutions to the problems to be overcome at various phases of the design process but also justify why a particular solution is proposed in the light of the constraints associated with a given problem. It may be considered that research into the development of expert CAD tools is a worthless venture, given the amount of progress made on silicon compilers; however, silicon compilers provide only a very constrained solution to a design problem and the more sophisticated compilers being developed do incorporate expert systems to some extent. Furthermore, there has evolved a growing community interested in implementing systems on silicon, but to whom the design process appears as a 'black art' and who subsequently require the assistance of an expert when certain design decisions have to be made; this assistance, however, cannot be obtained from the present range of CAD tools, which are inherently passive. Consequently, there is a need for a design system which can essentially 'think' and have the knowledge base of an expert so that when design decisions have to be made there is an expert knowledge base which can be interrogated to obtain an answer; the ability of the 'expert' CAD tool to divulge its reasons for making a given decision implies that it can also be used as a teaching tool.

To date, a number of 'expert' CAD tools for IC design have been developed, for example:

1. The *design automation assistant* [3], [4], which synthesizes algorithmic descriptions of VLSI systems into a list of technology independent registers, operators, datapaths and control signals;
2. *FLUTE* [5], which is a heuristic floor planner and operates within *CADRE* [6], an automatic synthesis tool which converts a hierarchical structural description into a VLSI layout;
3. *REDESIGN* [7], which analyses an existing design to suggest local modifications and evaluates their subsequent effect on the circuit, when a circuit is to be modified to improve its functionality;
4. *SOCRATES* [8], which optimizes a combinational logic design for a specific target technology by performing substitutions of equivalent gate configurations, thereby reducing the overall area and improving circuit speed;
5. *TALIB* [9], which is used in the cell layout phase of an integrated circuit design; it accepts a schematic of the cell as input together with a description of the cell layout boundary and generates the mask geometries required to realize the cell;
6. *WEAVER* [10], which is a knowledge-based routeing program which simultaneously considers various routeing metrics – for example complete routability, minimum area, wire length and the number of vias.

However, this chapter is concerned with the development of 'intelligent tools' used in testing and design for testability; before discussing these developments the basic components of an *expert* or *knowledge-based* system will be described.

11.2 BASIC COMPONENTS IN AN EXPERT SYSTEM [11] ,[12]

An expert system is a knowledge-intensive program capable of producing solutions to a range of non-numerical problems, which in the past required the knowledge of a human expert to provide the solution. An expert system differs from a conventional program in that it operates on knowledge rather than data which is passive; the procedures invoked tend to be heuristic rather than algorithmic and solutions are obtained by an inferential process rather than, say, some repetitive numerical process. In general, a conventional program does not contain an explicit knowledge of the problem to be solved, it simply contains a sequence of instructions. defined by the programmer, to solve a specific problem. In comparison to the human expert the expert system offers the following advantages [12]. In an expert system the knowledge is permanent, that is, facts are not forgotten through lack of use, whereas the human expert will forget facts if they are not used regularly; furthermore the 'reasoning' power of an expert system is not subject to emotional factors such as stress, which can make the performance of the human expert unpredictable. The knowledge contained in an expert system is also easily transferred or reproduced, which is not the case for the human expert. Finally the expert system is affordable, whereas the human expert can be very expensive. The question which subsequently arises is 'Do we need human experts?' At present the answer is 'yes' since current expert systems can neither emulate the creativity and adaptability of human experts nor take a global view of a problem to see how various peripheral aspects relate to the central issue; finally human experts have a vast reserve of commonsense knowledge which would be difficult to codify into the expert system.

Table 11.1 User/expert system interaction [13]

User	Expert system
Specify requirements	Request information when
Provide additional	required
information when requested	When sufficient information is
Make decisions on whether to	available search solution space
change requirements or	for an answer
objectives	Suggest solutions and ask for
Request system to make a	confirmation of acceptance
decision	Report on failures to find a
Ask for an explanation	solution and provide useful
Decide whether or not to	advice
accept suggestions made	Provide explanations
by the system	Make decisions when requested

In a design environment, however, the functions performed by the user and the system which produce the most effective synergistic relationship to provide satisfactory solutions to problems are summarized in Table 11.1; although there are expert systems used in some problem domains where the user may assume an inferior or superior role.

An expert system essentially comprises four modules, namely a knowledge base, an inference engine, a knowledge acquisition module and an explanatory/interface module.

The *knowledge base* contains all the information required by an expert system to make it act intelligently. The knowledge in the system comprises facts and rules, which use the facts to make decisions; 'certainty' factors may also be associated with the facts and rules to indicate the degree of confidence that a given fact or rule is true or valid. Some of the rules in the knowledge base may also be heuristic, that is, 'rule of thumb' which, for example, in the interest of efficiency will limit the area of search for a possible solution and hence will not guarantee to find a solution to a problem, although in the vast majority of cases an adequate solution or conclusion will be found. Within an expert system knowledge can be represented and organized in a number of ways. The majority of expert systems are knowledge based, that is the knowledge about the problem to which the system is to be applied is separate from the other modules in the system; this has the advantage of permitting fast prototyping of systems together with the ease of adding new facts or rules to the knowledge base. Within the knowledge base the rules and facts can be structured in a number of ways, two of the more common methods of representing knowledge being the rule-based and logic-based methods. In the rule-based method knowledge is represented by antecedent–consequent rules of the form 'IF<condition>THEN<action>'. The antecedents or conditions may be joined by logical connectives; the consequent or action in a rule is performed only if the conditions in the problem being solved match the antecedents in the rule. The logic-based technique for knowledge representation uses a logic programming language to write declarative clauses to describe facts, rules and goals. The general format of a clause is 'consequent: – antecedent 1, antecedent 2 . . . '. A clause without an antecedent is a fact, a clause without a consequent is a goal or an objective to be validated and a clause containing both consequent and antecedents is a rule. The most common language used in logic-based expert systems is Prolog [14]. Other methods of knowledge representation [12] include the frame-based method, the procedure-oriented, the object-oriented and the access-oriented methods; some expert systems support more than one method of knowledge representation.

Within the expert system the ability to 'reason' about facts in the knowledge base is achieved through the use of a program called an *inference engine*. The most common reasoning strategies used in inference engines are the 'forward' and 'backward' chaining techniques. The forward chaining technique, sometimes referred to as the data-driven technique, starts with given facts and attempts to find all the conclusions which can be derived from the facts. The backward chaining technique essentially starts by hypothesizing on a given consequence and attempting to find the facts to substantiate it. The technique used very much depends upon the problem to be solved. The inference engine may be either explicit or implicit to the expert system, trading off the flexibility and subsequent improvement in efficiency in choosing the most appropriate 'reasoning' technique to solve a problem against minimizing the amount of work required to produce an expert system. One of the disadvantages of using expert system shells, i.e. an expert system without the problem domain knowledge, is that the inference mechanism may not incorporate the most appropriate 'reasoning' strategy for the particular problem domain.

The *knowledge acquisition module* is one of the knowledge-based system building aids that help to acquire and represent the knowledge from a human expert in a given problem domain. The task of acquiring the knowledge from a human expert and choosing the most appropriate representation for it is the function of the knowledge engineer. Several techniques [12] are available for extracting knowledge from the human (domain) expert, which is one of the major bottlenecks in developing an expert system; some of the techniques used are on-site observation, problem discussion and problem analysis. The difficulties in producing the knowledge base are further compounded if more than one domain expert is consulted. Several tools have been developed to assist in producing the knowledge base: for example, in addition to basic editing facilities there are syntax checkers to verify the grammar used in the rules, consistency checkers to ensure that the semantics of the rules or data are not inconsistent with knowledge already existing in the system; there may also be a *knowledge extractor* which assists the end users of the system to add and modify rules, and is essentially a combined syntax and consistency checker together with prompting and explanation facilities should the end-user get into difficulties.

The *explanatory interface module* is the means by which the end user interacts with the system, although to date full natural language processing, except in some strict technical contexts, is unattainable. An important characteristic of expert systems is their ability to 'explain' their 'reasons' for coming to a given conclusion, which is not only useful in debugging prototype systems but also gives the end user a degree of confidence in the system. Most expert systems have a 'predictive' modelling capability [12], that is the user can interrogate the system to find out the effect of certain changes in a given situation or if the situation changes what caused the change.

Over the past few years a large number of expert systems have been developed in a diverse range of subjects ranging from medicine, to engineering, to law, to chemistry, to military science and space technology.

11.3 APPLICATIONS OF EXPERT SYSTEMS IN TESTING [15], [16]

A major issue in the design and subsequent maintenance of complex systems is the rising costs of testing not only individual VLSI circuits after fabrication, but also when these are mounted on printed circuit boards to build systems and when these systems are in their operational environment. Many developments have been made to improve the efficiency of test generation algorithms, which has also been enhanced by the introduction of design-for-testability techniques. Although these developments are reducing testing costs, the techniques used have many limitations – for example:

1. Adequate test generation algorithms have not been developed for highly sequential circuits;
2. Design for testability techniques used to reconfigure some sequential circuits for the purposes of testing – for example, scan path, increase test application times;
3. Patterns produced by test generation algorithms have no structure, that is, they essentially comprise bit streams, and hence cannot take full advantage of the facilities offered by the ATE equipment, again increasing application times or requiring excessive pin storage.

The fundamental problem with the present approach used in test generation programs is that tests are generated at too low a level of abstraction, namely gate level and in some instances transistor level, hence requiring a vast amount of CPU time. In comparison, the human test programmer can generate effective tests for complex systems without requiring astronomical amounts of time. The reason for his success is that he does not view the system as a collection of gates, but looks at it from a functional level, and when deriving tests he uses a wealth of knowledge obtained not only from the circuit but also from experience; for example, he 'knows' what tests will be effective in testing certain functions, or how to get the system into a given state or how to observe and control the main interfaces between high level function blocks. Furthermore tests are generated with the objective of detecting all faults of interest using as many of the normal circuit operations as possible. Consequently, it has been recognized that test programming is a highly knowledge intensive task, full of heuristics, and not a purely algorithmic exercise based on a gate level description of a circuit. It has thus been seen by some as an activity in which expert systems could be used to good advantage by capturing the 'knowledge' used by test programmers and incorporating it into test generation programs with the objective of making the test generation process more effective and less costly.

Several of the advances in developing expert systems for test generation and fault diagnosis are briefly described below.

11.3.1 The SUPERCAT system [17], [18]

Automatic Test Generation systems can be use efficiently on small circuits, but it is usually the problem of the designer to generate the global test program for complex circuits. This problem is solved in two stages: first, automatic test generation techniques are used on small functional blocks; second, these tests are then integrated to produce the overall test program for the circuit, and this later part is the more difficult. It is the objective of the

SUPERCAT system to give the designer some 'expert' assistance in generating the global test programs.

The knowledge base in SUPERCAT comprises:

1. Information on a range of automatic test generation programs categorized in terms of
   (a) The types of logic to which they can be applied, that is random logic (combinational or sequential), RAMs, PLAs with and without testability enhancements, microprocessors, etc.
   (b) The input description used by the programs, that is gate level or functional state table descriptions, or instruction sets for microprocessors, etc.
   (c) The types of faults covered, that is, stuck-at-1/0 faults, pattern sensitive faults, etc.
2. A library of commonly used test patterns, either structural or functional, which have been shown to be effective in testing given functions in the past;
3. Information on a range of design for testability techniques regarding their applicability to given functions and also the overheads incurred by their use.

The SUPERCAT system has been designed to operate on data driven and control driven circuits, each with its own form of abstract representation, namely a data flow and a control flow model respectively.

When SUPERCAT is processing a data driven circuit it makes use of the CATA system [19] to perform a number of functions, for example:

1. To determine the 'flows' in a system, that is the smallest group of functions that can be activated completely independently of the remainder of the circuit, so that the signal paths required to control or observe particular functions can be identified;
2. To determine, for a given test strategy, the test paths required amongst the set of flows (and the order functions have to be tested if a 'start small strategy' is used);
3. To perform a testability evaluation and suggest areas where testability enhancements should be inserted if it is required to diagnose faults to a more precise location.

When SUPERCAT is used on control circuits it attempts to generate tests based on the normal operation of the circuit. For the purposes of testing, the circuit is divided into two parts – the controller and controlled resources (functions). When creating tests for the controller a symbolic execution tree is generated: the set of paths through the tree represents sequences of states that the controller will pass during its normal operation, and which therefore must be activated when checking out the controller. The ability to observe the controller states is also determined, together with the capability of distinguishing between controller and resource faults; if necessary testability enhancements to overcome these problems will be suggested. When testing the resources, the major problem to be solved by the SUPERCAT system is finding the set of paths in the control flow graph which will permit the required patterns to be applied to a given function and the subsequent responses propagated to an observable output.

11.3.2 An intelligent test generator – HITEST [20], [22]

The major limitation of most automatic test generation programs is that they have a microscopic view of the circuit, usually in terms of interconnections of gates and flip-flops; although improved techniques are being developed for test generation the costs are still rising and are now an appreciable part of the overall design costs. However, test programmers can produce effective test sets without incurring excessive increases in test generation time as circuit complexity increases; the reason for this is that the test programmer and ATPG program are being asked to solve essentially different problems. The test programmer has a more macroscopic view of the circuit and hence the test generation problem is more tractable; however, if the programmer is given only a gate level description, test generation times would far exceed those required by the ATPG programs. It may be argued that high level ATPG programs have been developed, but these are still not as effective as the test programmer, because in addition to having a macroscopic view of the circuit the test programmer has a wealth of knowledge he can apply in testing a circuit which is not available to ATPG programs; in order to improve the efficiency of ATPG programs this knowledge base must be incorporated inside the test generation system; this has been the prime objective in developing the test generation system HITEST.

A block diagram of the HITEST system is shown in Fig. 11.1; the output from the system is a high level language 'test waveform' description file which can subsequently be translated [23] into a format suitable for a range of automatic test equipment. The contents, or more precisely a compiled version of the contents of the test waveform description file, is central to the control of simulator and test generator activities; whenever a test generator command is encountered in the waveform file, control is passed from the simulator to the test generator. Control is passed to the test generator via the TESTGEN command which has a 'task' as an argument; this results in a fragment of waveform source code being inserted and subsequently compiled in the waveform description file, before the original TESTGEN statement. The fragment of waveform source code may comprise statements: for example, to set nodes to given values or other calls for further test generation tasks. The statements in the waveform source code are subsequently executed with

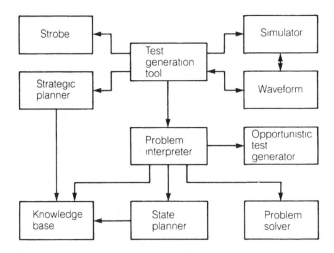

Fig. 11.1 HITEST system [21].

control being passed to either the simulator or test generator, depending upon the statement being executed; each return from the test generator usually results in another section of waveform source code being added to and subsequently compiled in the test waveform description file. Depending upon the task the call to TESTGEN may subsequently be deleted or the fragment of source code inserted above the call to TESTGEN, which will subsequently be executed at a later time. During the generation of the test waveform description file the test generator uses the simulator to determine the current state of the circuit for a given set of inputs in order to make decisions about the next step in the procedure.

Having briefly discussed the overall interaction between the waveform system, simulator and test generator in producing a test waveform file, which is generated incrementally, the function of the major blocks in the test generator will be described.

The 'strategy planner' is invoked by the test generator when a request is made to find a new test generation task; this consists of an 'outline waveform' comprising sections of waveform source code together with definitions of test generation problems (arguments in TESTGEN calls) which require to be solved in order to complete a given task. On returning from the test generator the 'outline waveform' is subsequently embedded into the test waveform file which is compiled and executed, subsequent calls to the test generator being activated by the TESTGEN statements. In these instances, however, when control is switched to the test generator the problem 'name' is used to identify a problem 'definition' in the knowledge base, which decides upon the algorithm or the actions necessary to solve the problem. The problem to be solved can be either setting up a specific set of conditions, for example, to apply a well known test sequence to a functional block, or calling the opportunistic test generator, which attempts to do something 'useful' given the present state of the circuit, such as propagating fault effects to a primary output or to the inputs of storage elements which are about to be clocked. The main objective of the opportunistic test generator, as the name implies, is to make best use of conditions in the circuit which provoke a lot of events, that is signal changes, to propagate and detect a vast number of fault conditions. This is in contrast to the problem solver block which is essentially a conventional automatic test generator based upon the PODEM algorithm (see Chapter 5), but modified to handle some of the peculiar characteristics found in MOS circuits, for example tristate and bidirectional components. The sorts of problems given to the problem solver are, for example, achieving the objective of setting a given node to a particular logic value, or creating a sensitized path to make a node observable or generating the necessary conditions to detect a fault and propagating its effect to an observable point. In solving these problems the inputs and outputs of storage elements are considered to be pseudo outputs and inputs to the circuit.

The problem interpreter and state planner blocks are used together to generate a scheme to change the output state of storage elements to values defined by the problem solver as essential to achieve some objective. The storage elements need not necessarily be included in a scan path, but may be any group of elements or an individual element which the test generator 'knows' can have its state(s) readily controlled or observed. In general, however, the function of the problem interpreter is to act upon test problem definitions and call upon the opportunistic test generator, the state planner or the problem solver as required.

Within the test waveform file, whenever a statement is executed which calls upon a primary input to be assigned a value, the logic simulator is called to evaluate the effect of the change on the circuit. If the subsequent statement to be executed is TESTGEN (strobe), this will cause the present primary output states to be written to the test waveform file; the fault simulator, which uses the PVL technique (see Chapter 4) is then passed these values and subsequently determines which faults can be detected; these faults are then removed from the list of undetected faults and a fault dictionary for the circuit is updated.

The knowledge base used in HITEST is a frame-based system [12] in which all the data is stored in named slots contained within a frame. The type of knowledge stored in the system comprises, for example, data about

1. parts of a circuit which are combinational and those which are sequential
2. techniques to control/observe well known signals
3. overall test strategies, i.e. whether to apply functional or algorithmically generated tests

4. waveform fragments, task and test generation problem definitions
5. for a given circuit the actual facilities used from multiple-function blocks extracted from a cell library, since it is only necessary to test those functions in the block which are used in a given circuit.

In general the knowledge contained in the system is directed at constraining the solutions proposed by the test generator to sequences which would be considered effective by the test programmer if he were performing the same task, and also at generating tests which can be easily translated into the appropriate format for a range of ATE equipment which makes full use of the pattern generation facilities offered by the equipment.

11.3.3 A Diagnostic Assistance Reference Tool (DART) [24]

DART is a test generation technique which uses a general inference procedure to identify faulty components and generate distinguishing test sequences. In performing these tasks DART works solely from a description of the intended structure and fault free behaviour of the device under test; this information is obtained, essentially, direct from the data files generated when the device was designed. Test generation techniques which use the structural description of a circuit are not novel – this is the essence of the path sensitization techniques; these are, however, very inefficient. DART overcomes this difficulty by exploiting the hierarchy which exists in most complex designs, in particular computer architectures. In diagnosing a fault in a system, DART first accepts a statement of the malfunction in a formal language; it subsequently suggests tests, analyses the results and isolates the failed unit. During the diagnosis procedure DART starts with a high level description of the system and attempts to isolate the major subcomponent containing the fault. This component is subsequently decomposed into its subcomponents and an attempt to isolate the fault to one of these subcomponents is made, and this procedure is repeated until the fault is isolated to the smallest replaceable unit. In this way the number of functions processed at any given time is small, keeping test generation costs and time to within acceptable limits.

In isolating a faulty component DART uses a deductive procedure in which a fault symptom is described in terms of a deviation from the expected behaviour of the circuit. From the fault free behaviour of the circuit DART reasons backwards to justify this behaviour; since the fault free behaviour was not observed all components involved in the justification process are considered to be potentially faulty. Subsequently, tasks must be generated which will isolate the faulty component from the list of potentially faulty components; this comprises seeking out paths to observable outputs and controllable inputs which only contain one of the suspected faulty devices, so that the faulty component is isolated when the actual and expected behaviours do not match. The basic flowchart of the DART procedure is shown in Fig. 11.2.

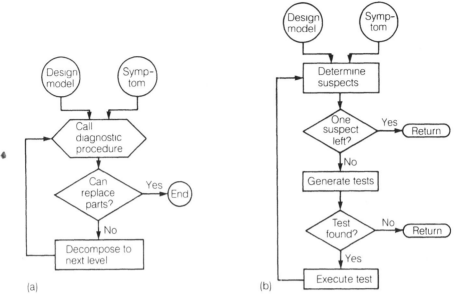

Fig. 11.2 DART algorithm [24]: (a) overall procedure; (b) diagnostic procedure.

Within DART the circuit structure and behaviour are described using a design description language called SUBTLE, whose syntax is similar to that for predicate calculus; the structure of the device is described in terms of its functions and interconnections, and each function can in turn be described in terms of its subfunctions and interconnections until the basic primitives are reached. The behaviour of a circuit is expressed in terms of the signal values on the inputs and outputs of the circuit; the most fundamental behavioural descriptions comprise a set of rules relating inputs to outputs. In order for DART to work on different levels of abstraction, a formal statement must also be made about the relationships between signals at the different levels of abstraction; this takes the form of a proposition which maps the inputs and outputs of a function at one level onto their counterparts at a lower level.

The main advantages offered by DART are that there are substantial savings in computational effort since the procedure exploits the hierarchy in the system being diagnosed. Also the procedure works from the design files; hence the knowledge base does not require a list of rules which relate symptoms to failure modes.

11.3.4 An Intelligent Diagnostic Tool (IDT) [25]

The diagnosis of faults in computer systems requires skilled personnel to reason about a large number of facts relating to the condition of the computer, obtained either by directly observing the operation of the machine or as a result of applying diagnostic test patterns to the machine. The personnel must also be able to realize when sufficient information has been accrued to identify the cause of the fault down to the smallest functional unit which can be replaced. IDT is a tool which can be run in either interactive or automatic mode to assist personnel in the identification of faulty components inside a computer system. IDT can switch between operating modes depending upon the skill of the technician using the system and his requirements. In fully automatic mode IDT selects and runs diagnostic tests and subsequently interprets the results; the selection of diagnostic tests is not sequential but is dependent upon the outcome of previous tests or on suggestions offered by the user; in this way the application of unnecessary tests is avoided. If a suggestion made by the user is found, as a result of the tests applied by IDT up

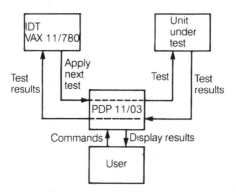

Fig. 11.3 IDT system [25].

to a given point in the test procedure, to be inconsistent with the information gathered, the user will be notified.

A block diagram of the IDT system is shown in Fig. 11.3; it comprises two computers in addition to the unit under test. The VAX 11/780 is a remote computer which contains the IDT system comprising knowledge base, inference engine and test strategies; the PDP 11/03 is a local computer which contains the diagnostic test sequences and drives the display software for the user interface. The PDP 11/03 plays a passive role in the IDT system and simply passes messages to the user, loads and runs tests when commanded by the IDT program. The IDT system is essentially menu driven, permitting the user to choose a range of options: for example

1. selecting the next test to be applied, this option only being used by competent operators who have a knowledge of the test sequences applied to the computer under test

2. entering an opinion as to the probable unit causing failure in the computer under test
3. requesting the current status of the test program in terms of the tests applied and their subsequent results
4. permitting the IDT system to choose the next test to be applied.

The display also shows a breakdown of the computer under test in terms of field replaceable units and their sub-units which can be tested by the diagnostic patterns. As the diagnostic tests proceed, the sub-units which have been shown to be fault free are removed from the display screen.

In diagnosing a fault the IDT must be capable of analysing test results and selecting the next test, depending upon the outcome of the previous test or an opinion supplied by the user or statistical data upon commonly occurring faults. Thus the IDT must have a mechanism [25] for interpreting the results, 'reasoning' from the interpreted results which sub-functions may be faulty, and a strategy for choosing the next test to be applied. The reasoning capability is achieved through the use of symbolic formulae which are used to represent the relationship between the test results and the faulty system. Test procedures are run under the single fault assumption. The IDT system has been used successfully by Digital Equipment Corporation to diagnose faults in the floppy disk subsystem in PDP 11/03s.

Some other systems which have been developed to assist in the diagnosis of faults in electronic equipment are

1. MIND [26], which is used to diagnose faults in VLSI testers. The system uses a hierarchical structural description of the tester together with 'expert' knowledge to reduce the down time of the testers, which incur a loss of approximately $10 000 for each hour that they are out of service in a production environment.
2. CRIB [27], which is used to assist engineers in the field to diagnose hardware/software faults in computer systems. The input to the expert system is an almost English Language description of the symptoms of a fault; these symptoms are subsequently matched with a known fault condition in a database. The diagnostic expertise is represented as 'symptom patterns' which suggest actions to be performed to obtain further information on the symptom of the fault.
3. FIS [28], which is used to assist technicians to diagnose faults in analogue electronic equipment; faults can be isolated down to the level of power supplies, amplifiers, etc. The system has a novel feature of being able to reason qualitatively from a functional model of the system, without recourse to numerical simulation. The techniques employed in FIS can also be used to diagnose faults in mechanical, hydraulic and optical systems.

11.4 APPLICATION OF EXPERT SYSTEMS IN DESIGN FOR TESTABILITY

Over the past decade a large number of design-for-testability methods have been developed in an attempt to reduce test generation costs. The techniques either enable the circuit to test itself or reconfigure the circuit so that the problem of generating test patterns and their subsequent application to the circuit is simplified. Although designers are aware that design-for-testability techniques will assist in reducing their testing problems they are probably not aware of all the techniques available, nor of the techniques most applicable to their circuits; since, in general, DFT techniques are not applicable to all classes of logic circuit, several different methods need to be incorporated into a design. Furthermore they must be aware of the implications of using particular DFT methods, not only with respect to their advantages but also their disadvantages.

Thus when choosing a DFT technique the designer must match the attributes of a given method against the constraints imposed by the design specification. The characteristics of this problem are again well suited to the use of expert system techniques, since:

1. There is a vast body of knowledge on DFT techniques, testing methods, constraints, etc. that the designer is probably unaware of but could use to his advantage;
2. The problem of choosing and incorporating a DFT technique into a given circuit requires a non-algorithmic solution since there is no unified theory about integrating DFT structures into a design or evaluating their trade-offs.

Several expert systems have been developed, recently, to assist designers to choose and subsequently integrate DFT structures into their designs, and these systems will be described briefly in the following sections.

11.4.1 Testable Design Expert System (TDES) [29]

The aim of the TDES system is to assist in the implementation of testable circuits. The input to the TDES system comprises a description of the circuit to be processed in terms of PLAs, RAMs, ROMs, registers, busses etc. together with the constraints imposed upon the circuit, for example, area overhead, pin-out limitations, ATE requirements, etc. The knowledge base within TDES contains information on a range of DFT techniques, their attributes, rules concerning their implementation and a method of evaluating the trade-offs between methods so that the method most suitable to the circuit being processed can be found. If necessary the system will incorporate additional hardware into a design to implement a given technique; if it cannot be realized with the existing hardware in the circuit.

The circuit to be processed is represented within TDES as a hierarchical graph model comprising nodes and arcs; the nodes represent structures and the arcs represent interconnections which can be either single wires or busses. Each node may have its own graph model depending upon the complexity of the structure it represents. A node is classified as *basic* if the structure it represents is a combinational logic function, a register or a RAM; otherwise it is classified as *complex*. Labels are associated with each node describing various attributes of the structure it represents; for example, the number of inputs/outputs, the number of bits if it is a register, the number of product terms if it is a PLA. The label also specifies the function of the structure, since TDES may be able to use the structure in implementing a given DFT technique; for combinational function blocks the label also specifies the design style, since this can affect the way it is tested. The labels also specify any global architectural features that TDES might utilize, for example bit slice or pipeline architectures. The arcs also have associated labels defining source, destination, width etc.

Since the information regarding DFT methods is well defined TDES uses a frame-based technique [12] to store information in its knowledge base relating to the structural, behavioural, quantitative and qualitative aspects of each DFT technique.

The objective of TDES is not simply to suggest and subsequently integrate a suitable DFT technique into a circuit, but to generate a 'total' test solution for a circuit; that is, TDES is not only concerned with producing a testable circuit but also producing a test plan for the circuit. This approach is referred to as a Testable Design Methodology (TDM).

Within the knowledge base the structural requirements for a given TDM are described by means of a 'template' which contains information on the type of function (kernel) to which the technique can be applied. The template also contains information, where applicable, on the hardware necessary to generate test patterns, capture circuit responses, evaluate the results, together with information on how to connect the test hardware to the kernel to be tested. In order to evaluate the 'goodness' of an implementation of a given DFT method, the template also contains information on the trade-offs associated with the method in terms of, for example, area overhead, fault coverage, test generation costs, test application time etc; these attributes of a DFT technique may be constants or variables which depend upon circuit size.

It was stated that TDES offers a 'total' test solution, in that part of the output from the system is the sequence of operations necessary to test a kernel using a given DFT method; the method of testing a circuit is stored in a *test plan* which is associated with each template. In essence the test plan describes the transfer of data along the arc of a graph and the processing of data at the nodes in a graph of the template.

For the purposes of testing TDES divides the circuit into kernels, which may or may not be functional blocks, but have the salient characteristic that they can be tested using some standard DFT technique; these kernels are identified by applying certain rules [29] to the graph model of the circuit being processed. To improve the efficiency of finding a 'total' test solution, TDES assigns weights to the kernels and subsequently processes the kernels with the highest weights first, the objective being that these kernels have been identified as difficult to test for various reasons; hence the number of TDMs applicable to test them will be small, so it can be readily discovered whether or not a complete solution can be obtained within the given constraints. The rules applied by TDES in deriving the weighting factors assigned to the kernels is contained within its knowledge base, and can subsequently be changed as experience is gained in using the system. The weighting factors are influenced by the size of the kernel to be tested, the degree of difficulty in gaining access to the inputs/outputs of the kernel, the number of TDMs applicable to the kernel, etc.

The process of structuring a DFT technique around a kernel is called *embedding*, and requires a subgraph of the kernel to be generated. Each node in the subgraph is assigned an *identity node*, which defines the method of transferring data through the node together with the necessary activation signals and the delays across the node. Similarly an *identity transfer path* is assigned to the arcs between the nodes, describing the activation path to transfer data along the path and the delay that the transfer is likely to cause. The process of embedding a TDM into a kernel comprises matching the subgraph of the TDM with that of the kernel; if a direct match cannot be made additional hardware is added. The embedding of the test structures is done interactively and at each stage the 'goodness' of a technique is evaluated and compared with the constraints imposed by the designer. Although a direct match with a TDM may exist, it may be advisable to consider techniques which require additional hardware, since the direct implementation may reduce the area overhead, but the testing time and fault coverage may be inferior to those techniques which cannot be implemented directly. Furthermore, local embeddings which require additional testability structures may incur a high overhead initially but may globally provide a better solution since other kernels may also be able to use these structures. When several possible TDMs exist to test a kernel the designer is presented with the solutions and their trade-offs and asked to select one or reject all solutions. The final output from the system defines the techniques to be used on each kernel, what overheads are incurred, if any, and a test plan for the circuit giving a 'total' test solution.

Recently the capabilities of TDES have been enhanced by incorporating procedures to optimize the execution time of the test plans. The optimization process is based on the theory of 'test plan execution overlap' [30] which uses the fact that in general test plans can be pipelined, thus reducing the testing time; however, it is necessary to identify steps in a test plan which cannot be executed in parallel since they may use the same piece of hardware or require conflicting control signals to be generated. The test schedule which is subsequently produced is unique to the circuit being processed.

11.4.2 PLA Expert System Synthesizer (PLA-ESS) [13]

PLA-ESS was developed to assist designers in their choice of DFT technique, specifically, for PLAs. A large number of DFT techniques have been developed for PLAs since these structures, which are commonly used in VLSI circuits, are not amenable to the standard test methods due to their high fanin/fanout characteristic and unusual fault modes – for example, cross-point faults. Consequently, when the designer has to decide upon the most effective way to test a PLA he must evaluate each method in the light of its advantages and disadvantages. This requires him to examine a multidimensional solution space in which global trade-offs have to be considered regarding fault coverage, area overhead, additional I/O pins, test application time, etc.; this type of problem, again, is amenable to solution using expert system techniques.

The input to PLA-ESS comprises a description of the PLA in terms of the number of inputs, outputs and product terms together with a requirements vector, which defines the constraints to be imposed upon the DFT technique with respect to its effect on the original design, the requirements of the test environment and the costs; a weighting factor can be assigned to each attribute in the requirements vector defining the importance of satisfying a given requirement in the proposed solution. Each DFT technique stored in the knowledge base of PLA-ESS has an associated attributes vector which specifies the characteristics of the method. If several methods can be applied to a given PLA an *evaluation matrix* will be generated, where the columns correspond to the techniques and the rows correspond to their attributes. In selecting the technique to be used from the evaluation matrix PLA-ESS generates an *evaluation function* which combines the attributes of each method into a numerical 'score' which reflects the 'goodness' of the technique in matching the designer's requirements. The evaluation function is calculated in such a way that the technique with the highest score will provide the closest match to the designer's requirements; in this way one or more possible solutions can be identified. If a suitable match cannot be found PLA-ESS invokes the 'reasoner' which uses the technique of '*reason analysis directed backtracking*' [13] to establish the reason for the failure, and when the cause is identified attempts are made to remedy it and the process of searching for a possible solution is restarted. Typical reasons for a PLA-ESS failing, initially, to find a match are that an entry in the requirements vector is too constraining or the system has been asked to attempt to satisfy two conflicting requirements. When PLA-ESS cannot find a solution under the conditions specified by the designer a dialogue is set up between the system and the designer in which the designer can ask for an explanation why a solution cannot be found, in which case the system will display the attributes of the requirements vector causing the problem together with the values which can be achieved by the system; the designer may then modify his requirements and the system will attempt to find a solution. If the requirements cannot be modified the system will declare that within the given constraints no solution can be found. The designer also has the capability of asking the system to display a trace of the 'design history' so that he can recall the modifications made to the requirements vector and the resultant effect it has had on the proposed solution. The designer is not obliged to accept suggestions made by the system ('last chance confirmation') but can make his own decisions and monitor the outcome. If a suggestion is rejected the system will prompt for a reason and then invoke the 'reasoner' to find a more acceptable solution if one exists. The designer can also ask the system to make a selection of a technique for him if there are several techniques which could be applied to his PLA. Once a satisfactory solution has been found, the PLA is subsequently modified so that the method can be implemented and an appropriate test set for the PLA is generated if it is required.

11.4.3 Automatic Design For Testability – ADFT system [31], [32]

ADFT is essentially an expert system which has been integrated into the SILC [33] SILicon Compiler so that the system, overall, guarantees not only 'correctness by construction' but also 'testability by construction'. In a similar way to TDES and PLA-ESS this system is also used to suggest suitable enhancements to be added to a circuit in order to improve its testability; however, it differs from the previous methods in that the hardware to enhance testability is synthesized at the outset with the normal functional logic of the circuit, thus reducing area overheads. The system has ben developed by GTE Laboratories for use in the design of VLSI chips for telecommunication applications.

The ADFT system essentially comprises:

1. A set of rules to ensure that the design complies with the requirements of the DFT techniques supported by the system, and an associated 'rules checker';
2. A testability evaluator which performs two functions. First, it determines the testability of a design based on the resulting effects of its implementation on the design, ATE requirements and costs. The testability evaluation is performed at three levels, namely at structural level, where the controllability/observability values for each node in the circuit are

calculated using a testability analyser; at path level, where a critical path-tracing algorithm is used to locate critical testing paths amongst the inter-connections between functions; finally, at functional block level, where an 'information flow approach' is used to identify potential testability bottle-necks. The second function of the evaluator is to identify parts of the circuit which will be difficult to test and require the expert system to make intelligent decisions about design modifications to overcome the problems;

3. A library of testable functions which may be used to replace equivalent functions in the circuit which have been declared by the testability eval-uator as difficult to test. The contents of the library comprise ROMs, RAMs, registers, PLAs, series/parallel converters, etc., together with the patterns necessary to test these functions.

4. The Testability EXPERT (TEXPERT) which is invoked to make intelli-gent decisions on how to solve testability problems within given constraints.

The ADFT system supports a modular design-for-testability approach, in that it does not simply apply a given DFT technique throughout the circuit, but implements the technique most suitable to a given function in its particular environment. When TEXPERT is invoked to make decisions upon where and what testability enhancement structures have to be integrated into a circuit it has access to the following information:

1. an estimate of how difficult a function is to test, from the testability evaluator
2. knowledge of how well a function conforms to the design-for-testability rules, from the testability rules checker
3. limitations on the automatic test generation algorithms to be used to produce test patterns for the circuit
4. general constraints imposed upon the solution by the designer
5. library of testable functions together with the overheads incurred in their use.

In making a decision TEXPERT uses information from the testability evaluator to prioritize its activities and uses the designer's constraints and the limitations of the automatic test generator algorithms to impose boundaries on the scope of possible solutions. When TEXPERT is presented with a function which has been declared 'difficult to test' it searches the library to determine if a testable version of the function exists. If a testable version exists the overheads incurred by its use are examined and if these are high the designer is asked if he can relax his constraints so that the testable function can be used; if the constraints cannot be relaxed an alternative DFT technique is examined, which may involve reconfiguring some of the functional hardware to perform a test function. TEXPERT also ensures that all testability enhancement structures can be connected onto the central bus system, OCTEBUS (On Chip Test Enhancement BUS) [31], which guarantees overall controllability and observability of all function blocks in the circuit regardless of the hierarchy. The final output from the ADFT system is a testable design which complies with the constraints imposed by the designer.

Logical programming languages have also been used to implement design-for-testability expert systems directly. One particular system developed by Horstmann [34], [35] uses the Prolog language and was designed to check the compliance of circuit to the design rules for LSSD; when a rule violation is detected the component/connection causing the violation is isolated and the circuit is subsequently modified, either automatically or manually, to comply with the design rule. In this system the circuit description is written as a set of Prolog clauses, and the expert knowledge of the system resides in the DFT, control sequence and transformation rules. The DFT rules define the criteria for a testable design; the control sequence rules are the data that the system 'thinks' is required by the automatic test generators and test equipment: for example, a definition of sensitization patterns for the clocks on storage elements for control and observation functions in the circuit. The transformation rules define local design modifications required to make a circuit testable. When the system is checking the circuit to determine if it complies with a given DFT rule, it essentially considers the circuit description as an 'input theory', and the DFT rules form the objectives or goals to be proved using the 'input theory'. When a rule fails the attempted proof, which is, essentially, the search tree of the Prolog clauses used so far, is sent to the analyser to identify the part of the circuit violating the design rule. Once the analysis is completed the system uses the transformation rules to determine the modifications required to make the design comply with the design rule; the required modifications are subsequently entered as clauses in the circuit description. The system has been used with some success in both identifying and correcting design rule violations in a four-bit binary up/down counter described at gate level, a DMA controller chip described at functional level and the M6800 using a bus level description.

The major obstacles to the pervasive use of expert systems, not only in IC design but also in other disciplines, are the problems of extracting knowledge from the domain expert and the start-up costs. Various techniques have been developed [12] to extract problem-specific knowledge from the expert but all depend, for their success, upon a firm commitment by the domain expert to fully co-operate on the project. The problem of start-up costs can be offset, to some extent, by using an expert system shell or using a logical programming language, e.g. Prolog, which has a built-in inference mechanism, so that a prototype system can produced relatively quickly and the knowledge base can be built up incrementally. In retrospect, start-up costs were also considered to be an obstacle in the development of CAD programs for IC design; these programs are now widely used and their development costs, although still considerable, have been accepted, primarily because these tools have been seen to produce a tangible benefit to the designer in terms of reduced effort and design time; hence if the development costs of expert systems are to be accepted they must also demonstrate that they can produce tangible benefits to the designer.

11.5 REFERENCES

1. Shrobe, H. E. (1983) *AI meets CAD, VLSI '83, VLSI Design of Digital Systems*, pp. 387–99, Elsevier Science Publishers.
2. Horstmann, P. W. and Stabler, E. P. (1984) Computer aided design (CAD) using logic programming. *21st Design Automation Conference Proceedings*, June, 144–51.
3. Kowalski, T. J. and Thomas, D. E. (1984) A VLSI design automation assistant: An IBM System/370 design. *IEEE Design and Test*, **1**(1), 60–9.
4. Kowalski, T. J., Geiger, D. J., Wolf, W. H. and Fitchner, W. (1985) The VLSI design automation assistant: From algorithms to silicon. *IEEE Design and Test*, **2**(4), 33–43.
5. Watanabe, H. and Ackland, B. (1987) FLUTE: An expert floorplanner for VLSI. *IEEE Design and Test*, **4**(1), 32–41.
6. Ackland, B., Dickenson, A., Ensor, R. *et al.* (1985) CADRE – A system of cooperating VLSI design experts. *Proc. International Conference Computer Design*, 99–104.
7. Steinberg, L. I. and Mitchell, T. M. (1985) The REDESIGN system: A knowledge-based approach to VLSI CAD. *IEEE Design and Test*, **2**(1), 45–54.
8. de Geus, A. J. and Cohen, W. (1985) A rule based expert system for optimising combinational logic. *IEEE Design and Test*, **2**(4), 22–32.
9. Kim, J. and McDermott, J. (1983) TALIB: An IC layout design assistant. *Proc. AAAI-83*, 197–201.
10. Joobbani, R. and Siewiorek, D. P. (1986) A knowledge base routing expert. *IEEE Design and Test*, **3**(1), 12–23.
11. Hayes-Roth, F. (1984) The knowledge-based expert system: A tutorial. *Computer*, **17**(9), 11–28.
12. Waterman, D. A. (1986) *A Guide to Expert Systems*, Addison-Wesley.
13. Breuer, M. A. and Zhu, X. (1985) A knowledge-based system for selecting a test methodology for a PLA. *22nd Design Automation Conference Proceedings*, June, 259–65.
14. Clocksin, W. F. and Mellish, C. S. (1981) *Programming in Prolog*. Springer-Verlag.
15. Robinson, G. D. (1984) Artificial intelligence and testing. *Digest of Papers, 1984 International Test Conference*, October, 200–3.
16. Angwin, J., Drake, P. and Reader, G. (1984) The need for real-time intelligence when testing VLSI. *Digest of Papers, 1984 International Test Conference*, 752–9.
17. Bellon, C., Robach, C. and Saucier, G. (1983a) An intelligent assistant for test program generation: *The SUPERCAT System*. *Digest of Technical Papers, International Conference Computer Aided Design*, September, 32–3.

18. Bellon, C., Robach, C. and Saucier, G. (1983b) VLSI test program generation: A system for intelligent assistance. *Proc. International Conference on Computer Design*, October, 49–52.

19. Robach, C., Malecha, P. and Michel, G. (1984) CATA: A computer-aided test analysis system. *IEEE Design and Test of Computers*, 1(2), 68–79.

20. Parry, P. S., Guyler, I. A. and Bayliss, J. S. (1986) An application of artificial intelligence techniques to VLSI test. *1986 Silicon Design Conference Proceedings*, July, 325–9.

21. Robinson, G. D. (1983) HITEST – intelligent test generation. *Digest of Papers, 1983 International Test Conference*, 311–23.

22. Bending, M. J. (1984) HITEST: A knowledge-based test generation system. *IEEE Design and Test of Computers*, 1(2), 83–92.

23. Maunder, C. (1983) HITEST test generation system – interfaces. *Digest of Papers, 1983 International Test Conferences*, 324–32.

24. Genesereth, M. R. (1982) Diagnosis using hierarchical design models. *Proc. AAAI-82*, 278–83.

25. Shubin, H. and Ulrich, J. W. (1982) IDT: an intelligent diagnostic tool. *Proc. AAAI-82*, 290–5.

26. Wilkinson, A. J. (1985) MIND: An inside look at an expert system for electronic diagnosis. *IEEE Design and Test of Computers*, 2(4), 69–77.

27. Hartley, R. T. (1984) CRIB: computer fault-finding through knowledge engineering. *Computer*, March, 76–83.

28. Pipitone, F. (1986) The FIS electronics troubleshooting system. *Computer*, July, 68–76.

29. Abadir, M. S. and Breuer, M. A. (1985) A knowledge based system for designing testable VLSI chips. *IEEE Design and Test of Computers*, 2(4), 56–68.

30. Abadir, M. S. and Breuer, M. A. (1986) Test schedules for VLSI circuits having built in test hardware. *IEEE Trans. Computers*, C-35(4), 361–7.

31. Fung, H. S., Hirschhorn, S. and Kulkarni, R. (1985) Design for testability in a silicon compilation environment. *22nd Design Automation Conference Proceedings*, June, 190–6.

32. Fung, H. S. and Hirschhorn, S. (1986) An automatic DFT system for the SILC silicon compiler. *IEEE Design and Test of Computers*, 3(1), 45–57.

33. Blackman, T., Fox, J. and Rosebrugh, C. (1985) The SILC silicon compiler: Language and features. *22nd Design Automation Conference Proceedings*, June, 232–7.

34. Horstmann, P. W. (1983) Design for testability using logic programming. *Digest of Papers, 1983 International Test Conference*, October, 706–13.

35. Horstmann, P. W. (1985) A knowledge-based system using design for testability rules. *Digest of Papers, 14th International Symposium on Fault-Tolerant Computing*, June, 278–84.

INDEX